Computational Thermodynamics: The Calphad Method

Phase diagrams are used in materials research and engineering to understand the interrelationship of composition, microstructure, and process conditions. Computational methods such as Calphad (calculation of phase diagrams), are employed to model thermodynamic properties for each phase and simulate multicomponent multi-phase behavior in complex systems. Written by recognized experts in the field, this is the first introductory guide to the Calphad method, providing a theoretical and practical approach. Building on core thermodynamic principles, this book applies crystallography, first principles methods and experimental data to computational phase properties modeling using the Calphad method. With a chapter dedicated to creating thermodynamic databases, the reader will be confident in assessing, optimizing, and validating complex thermodynamic systems alongside database construction and management. Several case studies put the methods into a practical context, making this suitable for use on advanced materials design and engineering courses and an invaluable reference to those using thermodynamic data in their research and simulations.

HANS LEO LUKAS received his Ph.D. from the Technische Hochschule Stuttgart (now the University Stuttgart) in 1960; he continued his research there from 1964 until he retired in 1995. He has developed the BINGSS software, which is the most used optimizer for binary systems worldwide. Among numerous responsibilities over the years, he was co-editor of the journal *Calphad* from 1979 until 2003. He received the Hume-Rothery Award from the Institute of Materials, Minerals and Mining in 1993.

SUZANA G. FRIES received her Ph.D. in Science from the Physics Institute at the Universidade Federal de Rio Grande do Sul in Porto Alegre, Brazil, in 1985, where she held a professorship. She worked as a researcher and was a guest professor participating in several national and European projects at, for example, the Max Planck Institute in Stuttgart, RWTH in Aachen, Université de Paris-Sud in Chatenay-Malabry and Université de Montpellier II. She is currently a freelance consultant and itinerant guest scientist.

BO SUNDMAN received his Ph.D. from the Royal Institute of Technology (KTH) of Sweden in Stockholm in 1981. He worked as researcher, lecturer, and professor at the KTH until 2006, when he became senior researcher at the Paul Sabatier University of Toulouse, France. He is a member of the editorial board of the journal *Calphad*, and an initiator and board member of the Foundation for Computational Thermodynamics (STT) and Thermo-Calc Software AB. He has received the Gibbs Triangle Award from the Calphad Society 2002 and the Hume-Rothery Award 2005.

Computational Thermodynamics

The Calphad Method

HANS LEO LUKAS, SUZANA G. FRIES AND BO SUNDMAN

CAMBRIDGE
UNIVERSITY PRESS

CAMBRIDGE UNIVERSITY PRESS
Cambridge, New York, Melbourne, Madrid, Cape Town,
Singapore, São Paulo, Delhi, Tokyo, Mexico City

Cambridge University Press
The Edinburgh Building, Cambridge CB2 8RU, UK

Published in the United States of America by
Cambridge University Press, New York

www.cambridge.org
Information on this title: www.cambridge.org/9780521868112

First published 2007

A catalogue record for this publication is available from the British Library

ISBN 978-0-521-86811-2 Hardback

Contents

Preface

The idea of this book came from Professor Petzow, during the PML *Betriebsausflug* in September 1991. In a very informal way Professor Petzow invited S. G. F. (who was ready to return to Brazil after two years working with H. L. Lukas) to help H. L. L. to collect all his ideas about, and experiences with, thermodynamic optimization and put them into a book. Work on optimizations has been going on at Stuttgart for a long time, and valuable experience has been accumulated. Dr Lukas' feeling for optimizations is very well defined and one can talk about a "Lukas school for optimizations."

Later the project was enriched by the cooperation with Professor Sundman, at that time Dr. Sundman, who brought his own large experience on computational thermodynamics as well as the Stockholm group's approach to the theme with all the formalisms so well developed by Professor Mats Hillert.

The three authors were very motivated by the idea, since the lack of such a book had always made it difficult to introduce students and researchers to this field. The knowledge necessary in order to obtain a better thermodynamic description of a system is very broad, requiring a judgment of the experimental data provided by the literature and also a wise selection of the model best able to describe the experimental evidence. This judgment is difficult, but the better "educated" the assessor, the greater his ability to judge well.

The three authors have never worked together in the same institute. When B. S. visited Stuttgart during 1994, S. G. F. was already at Aachen. They have found time, however, to work together and have been meeting for many years during coffee breaks, before and after conferences, and during the Schloss Ringberg Workshops, in order to make progress with the book.

Many conferences and five Ringberg workshops were necessary to achieve the objective. During this time we were able to incorporate the most recent theoretical achievements of first-principles calculations into our procedures. When the book collecting the thermodynamic-optimization experiences of Stuttgart and Stockholm was finally sent to the Press in July 2006, S. G. F. was working in Vienna, B. S. was working in Toulouse, and H. L. L. had retired and was living in Stuttgart – and it was in time for the 80th birthday of Professor Petzow! The modeling and software has developed significantly during the writing of the book and it has been a challenge to keep it updated. Continuous updates will be provided at the website of the book.

The authors discussing the book in the Hexenzimmer at Schloss Ringberg in 2005, during the Thermodynamic Modeling and First-Principles Calculations Workshop organized by the Max Planck Society.

Dedication

The authors wish to dedicate this book to Ibrahim Ansara, who always managed to put a smile on the face of thermodynamics, like in the phase diagram Fig. 5.6(k).

Acknowledgments

The authors are grateful to A. Fernández Guillermet, B. Hallstedt, M. Harmelin, M. Hillert, U. Kattner, Frau O. Kubaschewski, B. Legendre, and many others for inspiration, reading parts of the book, and giving valuable comments; and to B. Böttger, N. Dupin, F. Lechermann, P. Rogl, and N. Warken for providing figures from their ongoing research work.

The figures with crystal structures have been taken from the Crystal Lattice Structures web page, http://cst-www.nrl.navy.mil/lattice/, provided by the Center for Computational Materials Science of the United States Naval Research Laboratory.

The triangular symbol in the left-hand corner of some diagrams indicates that the figure has been calculated by the Thermo-Calc software (Thermo-Calc Software AB, http://www.thermocalc.com).

1 Introduction

The Calphad technique has reached maturity. It started from a vision of combining data from thermodynamics, phase diagrams, and atomistic properties such as magnetism into a unified and consistent model. It is now a powerful method in a wide field of applications where modeled Gibbs energies and derivatives thereof are used to calculate properties and simulate transformations of real multicomponent materials. Chemical potentials and the thermodynamic factor (second derivatives of the Gibbs energy) are used in diffusion simulations. The driving forces of the phases are used to simulate the evolution of microstructures on the basis of the Landau theory. In solidification simulations the fractions of solid phases and the segregation of components, as well as energies of metastable states, which are experimentally observed by carrying out rapid solidification, are used. Whenever the thermodynamic description of a system is required, the Calphad technique can be applied.

The successful use of Calphad in these applications relies on the development of multicomponent databases, which describe many different kinds of thermodynamic functions in a consistent way, all checked to be consistent with experimental data. The construction of these databases is still a very demanding task, requiring expertise and experience. There are many subjective factors involved in the decisions to be made when judging and selecting which among redundant experimental data are the most trustworthy. Even more subjective is the assessment of phases of which little or nothing is known, except perhaps in a narrow composition and temperature range. Furthermore, the growing range of applications of these databases increases the feedback and several corrections and modifications are required. The development of new models and the rapid advance of first-principles (from the Latin *ab initio*) calculations makes the assessment techniques very dynamic and challenging.

1.1 Computational thermodynamics

Thermodynamics describes the equilibrium state of a system. This is a necessary foundation for simulations of phase transitions and processes, since all systems try to reach this state. In computational thermodynamics (CT) the equilibrium state is described using thermodynamic functions that depend on temperature, pressure, and composition.

These functions can be extrapolated also outside the equilibrium state and thus, when they are included in a simulation model, provide information about values and gradients of the thermodynamic properties in space and time outside the range of stability of a phase.

The thermodynamic models used by CT contain adjustable parameters, which can be optimized in such a way that the models can reproduce many kinds of experimental data as well as theoretical models and first-principles data. Thus CT is more flexible and has much wider possibilities for realistic application calculations than merely using first-principles data. The quality of the results of CT is based on the fit to experimental data, which is also the criterion in judging the quality of results from first-principles calculations.

More fundamental theories are focused on understanding particular mechanisms and isolated properties, which they do very well, but they are not able to describe their coupling in complex systems. CT can make use of theoretical results as well as traditional experimental data and provides a unique framework of various types of information that can be obtained using rather simple models. Thus CT is able to provide consistent thermodynamic information with the accuracy required to describe multicomponent systems of technological interest. It is versatile enough to be extended to new applications and incorporate related types of information such as magnetism (see section 5.4.2) and viscosity (Seetharaman *et al.* 2005).

The important fact is that the thermodynamic information which can be extracted from CT can describe the equilibrium state as well as extrapolations from it. There are types of simulation software used today that depend on thermodynamic information such as heat capacities, partition coefficients, and latent heats, but for which the data have been collected from various sources and are inconsistent and cannot reproduce the real equilibrium state. The results from simulations using such software may be reasonable because the kinetic model has parameters that can be adjusted to compensate for the errors in the thermodynamic data, but the range of applicability of the simulation is poor. Of course, even with correct thermodynamics one will have to fit kinetic parameters in simulation models, but these will have a larger range of validity because they depend less on the thermodynamic properties of the system.

CT has shown its importance for calculating multicomponent phase diagrams and for process and phase-transformation simulations as mentioned above and in detail described in section 8.10. It is used by experimental researchers as a tool to test the compatibility between their results and for data found in the literature, as well as planning new experimental work. CT is also interesting to theoreticians, not only to use as a "reality" against which they can check their predictions from fundamental models but, also and more importantly, as a technique to improve the usefulness of their results by combining them with experimental data, i.e. the Calphad method.

Knowledge of phase equilibria is fundamental to all aspects of materials science, since the properties are determined by the microstructure, see for example Durand-Charre (2004), and the microstructure consists of several phases arranged in space as in Fig. 1.1. A practical way to obtain the phase equilibria in a multicomponent system is by calculations using assessed thermodynamic databases. Hence the generation of reliable and consistent computer-readable thermodynamic databases is very important.

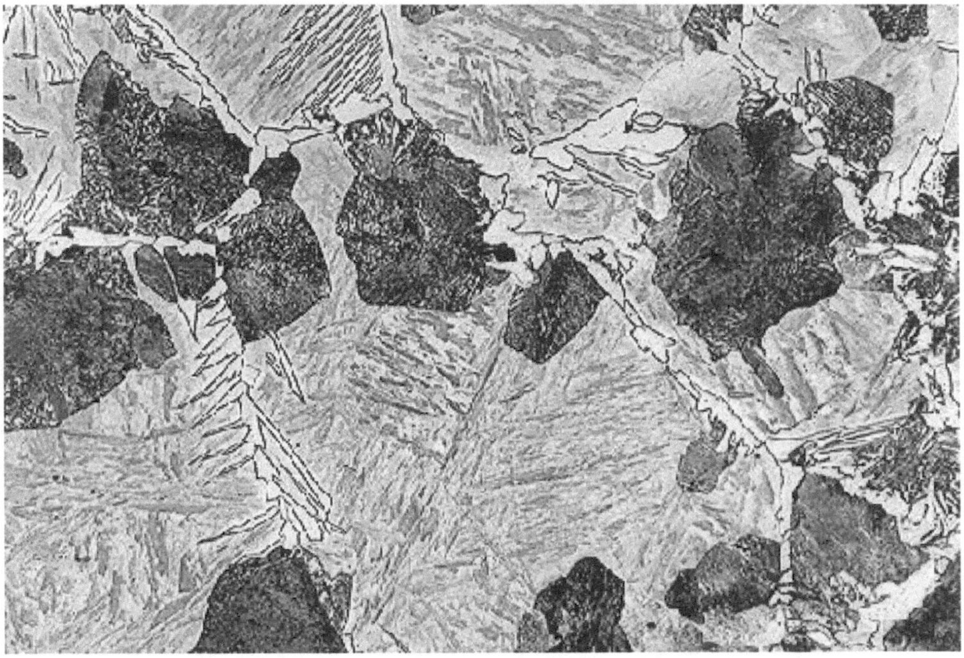

Figure 1.1 The microstructure of a low-carbon steel with many different stable and metastable phases formed during cooling. The properties of the material depend on the amounts and arrangements of these phases and their compositions. Computational thermodynamics together with application software can be used to simulate the development of such microstructures and predict the properties of materials. Courtesy of Mats Hillert.

New experiments are still very important, because the validated databases are based on the combination of theoretical and experimental data.

1.2 The past and present, the Calphad technique

Phase diagrams had been calculated from Gibbs-energy models by van Laar (1908a, 1908b) and many others, but the first general description of Calphad was written by Larry Kaufman in the book *Computer Calculations of Phase Diagrams* (Kaufman and Bernstein 1970), in which he developed the important concept of "lattice stability" which he had introduced earlier (Kaufman 1959). He explained clearly how parameters could be derived both from experimental phase diagrams and from the rudimentary first-principles techniques available at that time, and how they could be used to calculate phase diagrams. The concept of lattice stability was essential for the development of multicomponent thermodynamic databases, which was a very far-sighted goal because at that time it was a challenge to calculate even a ternary phase diagram.

The method of extrapolating solubility lines into the metastable range to obtain a thermodynamic property, such as the melting temperature of metastable fcc Cr, shows one of the important advantages of combining phase diagrams and thermodynamics.

Many such combinations have been used in exploring properties of other systems during the development of the Calphad techniques. Important contributions on how to use experimental and theoretical data in thermodynamic models were made by Kubaschewski *et al.* (1967), Hillert and Staffanson (1970), Hasebe and Nishizawa (1972), Ansara *et al.* (1973), and Lukas *et al.* (1977). For historic details, the recent paper by Kaufman (2002) is recommended.

Originally the term "Calphad" meant calculating phase diagrams from thermodynamic models with parameters adjusted to the available experimental data. The term "Calphad technique" has come to mean the technique of selecting models for phases that can be extrapolated both in composition and in temperature ranges, including also metastable ranges. A comprehensive description of this Calphad technique can be found in Kattner (1997). The "Calphad method" means the use of all available experimental and theoretical data to assess the parameters of the Gibbs energy models selected for each phase. That is the topic of this book. To describe the use of these models and parameters stored in thermodynamic databases for various applications, the term "computational thermodynamics" has been adopted.

It may be instructive to mention here that in the early days of Calphad there was a heated argument among material scientists about how to model a dilute solution. The thermodynamic properties of a dilute solution of B in a solvent A can be modeled with a simple Henrian coefficient as described in section 5.5.9. With the Calphad technique one needed two parameters, one representing the solvent phase consisting of pure B, i.e. a "lattice stability," and one interaction parameter between A and B. In the dilute range the sum of these two parameters is the only important quantity and is in fact identical to the Henrian coefficient. The fact that the Calphad technique needed more parameters, and that the parameter representing pure B in the same phase as A often was an unstable phase (like pure fcc Cr), was taken to be a severe drawback of the Calphad technique.

However, the drawback of dilute models is more severe because there are many cases, for example at solidification, for which one must find a way to describe how the thermodynamic properties vary when the solvent phase changes. Upon investigating this problem in more detail, one finds that the Calphad technique is the simplest and most consistent way to handle a multicomponent thermodynamic system with many "solvent" phases. Similar modeling problems in which one has to evaluate "lattice stabilities" of phases with more or less limited solubilities occur often in Calphad assessments. This is discussed in chapter 6.

1.3 The future development of databases and software applications

As computers become faster, models and techniques of greater sophistication and accuracy are being developed within CT. However, there is an important "inertia" present in the thermodynamic databases, which are compiled from a large number of both independent and inter-dependent assessments. In a reference book like Hultgren *et al.* (1973) one may replace the data for one element without changing anything else, but in a database the binary assessments depend on the unary data for pure elements and on the models selected for the phases. Ternary and higher-order assessments depend on the selected assessments of the corresponding lower-order systems. This means that, in order to introduce a new value for a pure element, a new assessment of a binary system, or a new model for a phase present in a thermodynamic database one must revise all assessments depending on this value, assessment, or model. The current commercial databases depend on a number of decisions taken for pure elements and models made more than 20 years ago by the Scientific Group Thermodata Europe (SGTE, http://www.sgte.org). It is a great credit to the scientists involved in these decisions that there have been so few problems with using these data to build multicomponent databases for so many years. Nonetheless, there is a need to revise the set of unary and binary data as well as models continuously and maybe every 10 years one should start creating improved versions of the databases from a new revised set of data. The maintaining of databases is briefly discussed in chapter 8.

The number of publications on thermodynamically assessed systems is increasing rapidly and most of them use SGTE unary data (Dinsdale 1991), but many important binaries have been assessed several times using different models for some of the phases, which are not always compatible. When selecting which assessed version of a system should be incorporated into a database, the thermodynamic description should, at least ideally, be the best that can be obtained at present, taking into account theoretical approaches and technical possibilities. Of course, what should be judged the "best" assessment is not easy to define, but some rules can be established, as will be discussed in this book.

A series of thermodynamic and modeling workshops organized by the Max Planck Institute in Stuttgart and held in the conference center at Schloss Ringberg was initiated in 1995. The aim of these workshops was to build the foundation for the next generation of thermodynamic databases and software. They had a unique organization with from five to seven groups, with from seven to nine participants in each. The participants in each of the groups were expected to write a paper together on a specific topic during the workshop, with some time to complete it allowed afterwards. The participants had quite different opinions initially, but during the workshop many new ideas on how to resolve the differences or find ways to resolve them appeared. The first workshop was dedicated to pure elements and compounds and was published as a special issue of *Calphad* (Aldinger *et al.* 1995). The second workshop was dedicated to modeling of solutions and was published in *Calphad* (Aldinger *et al.* 1997). The third was dedicated to applications and published in *Calphad* (Aldinger *et al.* 2000) and the *Zeitschrift für Metallkunde* (Burton *et al.* 2001). The fourth was about applications and modeling of special phases such as oxides and was also published in the *Zeitschrift für Metallkunde* (Aldinger *et al.* 2001). The fifth workshop was about Calphad and *ab initio* techniques and was published in *Calphad* (Aldinger *et al.* 2007).

1.4 The structure of the book

This book is intended to be an introductory text as well as a reference book for optimization of thermodynamic descriptions, but it is not intended to be read from beginning to end. This first chapter gives some introduction to the scope of the book. The second chapter is for reference, since its content should be well known to any student of physics, chemistry, or materials science, and the notation for thermodynamic quantities used in the book is explained.

Chapter 3 gives a short introduction to *ab initio* calculations, in particular how results from such calculations can be used for thermodynamic modeling and assessments. In the fourth chapter various experimental techniques that provide the data necessary for assessments are described. The fifth chapter gives a detailed description of most of the models currently used to describe thermodynamic functions of phases. This is mainly intended to be a reference for the sixth chapter, where the selection of models for phases is discussed in terms of their properties.

Chapter 6 is the central part of the book, where the experience from many assessments has been condensed into a few rules of practice. As usual, any good rule has many exceptions; many of these exceptions will be described in chapter 9, which deals with case studies. The reader may find that some topics are repeated several times; that is usually because they are important.

Chapter 7 describes two of the most-used software systems for thermodynamic assessments, BINGSS, developed by H. L. Lukas (Lukas *et al.* 1977, Lukas and Fries 1992) and PARROT, developed by Dr. B. Jansson (Jansson 1983) as a part of the Thermo-Calc software (Sundman *et al.* 1985, Andersson *et al.* 2002). The emphasis is on the main features of these items of software; many of the peculiarities will be explained only in the case studies in chapter 9. Chapter 8 deals with the creation of databases from separate assessments and how to maintain a database.

Chapter 9 is again a crucial part of the book. The beginner should try to follow some of the case studies in order to learn the technique from a known system. A careful reading of the case studies is recommended, since they give many hints on how to use experimental data of various types. The reader is also advised to look at the website for the book, because more case studies will be available there. References to existing software and databases can be found in a special issue of *Calphad* (2002, **26**, pp. 141–312).

2 Basis

2.1 Thermodynamics

A short overview on the rules of thermodynamics shall be given here, with special emphasis on their consequences for computer calculations. This part will not replace a textbook on thermodynamics, but shall help the reader to remember its rules and maybe present them in a more practically useful way, which facilitates the understanding of thermodynamic calculations.

Thermodynamics deals with energy and the transformation of energy in various ways. In thermodynamics all rules are deduced from three principal laws, two of which are based on axioms expressing everyday experiences. Even though these laws are very simple, they have many important consequences.

Thermodynamics can strictly be applied only to systems that are at equilibrium, i.e. in a state that does not change with time. As noted in the introduction, the thermodynamic models can be extrapolated outside the equilibrium state and provide essential information on thermodynamic quantities for simulations of time-dependent transformations of systems.

2.1.1 *The equation of state*

The concept of thermodynamic state must be introduced before beginning with the principal laws. This can be done by invoking the principle of the "equation of state." This is connected with the introduction of temperature as a measurable quantity. If pressure–volume work is the only work considered, then one can state that *in a homogeneous unary system the state is defined by two variables*. Of the three variables temperature T, pressure p, and volume V in a unary system, only two are independent, i.e. there exists a condition

$$F(T, p, V) = 0 \tag{2.1}$$

which means that one of the three variables is determined by the other two:

$$V = f(T, p); \qquad T = f(V, p); \qquad \text{or} \qquad p = f(T, V) \tag{2.2}$$

A unary system consists of a single component, either an element from the periodic table or a molecule that will not form any other molecules under the conditions considered. For the definitions of the terms "homogeneous," "system," etc., any general textbook on thermodynamics, for example Hillert *et al.* (1998), can be consulted.

For each additional kind of work considered, e.g. magnetic or electric, an additional variable is necessary to define the state. Nevertheless, the above consideration is very useful and, if the pressure dependence can be neglected, the state functions of pure substances (unary phases) are functions of the temperature only.

For practical reasons a "state" may be defined for inhomogeneous systems, but that can be taken as a simplified notation for the sum or integral of (eventually infinitesimally small) homogeneous systems, for each of which the state has to be defined separately.

To define the state of non-unary (binary, ternary, . . .) systems, one needs additional variables, for example the amounts of the components. These may be replaced by mole- or mass-fractions together with the total amount.

2.1.2 The first law of thermodynamics

This law is derived from the axiom of conservation of energy. A formulation well suited for our purpose is the following: *the sum of the heat and work transferred to an otherwise closed system defines a function not depending on the way in which this transfer took place*. The function defined in this way is called the internal energy U. A "closed system" means a system that does not exchange any heat, work, or matter with its surrounding. Besides constant internal energy U, it has constant volume V and a constant amount of matter (expressed as constant amounts N_i of different components i). All these quantities depend only on the "state" of the system and they are called state variables or state functions. The concept of state functions is very important in thermodynamics. A feature of the internal energy U, which must be kept in mind for numerical calculations, is that only differences between the values of this function for two well-defined states have a physical meaning. No absolute value of U can be defined.

If the system is opened and either *heat*, q, or *work*, w, is transferred to it from the surroundings, the above rule can be formulated in terms of a change of the internal energy:

$$\Delta U = q + w \tag{2.3}$$

Neither q nor w is itself a state variable. Transferring either only heat or only work may be different ways of bringing about the same change of state.

2.1.3 The second law of thermodynamics

This law is derived from the axiom that a complete conversion of heat to work is not possible. It may be formulated as follows: *a function of state, called entropy and denoted S, can be defined, which can increase, but never decrease, in a closed system.* The state in which the entropy of the closed system has its maximum is the equilibrium state. No further change of state can happen in this system as long as it remains closed.

The difference between the entropies of two well-defined states of a system (that is not closed) is defined by the integral

$$S_2 - S_1 = \int_{\text{state 1}}^{\text{state 2}} \left(\frac{\mathrm{d}Q}{T} \right)_{\text{rev}} \tag{2.4}$$

$\mathrm{d}Q$ is the infinitesimal amount of heat transferred to the system on the way from state 1 to state 2. The subscript "rev" indicates that the path from state 1 to state 2 must be reversible, which means that it is restricted to a sequence of equilibrium states. On going in non-reversible ways from state 1 to state 2, some work is added instead of heat. All real changes of state are irreversible, but a reversible change of state may be simulated by a *Gedankenexperiment*, i.e., an experiment that one may think of doing, but which it is impossible to do in reality.

2.1.4 The third law of thermodynamics

This law is derived from the axiom that it is impossible to reach the temperature of 0 K. This temperature can be approached only asymptotically. A consequence of this axiom is that *the change in entropy of a reversible reaction approaches 0 when the reaction temperature approaches 0 K*. By virtue of this law an absolute value can be defined for the state function entropy, S, in contrast to the internal energy U. By convention S is set to zero at 0 K.

2.1.5 Definition of some terms and symbols

A number of symbols will be used and the most important are summarized here. In most cases they refer to just one phase. Whenever more than one phase is involved, a superscript like α or β will be used to distinguish the phases involved. A superscript tot will be used when it is emphasized that the symbol refers to the total value for the system, summed over all phases. A phase is distinguished by its crystal structure (see section 2.2), and at equilibrium its composition is homogeneous in space. The gas, liquid, and the amorphous or glass phase are also phases, even though they have no crystal structure. Names of phases are discussed in section 8.4.

The term "constituent" means any species that can be treated as an independent entity to describe the constitution of a phase, for example molecules in a gas or a defect in a crystalline phase; see section 5.3.

- T the absolute temperature.
- R the gas constant, 8.31451 J mol^{-1} K^{-1}.
- p the pressure.
- V the volume.
- Q the heat.
- N_i the number of moles of component i.

N the total number of moles, $N = \sum_{i=1}^{n} N_i$.
N_A Avogadro's number, 6.023×10^{23} atoms per mole.
x_i the mole fraction of component i, $x_i = N_i/N$.
y_i the fraction of constituent i. The sum of the constituent fractions is unity; however, since y_i depends on the solution model for the phase, there may be more constituents than components, with the consequence that there is no general relation between the constituent fractions and the mole fractions.
Y short for denoting the constitution, i.e., all constituent fractions in a phase.
G the total Gibbs energy of a system.
G_m the Gibbs energy per mole of components of a system.
G_m^θ the Gibbs energy per mole of components of phase θ.
G_i^θ the partial Gibbs energy of component i in phase θ.
${}^\circ G_i^\theta$ the Gibbs energy for the pure component i in phase θ.
μ_i the chemical potential of component i.
a_i the activity of component i, $a_i = \exp[\mu_i/(RT)]$.

There are often several specifications to these symbols placed as superscripts and subscripts. The superscript is reserved for the phase, power, or sublattice indication. As subscript one can have the normalization "m" or the specification of a component or other things. The "pre"-superscript is used to specify that the symbol is for a pure element, "$^\circ$", an excess quantity "^{E}" or anything else that does not fit as a subscript.

2.1.6 *Equilibrium conditions and characteristic features*

A first condition of equilibrium is that all parts of a system have the same temperature and the same pressure. Inhomogeneities at equilibrium may occur only on having different "phases," each of which, however, must be homogeneous in itself. This is a direct consequence of the second law. A system with a temperature gradient may be simplified by dividing it into hotter and colder parts. By transferring heat from a hotter part to a colder part, according to Eq. (2.4) the entropy loss in the hotter part is less than the entropy gain in the colder part, and thus the total entropy of the system is increased. Similarly, equilibration of composition gradients increases entropy. This can be shown by "van 't Hoff's equilibrium box," as is explained in most textbooks on thermodynamics.

Experimentally, pressure differences can be maintained for quasi-infinite times and thus equilibrium with different pressures inside a system would seem possible. However, such a pressure barrier means that the system is divided into parts forming independent closed systems. Considering it as a single closed system implies that the pressure barrier can be opened.

Systems in equilibrium have single scalar values for temperature and pressure, the temperature and pressure of the system. A general system, in contrast, may have a temperature and a pressure field, i.e., temperature and pressure may vary with coordinates in space and the expression "temperature of the system" would be meaningless.

In an open system the differential of the internal energy can be expressed as a sum of products of an intensive variable and the differential of its "conjugate" extensive variable:

$$dU = T\,dS - p\,dV + \sum_{i=1}^{n} \mu_i\,dN_i \tag{2.5}$$

The conjugate pairs of an intensive and an extensive variable are thus temperature and entropy, T and S, negative pressure and volume, $-p$ and V, as well as the chemical potential and amount of component i, μ_i and N_i.

Dividing Eq. (2.5) by T and exchanging the left-hand side with the first term on the right-hand side gives another series of conjugate pairs of variables:

$$dS = \frac{1}{T}\,dU + \frac{p}{T}\,dV - \sum_{i=1}^{n} \frac{\mu_i}{T}\,dN_i \tag{2.6}$$

with the conjugate pairs $1/T$, U; p/T, V; and μ_i/T, N_i.

From section 2.1.3, the equilibrium condition can be formulated as follows: *in a closed system equilibrium is reached when the entropy reaches its maximal value.* This formulation is difficult to use practically because a "closed system" is very difficult to realize experimentally, since it must have constant internal energy, constant volume, and constant amounts of components.

However, under other conditions (in non-closed systems), one or more intensive variables may be kept constant. If the temperature is kept constant, then the internal energy U is no longer constant, since heat or work has to be exchanged with the surroundings in order to keep the temperature constant. The role of being kept constant is thus taken away from U and given to its "conjugate intensive variable," which is $1/T$. Then the function which has a maximum at equilibrium is the state function derived by a Laplace transformation from the entropy by subtracting the product of these conjugated extensive and intensive variables, see for example Callen (1985) or Hillert *et al.* (1998). This leads to Massieu's and Planck's function (Table 2.1).

In practice another series of state functions, starting with the internal energy U, is preferred. The other members of this series are formed by Laplace transformations using the conjugate pairs of Eq. (2.5). For U the equilibrium condition is formulated as follows: *at constant values of entropy, volume, and amounts of the components, equilibrium is characterized by a minimum of the internal energy.* Because the sign of this series of functions is defined to be opposite to that of S, Ψ, or Φ, these state functions have a minimum at the equilibrium state. The state variables which have to be kept constant in searching for the minimum are called characteristic state variables. They are given in the last column of Table 2.1.

A state function in which all the characteristic variables are intensive variables must always be zero. It is called the *Gibbs–Duhem equation.* At least one extensive variable is necessary in order for us to be able to distinguish the "system" from its surroundings or, in other words, to define the size of the system.

Table 2.1 *State functions having an extremum at equilibrium, if the characteristic state variables are kept constant*

Name	Symbol or definition	Characteristic state variables		
Entropy	S	U	V	N_i
Massieu's function	$\Psi = S - U/T$	$1/T$	V	N_i
Planck's function	$\Phi = S - U/T - p \cdot V/T$	$1/T$	p/T	N_i
Internal energy	U	S	V	N_i
Enthalpy	$H = U + p \cdot V$	S	$-p$	N_i
Helmholtz free energy	$F = U - T \cdot S$	T	V	N_i
Gibbs energy	$G = U + p \cdot V - T \cdot S$	T	$-p$	N_i
Grand potential	$-p \cdot V = U - T \cdot S - \sum_i \mu_i \cdot N_i$	T	V	μ_i
Gibbs–Duhem equation	$0 = U + p \cdot V - T \cdot S - \sum_i \mu_i \cdot N_i$	T	p	μ_i

The most commonly used of these functions is the Gibbs energy, G, and, using it, the equilibrium conditions are explicitly specified: *in an isothermal isobaric system with constant amounts of all components, the equilibrium is reached when the Gibbs energy reaches its minimal value.*

The function enthalpy, H, is often used in evaluating calorimetric measurements because its change under isobaric conditions is, with few exceptions, equal to the heat received, ΔQ, if this change is performed in a reversible way. Thus, for isobaric changes of state Eq. (2.4) results in

$$S_2 - S_1 = \int_{\text{state 1}}^{\text{state 2}} \left(\frac{\mathrm{d}H}{T}\right) \qquad p = \text{constant} \tag{2.7}$$

2.1.7 *Functions of state*

In the preceding section several functions were mentioned. The values of all these functions are defined if the "state" of a system is defined. Therefore they are called state functions or state variables. There are two classes of state variables, the "extensive" ones, the values of which are proportional to the size of the system, and the "intensive" ones, which at equilibrium have the same value everywhere inside a system. Extensive state variables include the volume, V, the number of moles, N, the number of moles of component i, N_i, the internal energy, U, the entropy, S, the Gibbs energy, G, and Planck's function, Φ. Some intensive state variables are the pressure, p, the chemical potential of component i, μ_i, the temperature, T, and the variable μ_i/T.

Extensive variables depend on the size of a system; they must be distinguished from the "integral molar" functions, which are examples of a more general class of functions, namely the "quotients of extensive variables." These are independent of the size of the system, but, in contrast to the intensive variables, have different values in different phases of the same equilibrium. Examples of this type of state variables or state functions are the molar volume, V/N, the molar Gibbs energy, G/N, the mole fractions $x_i = N_i/N$, the molar entropy, S/N, and the Gibbs energy per volume, G/V. The symbol G initially is reserved for the extensive variable "Gibbs energy of the system" (measured in joules, J).

In the literature, however, very often it is used instead of G/N or G_m for the "integral molar Gibbs energy," which is a quotient of extensive variables and is measured in J mol^{-1}.

Besides the "integral molar" quantities Z/N, there are "partial molar" quantities defined as $\partial Z/\partial N_i$, where Z represents any extensive function of state and the partial derivative is taken at constant values of the other characteristic variables. The partial Gibbs energy thus defined as the partial derivative of the total Gibbs energy with respect to the amount of component i, $(\partial G/\partial N_i)_{N_j,T,p}$, is identical to the chemical potential μ_i, which is the intensive state variable conjugate to the amount of component i. μ_i can also be obtained as the partial derivative of the total enthalpy H with respect to the amount of component i, $(\partial H/\partial N_i)_{N_j,S,p}$, but taken at constant entropy S, p, and N_j $(j \neq i)$. This, however, is less interesting, because it is difficult to perform an experiment at constant S, p, and N_j.

2.1.8 Gibbs' phase rule

The number of independent state variables in a system determines the maximum number of phases that can be stable simultaneously. Gibbs was the first to formulate this and it is known as Gibbs' phase rule:

$$f = c + 2 - p \tag{2.8}$$

where f is the degree of freedom, c is the number of components, and p is the number of stable phases. The number 2 represents the state variables T and p, since they appear as state variables in all thermodynamic systems. If there are types of work other than pV work, for example electric or magnetic work, there are additional state variables and the degree of freedom increases by one for each additional type of work.

If either T or p is fixed, the 2 is reduced to 1. If both T and p are fixed, the 2 is reduced to zero. Thus, in a system with one component at fixed T and p, just a single stable phase can exist. If T or p is variable, one or two stable phases can exist, and if there are two phases stable the degree of freedom is zero, i.e., this can occur only at a single value of T or p. If both T and p are variable in a single-component system, then at most three phases can be stable and this can happen only at a specific value of each of T and p, and thus is called an invariant equilibrium. For example, the triple point of H_2O, at which ice, water, and vapor are stable simultaneously, is an invariant equilibrium. If there are two phases stable in a unary system, the degree of freedom is 1 and it is called a monovariant equilibrium. Here T can be expressed as a function of p or vice versa, represented as a one-dimensional curve in a diagram. The T–p diagram for pure Fe is shown in Fig. 2.6(a) later.

Most of the diagrams of binary or higher-order systems discussed below are drawn for a fixed value of pressure and thus the maximum number of stable phases is three in a binary system, four in a ternary system, etc. In an isothermal section of a ternary system, i.e., at constant pressure and temperature, the maximum number of phases is again three. In a phase diagram with potentials (intensive state variables) as axes the degree of freedom defines the number of conditions that can be changed simultaneously (at least by a small amount). Thus, for invariant equilibria no condition can be changed, for monovariant equilibria one condition can be changed. In a binary system at fixed pressure the degree of freedom is two only in single-phase regions. In a binary diagram with a potential plotted versus temperature single-phase regions appear as areas and two-phase regions as lines, see Fig. 2.6(b) later. In a composition–temperature diagram the two-phase regions also appear as areas, but the degree of freedom is only one there, as shown in Fig. 2.6(c). This reflects the fact that phase compositions but not phase amounts are counted as variables in Gibbs' phase rule.

The degree of freedom is an important quantity when determining the number of conditions necessary to define the equilibrium, as will be discussed in section 2.3.2.

2.1.9 Statistical thermodynamics

In statistical thermodynamics the rules of phenomenological thermodynamics are explained as resulting from the mechanics of all the atoms or molecules present in matter. The main topic not included in mechanics itself is the statistical explanation of the entropy. The entropy was explained by Boltzmann as being given by

$$S = k \cdot \ln(W) \tag{2.9}$$

where $k = R/N_A$ is Boltzmann's constant and W is the number of different microscopic states leading to the same macroscopic state. The assumptions necessary in order to enumerate W were more clearly specified by quantum mechanics.

Statistical thermodynamics is the physical background of modeling expressions to describe the Gibbs energy of a phase. The energy terms of the models are usually defined by energies of formation of molecules or building units of crystal lattices. The entropy terms are divided into vibrational terms, which are treated similarly to the energy terms, and entropies of mixing.

Statistical treatment of vibrational entropy has been used successfully for two topics, the calculation of the entropy of molecules in the gas phase, where the possible states of rotation, vibration, and excitation can be related to spectroscopical experimental data, and to low-temperature heat capacities of solids. The latter are normally not explicitly used in the assessments described here, but appear in integrated form as entropies at the "standard temperature" (298.15 K).

The entropy of mixing in many of the models in chapter 5 is derived from the following question: what is the increase in entropy due to Boltzmann's law, if two or more kinds of different atoms are randomly mixed at N fixed places? Let N_A and N_B be the numbers of A and B species randomly mixed at $N = N_A + N_B$ places. Then the number W of possible different configurations is

$$W = \frac{N!}{N_A! \cdot N_B!} \tag{2.10}$$

After using Stirling's formula $\ln(N!) \approx N \cdot \ln(N) - N$, neglecting $-N$ for large N, identifying N_A/N and N_B/N with the mole fractions x_A and x_B and taking Avogadro's number for N, $kN = R$, it follows that

$$k \cdot \ln(W) \approx k \cdot [N \cdot \ln(N) - N_A \cdot \ln(N_A) - N_B \cdot \ln(N_B)] = -R \cdot [x_A \cdot \ln(x_A) + x_B \cdot \ln(x_B)] \tag{2.11}$$

This formula is used extensively in chapter 5. Since chapter 5 deals with applied statistical thermodynamics, no further details are mentioned here. For a better understanding of other features of statistical thermodynamics used in the modeling, textbooks on statistical thermodynamics may be consulted.

2.1.10 Important thermodynamic relations

Only very simple thermodynamics is needed in order to describe the models. Since most thermodynamic data are measured at known temperature, pressure, and composition, it is convenient to choose the Gibbs energy, denoted G, as the basic modeling function. If the Gibbs energy is known, one may derive other quantities from this in the following way:

$$\begin{aligned}
&\text{Gibbs energy} && G = G(T, p, N_i) \\
&\text{entropy} && S = -\left(\frac{\partial G_m}{\partial T}\right)_{p,N_i} \\
&\text{enthalpy} && H = G + TS = G - T\left(\frac{\partial G}{\partial T}\right)_{p,N_i} \\
&\text{volume} && V = \left(\frac{\partial G}{\partial p}\right)_{T,N_i} \\
&\text{chemical potential of component } i && \mu_i = \left(\frac{\partial G}{\partial N_i}\right)_{T,N_{j\neq i}} \\
&\text{heat capacity} && C_p = -T\left(\frac{\partial^2 G}{\partial T^2}\right)_{p,N_i} \\
&\text{thermal expansion} && \alpha = \frac{1}{V}\left(\frac{\partial^2 G}{\partial p\, \partial T}\right)_{N_i} \\
&\text{isothermal compressibility} && \kappa = -\frac{1}{V}\left(\frac{\partial^2 G}{\partial p^2}\right)_{T,N_i} \\
&\text{bulk modulus} && B = 1/\kappa
\end{aligned} \tag{2.12}$$

All of the above quantities are valid for all thermodynamic systems. The partial derivatives with respect to the characteristic variables of G are taken at constant values of all the other characteristic variables. In solutions one may additionally define the partial Gibbs energy for component i:

$$G_i = \left(\frac{\partial G}{\partial N_i}\right)_{T,P,N_j} \qquad (j \neq i) \tag{2.13}$$

In solution modeling we will use the Gibbs energy per mole of formula units,

$$G_{\mathrm{m}} = G/N \tag{2.14}$$

To use the above relations for molar functions, the derivatives are taken from the molar Gibbs energy G_{m} at constant values of the mole fractions $x_i = N_i/N$ and constant total amount $N = 1$ mol.

In modeling the thermodynamic properties of a system, one must model each phase in the system separately. The properties of the system at equilibrium are then functions of the properties of the individual phases, if surface effects can be neglected.

Since many models yield expressions for the molar Gibbs energy as a function of mole fractions, it is useful to transform Eq. (2.13) into a function of these quantities:

$$G_i = G_{\mathrm{m}} + \left(\frac{\partial G_{\mathrm{m}}}{\partial x_i}\right)_{T,p,x_k} - \sum_j x_j \left(\frac{\partial G_{\mathrm{m}}}{\partial x_j}\right)_{T,p,x_l} \qquad (k \neq i,\ l \neq j) \tag{2.15}$$

The molar Gibbs energy of a phase is often formulated not as a function of the mole fractions x_i, but rather, for example, as a function of the constituent fractions y_j. In this case derivatives with respect to temperature used in the above relations must be calculated as

$$\left(\frac{\partial G_{\mathrm{m}}}{\partial T}\right)_{p,x_i} = \left(\frac{\partial G_{\mathrm{m}}}{\partial T}\right)_{p,y_j} + \sum_j \left(\frac{\partial G_{\mathrm{m}}}{\partial y_j}\right)_{T,p,y_k} \cdot \left(\frac{\mathrm{d}y_j}{\mathrm{d}T}\right)_{p,y_k} \qquad (k \neq j) \tag{2.16}$$

This equation means that the entropy and heat capacity have a contribution from the speciation of the phase; see Fig. 9.8(b) later.

The second derivative of the Gibbs energy with respect to mole fractions is

$$\Omega_{ij} = \left(\frac{\partial^2 G}{\partial x_i\, \partial x_j}\right)_{T,p,x_k} \qquad (k \neq i,\ k \neq j) \tag{2.17}$$

The second derivatives are used for the thermodynamic factor in diffusion calculations and they are also used to calculate the stability function. A phase is stable against fluctuations in its compositions if $\det|\Omega| > 0$. In some systems the stability function changes sign and the locus of $\det|\Omega| = 0$ is called the spinodal. If the phase composition is inside this spinodal, i.e., $\det|\Omega| < 0$, it can lower its Gibbs energy spontaneously by decomposing into two phases with the same structure but different compositions across a miscibility gap. A "composition-set number" is commonly used as a suffix, e.g., bcc#2, to a phase that can be stable with two different compositions at the same equilibrium. In Figs. 2.1(a)–(c) the Gibbs energy, chemical potential, and stability function as functions of the mole fraction of Cr are shown for the miscibility gap in the Cr–Fe system at 600 K, using data from Andersson and Sundman (1987). The phase diagram for Cr–Fe is shown in Fig. 5.4(a) later.

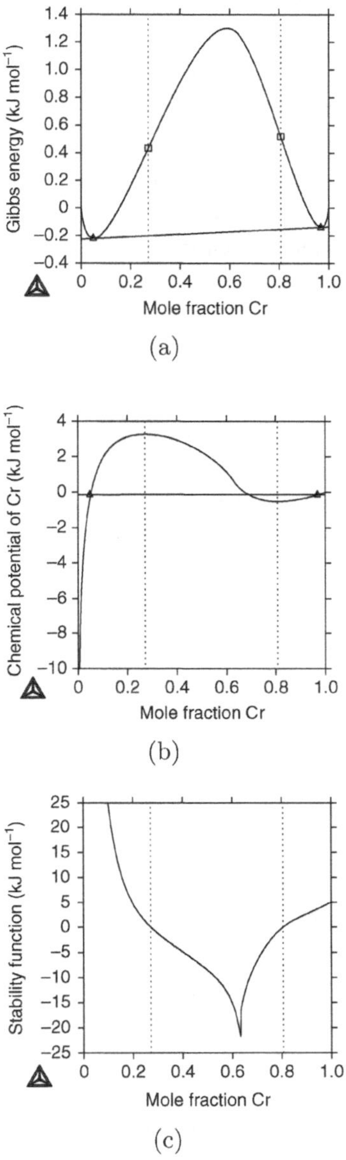

The Gibbs-energy curve for the bcc phase in the Cr–Fe system at 600 K. The stable equilibrium for compositions between the △ symbols (at the common tangent) consists of two bcc phases, one with mainly Fe and the other with mainly Cr. The spinodal points where the stability function is zero are marked with □ symbols on the vertical dashed lines. For compositions between the dashed lines, the bcc phase may decompose spinodally, i.e. without nucleation, to the two equilibrium compositions.

The chemical potential of Cr for the Gibbs-energy curve in (a). The equilibrium compositions are marked with △ symbols and have the same chemical potential for both compositions. The spinodal compositions are indicated by the dashed lines and coincide with the maxima and minima of the chemical-potential curve.

The stability function $\det|\Omega|$ for the Gibbs-energy curve in (a). This curve passes through zero at the spinodal compositions, which are indicated by the dashed lines. The second derivatives of the Gibbs energy are also the "thermodynamic factor" in the diffusion coefficient. The strong cusp in the stability function originates from the ferromagnetic transition.

Figure 2.1 The Gibbs-energy curve for the bcc phase in Cr-Fe at 600 K and its first and second derivatives with respect to composition.

2.2 Crystallography

2.2.1 Connection with thermodynamics

The modeling of thermodynamic functions of solid phases must be done in a manner closely related to the crystal structure, because thermodynamic modeling is the application of statistical thermodynamics and the crystal structure is one of the bases which must be considered in statistical treatment of solid phases. In particular, the sublattices defined in the compound-energy formalism in section 5.5.1 have to be related to crystallographic sublattices. The occupation of crystallographic positions by constituents defined in this formalism can have some physical reality only if it has crystallographic reality.

The terminology of crystallography, insofar as it is useful for the modeling of thermodynamic functions, shall be summarized here. It is used in section 6.2.5.5 and chapter 9. For a deeper review of crystallography, the comprehensive treatment by Ferro and Saccone (1996) is recommended.

2.2.2 Crystal symmetry

In crystalline solids the atoms are arranged in a "**lattice**," which exhibits the same pattern of atoms periodically in three dimensions. Three nonplanar vectors selected in accord with the spatial periodicity define the "**unit cell**." Since sums and differences of these vectors are also periods of the lattice, the unit cell is not uniquely defined. The smallest possible unit cell is called the "**primitive unit cell**."

The primitive unit cell is not uniquely defined either, since a parallelepiped with the vectors $(\vec{a}+\vec{b})$, $\vec{b}$, and $\vec{c}$ has the same volume as a parallelepiped with the vectors $\vec{a}$, $\vec{b}$, and $\vec{c}$. The commonly used unit cell is selected according to symmetry. The periodic lattice usually contains a periodic set of "**symmetry elements**" (rotation axes, screw axes, mirror planes, glide mirror planes, centers of symmetry).

The number of possible combinations of symmetry elements in a periodic lattice is limited to 230 different "**space groups**." The space groups are ordered according to the "**crystal systems**," which are defined in terms of the "highest" symmetry element. Also the coordinate system of the unit cell is chosen according to the crystal system (Table 2.2).

The coordinates given in Table 2.2 do not always allow one to describe a primitive unit cell. In the cubic coordinate system, for example, the vectors $(\vec{a}+\vec{b})/2$, $(\vec{a}+\vec{c})/2$, and $(\vec{b}+\vec{c})/2$ may define a primitive unit cell with one quarter of the volume of the unit cell with three perpendicular equal vectors $\vec{a}$, $\vec{b}$, and $\vec{c}$. This case is called "face-centered cubic" (fcc). After these differences, the 14 "**Bravais lattices**" are distinguished: primitive cubic, body-centered cubic (bcc), face-centered cubic (fcc), primitive tetragonal, body-centered tetragonal (bct), primitive hexagonal, rhombohedral, primitive orthorhombic, body-centered orthorhombic, one-face-centered orthorhombic, all-faces-centered orthorhombic, primitive monoclinic, face-centered monoclinic, and (primitive) triclinic.

The position of an atom inside the unit cell is given by three parameters, x, y, and z, giving fractions of the unit cell parameters a, b, and c, respectively. The symmetry elements of the space group for any position x, y, z produce images at several other positions, called "**equipoints**." For special positions, for example on a mirror plane or on a rotation axis, several of the equipoints coincide. These special positions are distinguished by the Wyckoff notation. The coordinates of all equivalent general and special positions of all space groups are given in the *International Tables for X-Ray Crystallography* (Henry and Lonsdale, 1965) or can be obtained from the website of the International

Table 2.2 *Highest symmetry elements and coordinate systems for the crystal systems*

Crystal system	Symmetry element	Coordinate system
Cubic	4- and 3-fold axes	Three perpendicular equal axes
Tetragonal	One 4-fold axis	Three perpendicular axes, two equal
Hexagonal	One 3- or 6-fold axis	Two equal axes at 120° and a third one perpendicular to them
Orthorhombic	Three perpendicular 2-fold axes	Three unequal perpendicular axes
Monoclinic	One 2-fold axis	Two oblique unequal axes and one perpendicular to them
Triclinic	No symmetry axis	Three unequal oblique axes

Table 2.3 *Example of a space group description, ($C\ 2/m\ 2/c\ 2_1/a$), Cmca, D_{2h}^{18}*

Number of sites	Wyckoff notation	Point symmetry	Equivalent positions 0 0 0; $\frac{1}{2}\,\frac{1}{2}\,0+$			
16	(*g*)	1	$x, y, z;$	$x, \bar{y}, \bar{z};$	$x, \frac{1}{2}-y, \frac{1}{2}+z;$	$x, \frac{1}{2}+y, \frac{1}{2}-z;$
			$\bar{x}, \bar{y}, \bar{z};$	$\bar{x}, y, z;$	$\bar{x}, \frac{1}{2}+y, \frac{1}{2}-z;$	$\bar{x}, \frac{1}{2}-y, \frac{1}{2}+z$
8	(*f*)	*m*	$0, y, z;$	$0, \bar{y}, \bar{z};$	$\frac{1}{2}, y, \frac{1}{2}-z;$	$\frac{1}{2}, \bar{y}, \frac{1}{2}+z$
8	(*e*)	2	$\frac{1}{4}, y, \frac{1}{4};$	$\frac{3}{4}, \bar{y}, \frac{3}{4};$	$\frac{3}{4}, y, \frac{1}{4};$	$\frac{1}{4}, \bar{y}, \frac{3}{4}$
8	(*d*)	2	$x, 0, 0;$	$\bar{x}, 0, 0;$	$x, \frac{1}{2}, \frac{1}{2};$	$\bar{x}, \frac{1}{2}, \frac{1}{2}$
8	(*c*)	$\bar{1}$	$\frac{1}{4}, \frac{1}{4}, 0;$	$\frac{1}{4}, \frac{3}{4}, 0;$	$\frac{1}{4}, \frac{1}{4}, \frac{1}{2};$	$\frac{1}{4}, \frac{3}{4}, \frac{1}{2}$
4	(*b*)	$2/m$	$\frac{1}{2}, 0, 0;$	$\frac{1}{2}, \frac{1}{2}, \frac{1}{2}$		
4	(*a*)	$2/m$	$0, 0, 0;$	$0, \frac{1}{2}, \frac{1}{2}$		

Union of Crystallography (http://www.iucr.org). Another useful website is that of the Bilbao crystallographic server (http://www.cryst.ehu.es).

In Table 2.3, as an example, all the symmetry elements and equipoints of space group Cmca, D_{2h}^{18}, with their Wyckoff notations are given. $\bar{x}$ is used as an abbreviation of $1-x$. There are three different notations for the same space group: the "full Herrmann–Mauguin notation," $C\ 2/m\ 2/c\ 2_1/a$, shows most of the symmetry elements: *C*-centered Bravais lattice, two-fold axes (2) in three directions, one of them a screw axis (2_1), mirror planes (m) and glide planes (c, a) with glide vectors $\vec{c}/2$ and $\vec{a}/2$ perpendicular to the three axes, respectively. The "abbreviated Herrmann–Mauguin notation," Cmca, shows all symmetry elements necessary to identify the space group. The Schönflies symbol, D_{2h}^{18}, is based on the macroscopic symmetry or "point symmetry," D_{2h}, with the counter "18." In the "point symmetry" mirror planes are not distinguished from glide planes, and screw axes and rotation axes are treated as equivalent, i.e. only the macroscopic symmetry is taken into account, neglecting details on the microscopic scale of atomic distances.

The symmetry elements in Henry and Lonsdale (1965) are given by figures and in Table 2.4 all these symmetry elements are tabulated for the space group Cmca.

Table 2.4 *Symmetry elements in space group Cmca,* D_{2h}^{18}

Bravais lattice	Orthorhombic, *C*-centered							
Symmetry-center	$0\,0\,0$	$\frac{1}{2}\,\frac{1}{2}\,0$	$\frac{1}{2}\,0\,\frac{1}{2}$	$0\,\frac{1}{2}\,\frac{1}{2}$	$\frac{1}{2}\,\frac{1}{2}\,\frac{1}{2}$	$0\,0\,\frac{1}{2}$	$0\,\frac{1}{2}\,0$	$\frac{1}{2}\,0\,0$
at $x\ y\ z=$	$\frac{1}{4}\,\frac{1}{4}\,0$	$\frac{3}{4}\,\frac{3}{4}\,0$	$\frac{1}{4}\,\frac{3}{4}\,\frac{1}{2}$	$\frac{3}{4}\,\frac{1}{4}\,\frac{1}{2}$	$\frac{3}{4}\,\frac{3}{4}\,\frac{1}{2}$	$\frac{1}{4}\,\frac{1}{4}\,\frac{1}{2}$	$\frac{3}{4}\,\frac{1}{4}\,0$	$\frac{1}{4}\,\frac{3}{4}\,0$
Two-fold rotational axis, $y\ z=$	$0\,0$	$\frac{1}{2}\,0$	$0\,\frac{1}{2}$	$\frac{1}{2}\,\frac{1}{2}$				
Mirror plane, $x=$	0	$\frac{1}{2}$						
Glide plane, $\vec{b}/2,\ x=$	$\frac{1}{4}$	$\frac{3}{4}$						
Two-fold rotational axis, $x\ z=$	$\frac{1}{4}\,\frac{1}{4}$	$\frac{1}{4}\,\frac{3}{4}$	$\frac{3}{4}\,\frac{1}{4}$	$\frac{3}{4}\,\frac{3}{4}$				
Glide plane, $\vec{c}/2,\ y=$	$\frac{1}{4}$	$\frac{3}{4}$						
Glide plane, $(\vec{a}+\vec{c})/2,\ y=$	0	$\frac{1}{2}$						
Two-fold screw axis, $x\ y=$	$\frac{1}{4}\,0$	$\frac{3}{4}\,0$	$\frac{1}{4}\,\frac{1}{2}$	$\frac{3}{4}\,\frac{1}{2}$	$0\,\frac{1}{4}$	$0\,\frac{3}{4}$	$\frac{1}{2}\,\frac{1}{4}$	$\frac{1}{2}\,\frac{3}{4}$
Glide plane, $\vec{a}/2,\ z=$	$\frac{1}{4}$	$\frac{3}{4}$						
Glide plane, $\vec{b}/2,\ z=$	$\frac{1}{4}$	$\frac{3}{4}$						

2.2.3 *Crystal structures*

2.2.3.1 Definition

A "**crystal structure**" is described by giving the space group, the lattice parameters (lengths of axes of the unit cell and angles between them), the number of constituents per unit cell, the coordinates of one site of each set of equivalent positions, and the type of constituent occupying each such site. An example is given in Table 2.5.

The crystal structures form families of (more or less closely) related structures. The main argument of relationship is similarity of the geometry of the coordination of all the atoms by other (like and unlike) atoms. For binary metallic phases a good classification is given in Schubert (1964). A more recent review of crystal-structure classification is the series of three papers by Daams *et al.* (1992) and Daams and Villars (1993, 1994). It has to be mentioned, however, that the symmetry (space group, Bravais lattice, etc.) does *not* show these relationships. It may be different for closely related structures and identical for very different structures.

There exists computer software that can visualize the structures and make understanding of the crystal structure and its use for thermodynamic modeling much easier.

Table 2.5 *An example of a crystal-structure description*

$PdSn_3$-type (Schubert *et al.* 1959)		Pearson symbol: oC32	
Space group:	Cmca	(equivalent to Bbam)	
Lattice parameters:	$a=1717\,\text{pm}$,	$b=646\,\text{pm}$,	$c=649\,\text{pm}$
8 Pd in 8(d)	$x\,y\,z=0.084$	0	0
8 Sn in 8(f)	$x\,y\,z=0$	0.17	0.33
16 Sn in 16(g)	$x\,y\,z=0.168$	0.33	0.17

2.2.3.2 Nomenclature of crystal structures

Several nomenclature systems for the identification of crystal structures are in use. The "**prototype**" is the name of the phase for which this crystal structure was first determined. The "**Strukturbericht**" notation gave letters and counters to crystal structures in the sequence in which they were classified. This classification stopped around 1950 and thus this nomenclature does not cover all the crystal structures determined later.

The "**Pearson symbol**" is a short description of the structure, denoting the Bravais lattice by two letters (c, t, h, o, m, a for cubic, tetragonal, hexagonal, orthorhombic, monoclinic or triclinic; P, I, F, C, R for primitive, body-centered, face-centered, *C*-centered or rhombohedral) and the number of atoms in the unit cell. Since these two specifications may be the same for different crystal structures, this nomenclature is not unique, but reflects explicitly the most important details of the crystal structure.

Mineral names very often are uniquely connected with a crystal structure and used as a name for this structure type, like "spinel."

In some thermodynamic databases the phase names used for several important crystal structures do not belong to any of the above nomenclature systems. Some of these names are ambiguous. The names fcc and bcc were initially designated for Bravais lattices (face-centered cubic and body-centered cubic), but they are very commonly used as names for the crystal structures denoted by prototype, Strukturbericht, and Pearson symbols as Cu-type, A1, cF4, and W-type, A2, cI2, respectively. Even derivatives of these structures are sometimes denoted by these two names, although they no longer have the corresponding Bravais lattice. For example "ordered bcc" is used for the CsCl-type (B2, cP2) and $FeAl_3$-type (cF16) structures having cubic primitive or face-centered cubic Bravais lattices, respectively; "ordered fcc" is in use for the CuAu-type ($L1_0$, tP2) and Cu_3Au-type ($L1_2$, cP4) structures having tetragonal or cubic primitive Bravais lattices respectively.

The nomenclature systems are compared for a few important crystal structures in Table 2.6. A more detailed table is given in Westbrook and Fleischer (1995).

2.2.3.3 Sublattice modeling

The crystal structure of a phase is very important for modeling its Gibbs energy by statistical thermodynamics, for example in the compound energy formalism (see section 5.5.1). All atoms in equivalent positions have the same coordination of neighboring atoms. Therefore each set of equivalent positions (sites belonging to the same Wyckoff position) may be treated as a "**sublattice**." This means that, if one of these atoms can be substituted by another one, the whole set of equivalent positions can be randomly substituted. To simplify the thermodynamic modeling, several sets of equivalent positions with similar coordination may be combined and treated as a single sublattice. The contrary, however, namely treating atoms of the same set of equivalent positions differently in the compound energy formalism, is usually not allowed. The chemical ordering, to be described in section 2.2.3.4, does not contradict this rule, since in the ordered

Table 2.6 *Examples for the nomenclature of crystal structures*

Prototype	Struktur-bericht	Pearson symbol	Mineral names	Other names
Cu	A1	cF4		fcc (austenite)
W	A2	cI2		bcc (ferrite)
Mg	A3	hP2		hcp
NaCl	B1	cF8	Halite, Periclase (MgO)	
CsCl	B2	cP2		Ordered bcc
Cu_3Au	$L1_2$	cP4		Ordered fcc
CuAu	$L1_0$	tP2		Ordered fcc
Ni_3Sn	$D0_{19}$	hP8		Ordered hcp
CaF_2	C1	cF12	Fluorite	
$MgCu_2$	C15	cF24		(Cubic) Laves phase
$MgZn_2$	C14	hP12		(Hexagonal) Laves phase
$MgNi_2$	C36	hP24		Laves phase
Diamond	A4	cF8	Diamond	

structure the symmetry is diminished, splitting the set of equivalent positions into different ones having also different Wyckoff notations in the space group of the ordered structure. Inside these Wyckoff positions the substitution has to be treated as identical for all sites.

There is only one exception to this rule: if several atoms of the set belong to a molecule, substituting one of them changes the molecule and may prohibit the substitution of the remaining atoms. For example, if the four O atoms of a SO_4^{2-} ion in a sulfate crystal structure are on equivalent positions, substituting one of them by S gives a thiosulfate ion SO_3S^{2-}. Random substitution of the O atoms by S atoms would yield ions with two O atoms substituted, $S(S_2O_2)^{2-}$, which do not exist. To describe a solid solution of sulfate with thiosulfate in the compound energy method, the positions of the whole anions SO_4^{2-} and $S(SO_3)^{2-}$ should be treated as single positions forming one sublattice. Similar situations may occur, but less obviously, in other crystal structures with homopolar bondings, such as silicates, carbides, nitrides etc.

2.2.3.4 Chemical ordering

Another feature of crystallography needed in modeling of Gibbs energies is the description of order–disorder transitions, also called superstructure formation. Ordering means in general that a set of equivalent positions, occupied randomly by different atoms, splits into two or several different sets of positions, each preferably occupied by one kind of the different atoms. This is possible only with loss of some of the symmetry elements of the space group that in the disordered state provide equivalence of the positions. This loss of symmetry elements may happen in the following ways.

- Without changing the unit cell, just losing sets of mirror planes or glide planes. An example is hcp with interstitials, space group $P6_3/mmc$, with two metal atoms in Wyckoff position 2(c), and two octahedral voids in position 2(a), which can be occupied by interstitial atoms. Neither position has any variable parameter. By omitting a set of mirror planes and a set of glide planes the symmetry diminishes to space group P$\bar{3}$m1, whereby the position of the octahedral voids splits into the two independent positions 1(a) and 1(b). The position of the metal atoms in this space group has Wyckoff notation 2(d) with a variable parameter x. If $x > 0.25$, the voids in position 1(a) become larger than those in 1(b) and may be preferably occupied by the interstitial atoms, thus reducing the strain energy. The structure now is called Cr_2C-type.

- Diminishing the symmetry may be accompanied by going to a lower crystal system (cubic → tetragonal, hexagonal → orthorhombic, hexagonal → monoclinic, . . .). Examples are the CuAu structure and martensite, in which the ordered phase is tetragonal, but derived from a cubic disordered phase.
- The primitive unit cell of the ordered phase is composed of several primitive unit cells of the disordered phase. The simplest example is the CsCl-type ordering, in which the cubic unit cell is identical with the primitive unit cell of the ordered phase, but contains two primitive unit cells of the disordered W-type (bcc) phase, although the commonly used cubic unit cell is the same in the ordered and disordered states. A second step of ordering of this type gives the $MnCu_2Al$-type (Heussler phase) or Fe_3Al-type, in which the commonly used cubic unit cell (fcc) contains eight unit cells of the CsCl-type, i.e., the primitive unit cell is twice as large as that in the CsCl-type and four times that of the W-type.

A modeling of chemical order, using some simplifications proposed by Bragg and Williams (1934), is described in section 5.8.4.

2.3 Equilibrium calculations

2.3.1 *Minimizing the Gibbs energy*

In section 2.1.6 various equilibrium conditions were described. Each of these conditions can be used to calculate an equilibrium, but the most commonly used method is to find the minimum of the Gibbs energy G at constant values of temperature T, pressure p, and amounts of components N_i. Instead of amounts of the components one may use the mole fractions, $x_i = N_i/N$, of all but the major component and the total amount $N = \sum N_i$. The usage of these conditions for a calculation needs an analytical expression of the molar Gibbs energy for each phase α, G_{m}^{α}, as a function of these and sometimes also other internal variables such as the fraction of molecules or site occupancies, generally called site fractions or constituent fractions.

The Gibbs energy may have several minima compatible with these conditions. That with the most negative value of G is the "global minimum" and corresponds to the "stable equilibrium"; the other ones are "local minima" and correspond to "metastable equilibria." They differ in terms of the phases and their compositions present at equilibrium, sometimes only by the compositions of the same set of phases.

The above conditions do not yet specify which and how many phases are present. Analytical expressions of Gibbs energies, however, exist only for phases and thus an equilibrium calculation can be performed only for selected sets of phases. Thermo-Calc uses the "driving force," see section 2.3.6, to search automatically to determine whether another set of phases has a minimum with a lower value of the Gibbs energy. In BINGSS it is assumed that experimental equilibrium data are reliable only if the phases present are known. Thus, the calculated value to be compared with an experimental one must be an equilibrium involving just the phases present in the experiment. Of course, finally it must be checked whether, by extrapolation of a phase description, this phase appears to be stable in areas where experimentally it was found not to be stable.

In chapter 5 analytical expressions for integral molar Gibbs energies of phases, G_{m}^{α}, are given. The total Gibbs energy is derived from these expressions by summing them up, multiplied by the fractions or amounts of the phases, m^{α}:

$$G = \sum_{\alpha} m^{\alpha} \cdot G_{\mathrm{m}}^{\alpha} \tag{2.18}$$

with the constraint $m^{\alpha} \geq 0$.

The analytical expressions of the molar Gibbs energies of the phases are expressed as functions of T, p, and either the mole fractions x_i^{α} of the components i in this phase α, or the fractions $y_k^{(l,\alpha)}$ of constituents k on sublattices l of phase α (constituent fractions). The mole fractions x_i^{α} are definite functions of the site fractions $y_k^{(l,\alpha)}$, but the inverse relation may contain additional variables. The amounts of phases, m^{α}, are definite functions of the mole fractions x_i^{α} and the amounts of components $N_i = N \cdot x_i^0$, expressed by the (generalized) lever rule. The lever rule states that the amount of each stable phase in the equilibrium multiplied by the mole fraction of the component in that phase, summed over all stable phases, must be equal to the overall fraction of the component:

$$x_i = \sum_{\alpha} m^{\alpha} x_i^{\alpha} \tag{2.19}$$

The equilibrium condition may now be written

$$\min(G) = \min\left(\sum_{\alpha} m^{\alpha} G_{\mathrm{m}}^{\alpha}(T, p, x_i^{\alpha} \text{ or } y_k^{(l,\alpha)})\right) \tag{2.20}$$

The variables m^{α} and x_i^{α} or $y_k^{(l,\alpha)}$ are usually the unknowns, the values of which have to be calculated from the conditions of the energetic minimum.

2.3.2 *Equilibrium conditions as a set of equations*

The minimum of Eq. (2.20) is found either by hill-climbing methods or by setting the derivatives with respect to the unknowns to zero, but keeping in mind the constraints interrelating these variables. This set of nonlinear equations is solved for the unknowns by an appropriate iteration algorithm.

For the latter method Gibbs gave a formulation in which the constraints relating m^α, x_i^α, and N_i were used to eliminate the variables m^α and N_i:

$$G_i^\alpha(T, p, x_i^\alpha) = G_i^\beta(T, p, x_i^\beta) \qquad (i = 1, \ldots, c,\ \alpha = 1, \ldots, p-1,\ \beta = \alpha+1, \ldots, p) \tag{2.21}$$

In equilibrium the partial Gibbs energies, defined in Eq. (2.13), of all components i in each pair of different phases α and β are equal. A variation of this formulation of the equilibrium conditions is that the partial Gibbs energies G_i^α of all phases are identical to the equilibrium chemical potentials μ_i, which are defined as the intensive variables conjugated to the extensive variables N_i. By their definition as derivatives of the total Gibbs energy with respect to the N_i, they are defined as partial Gibbs energies of the whole system:

$$G_i^\alpha(T, p, x_i^\alpha) = \mu_i \qquad (i = 1, \ldots, c,\ \ \alpha = 1, \ldots, p) \tag{2.22}$$

Besides the x_i^α in Eq. (2.22), also the μ_i now belong to the unknowns to be calculated from the minimization of the total Gibbs energy.

The formulations Eq. (2.21) and, especially, Eq. (2.22) are very useful, if the G^α are given as functions of the x_i^α, because then all relevant constraints are already included in the equations. For a stoichiometric phase the equations concerning this phase must be modified, which will be discussed later.

If the G_m^α are given as functions of site fractions $y_k^{(l,\alpha)}$ instead of mole fractions x_i^α, however, Eqs. (2.22) are difficult to use, because then the constraints interrelating the x_i^α and the $y_k^{(l,\alpha)}$ have to be considered. Eriksson (1971) and Hillert (1981) used the Lagrange-multiplier method to derive a formulation whereby all the variables in Eq. (2.20) are treated as independent. There are three types of constraints: (1) the total amount of each component, N_i, has to be kept constant; (2) the site fractions in each sublattice sum up to unity; and (3) the charges of ionic species sum up to zero in each phase:

$$\sum_\alpha m^\alpha \sum_l a^{(l)} \sum_k b_{k,i}^{(\alpha,l)} \cdot y_k^{(\alpha,l)} - N_i = 0 \qquad \text{(all components)} \tag{2.23}$$

$$\sum_k y_k^{(\alpha,l)} - 1 = 0 \qquad \text{(all sublattices)} \tag{2.24}$$

$$\sum_l a^{(l)} \sum_k \nu_k^{(\alpha,l)} \cdot y_k^{(\alpha,l)} = 0 \qquad \text{(all phases with charged species)} \tag{2.25}$$

Each constraint is multiplied by a "Lagrange multiplier," and added to the total Gibbs energy in Eq. (2.20) to get a sum L. If all the constraints are satisfied, L is equal to G and a minimum of L is equivalent to a minimum of the total Gibbs energy G. The $b_{k,i}^{(\alpha,l)}$ are the stoichiometric numbers of component i in species k on sublattice l of phase α, the $a^{(l)}$ are the fractions of sites of sublattice l referred to all sites of the phase, and the $\nu_k^{(\alpha,l)}$ are the charges of ionized species k on sublattice l of phase α. The Lagrange multipliers are treated as additional unknowns. The derivatives of L with respect to the Lagrange multipliers yield the constraints, but the derivatives of L with respect to the m^α and $y_k^{(l,\alpha)}$ are now all independent. The Lagrange multipliers of constraints in Eq. (2.23) were shown to be identical to the chemical potentials μ_i (Hillert 1981). The Lagrange multipliers of Eq. (2.24) and Eq. (2.25) are called $\lambda^{(\alpha,l)}$ and λ_e^α, respectively. Some of the resulting equations may be simplified by replacing the unknowns $\lambda^{(\alpha,l)}$ and λ_e^α by $\pi^{(\alpha,l)} = \lambda^{(\alpha,l)}/m^\alpha$ and $\pi_e^\alpha = \lambda_e^\alpha/m^\alpha$, respectively. It was shown that these quotients remain finite, even when m^α approaches zero (Lukas *et al.* 1982).

Setting the derivatives of L with respect to the unknowns to zero yields the following set of equations as equilibrium conditions additionally to Eqs. (2.23), (2.24), and (2.25):

$$\frac{\partial G_m^\alpha}{\partial y_k^{(\alpha,l)}} + \sum_{i=1}^{c} \mu_i \cdot a^{(l)} \cdot b_{k,i}^{(\alpha,l)} + \pi^{(\alpha,l)} + \pi_e^\alpha \cdot a^{(l)} \cdot \nu_k^{(\alpha,l)} = 0 \quad \text{(all species)} \tag{2.26}$$

$$G_m^\alpha - \sum_i \mu_i \sum_l a^{(l)} \sum_k b_{k,i}^{(\alpha,l)} \cdot y_k^{(\alpha,l)} = 0 \quad \text{(all phases)} \tag{2.27}$$

The set of equations (2.23)–(2.27) is a set of nonlinear equations with the unknowns m^α, $y_k^{(l,\alpha)}$, and the Lagrange multipliers μ_i, $\pi^{(\alpha,l)}$, and π_e^α as well as the variables T, p, and N_i. The latter values have to be kept constant during searching for the minimum, meaning that no derivatives of the total Gibbs energy or of the function L with respect to these variables are formed, but nevertheless they may be treated as unknowns according to the question "at which (constant) values of T, p, and N_i does there exist a solution for the equilibrium conditions?" The number of independent equations must equal the number of unknowns. Thus, if any of T, p, and N_i are treated as unknowns, some other conditions must be set, which are explained in section 2.3.5.

These equilibrium conditions are used in Thermo-Calc. In the following text they will be referred to as "Hillert's equilibrium conditions."

If the molar Gibbs energies of the phases are expressed as functions of the mole fractions x_i^α without considering constituent fractions, as for example in substitutional solutions, Hillert's equilibrium conditions reduce to

$$\frac{\partial G_m^\alpha}{\partial x_i^\alpha} + \mu_i + \pi^\alpha = 0 \quad \text{(all components in each phase)} \tag{2.28}$$

$$G_m^\alpha - \sum_i \mu_i \cdot x_i^\alpha = 0 \quad \text{(all phases)} \tag{2.29}$$

$$\sum_i m^\alpha \cdot x_j^\alpha - N_i = 0 \quad \text{(all components)} \tag{2.30}$$

$$\sum_i x_i^\alpha - 1 = 0 \quad \text{(all phases)} \tag{2.31}$$

Multiplying the i Eqs. (2.28) for phase α by x_i^α and subtracting them from Eq. (2.29) results (keeping in mind that $\sum_i x_i^\alpha = 1$) in

$$G_m^\alpha - \sum_j x_i^\alpha \cdot \frac{\partial G_m^\alpha}{\partial x_i^\alpha} - \pi^\alpha = 0 \quad \text{(all phases)} \tag{2.32}$$

This relation identifies the variable π^α with $G_m^\alpha - \sum_i x_i^\alpha \cdot (\partial G_m^\alpha / \partial x_i^\alpha)$ (Hillert 1981). On replacing π^α in Eq. (2.28) by this expression and using Eq. (2.15), one can obtain Eq. (2.22).

The set of equations (2.22), (2.30), and (2.31) is a set of nonlinear equations alternative to "Hillert's equilibrium conditions" if the molar Gibbs energies are expressed as functions of the mole fractions x_i^α. In the following, this set will be referred to as "modified Gibbs equilibrium conditions." They are used in BINGSS, BINFKT, TERGSS, and TERFKT.

The further modification of this set in the case of a stoichiometric phase α is now easy. Since the mole fractions of a stoichiometric phase are not unknown, Eqs. (2.28) are not used and the equation for phase α from the set (2.29) is used, because here it is not used up to eliminate π^α.

For semi-stoichiometric phases, equations intermediate between Eq. (2.22) and Eq. (2.29) may be formulated. As an example the two equations replacing Eqs. (2.28) and (2.29) for a line compound with constant mole fraction of A in a ternary system are

$$G_\mathrm{m}^\alpha - x_\mathrm{C}^\alpha \cdot \left(\frac{\partial G_\mathrm{m}^\alpha}{\partial x_\mathrm{B}^\alpha} - \frac{\partial G_m^\alpha}{\partial x_\mathrm{C}^\alpha} \right) - (x_\mathrm{B}^\alpha + x_\mathrm{C}^\alpha) \cdot \mu_\mathrm{B} - x_\mathrm{A}^\alpha \cdot \mu_\mathrm{A} = 0 \tag{2.33}$$

$$G_\mathrm{m}^\alpha - x_\mathrm{B}^\alpha \cdot \left(\frac{\partial G_\mathrm{m}^\alpha}{\partial x_\mathrm{C}^\alpha} - \frac{\partial G_\mathrm{m}^\alpha}{\partial x_\mathrm{B}^\alpha} \right) - (x_\mathrm{C}^\alpha + x_\mathrm{B}^\alpha) \cdot \mu_\mathrm{C} - x_\mathrm{A}^\alpha \cdot \mu_\mathrm{A} = 0 \tag{2.34}$$

In Thermo-Calc the equations are not modified with respect to the model of the phase, and, for example, if a quasibinary section of a ternary system is considered, the "extra" degree of freedom must be removed by assigning an arbitrary activity to the component that is outside the quasibinary section. For example, in the ternary system Ca–Si–O, CaO–SiO_2 is a quasibinary system and, with the components CaO and SiO_2, the activity of O_2 can be set to an arbitrary value, since all phases in the quasibinary system have their oxygen contents determined already by the Ca : Si ratio.

In "Hillert's conditions" as well as in the "modified Gibbs equilibrium conditions" the number of unknowns is larger than the number of equations by $c+2$, where c is the number of components, because, for each unknown, except the $c+2$ ones N_i, T, and p, there is just one equation. This means that the number of degrees of freedom is $c+2$, which looks like a contradiction to Gibbs' phase rule insofar as the number of phases is not mentioned. The explanation is that the phase rule is derived from the set of equations (2.21), from which the variables m^α and N_i have been eliminated and a free change of these variables is not counted as a degree of freedom. For example, a two-phase field in an isobaric binary system is treated in the phase rule as monovariant. Nevertheless, after fixing the single variable T, the N_i still may be independently changed, yielding different m^α. The success of the phase rule shows that ignoring the variables m^α and N_i in counting the number of degrees of freedom has its advantage. In the above sets this can be done by ignoring the c equations (2.23) or (2.30), respectively, which are the only ones containing the $p+c$ variables m^α and N_i. On losing $c+p$ variables and c equations, the number of degrees of freedom changes to $c-p+2$, in agreement with the phase rule. This is also a reason why in Eq. (2.26), contrary to Hillert (1981), $\lambda^{(i,l)}$ and $\lambda_\mathrm{e}^\alpha$ are replaced by $\pi^{(i,l)}$ and π_e^α in order to eliminate m^α from these equations.

To calculate a state of equilibrium, for each degree of freedom an independent condition must be added to the set of equations, reducing the number of degrees of freedom to zero. These additional conditions and their purposes will be discussed in section 2.3.5. In section 2.3.3 the solution of these sets of equations is described.

2.3.3 The Newton–Raphson method

Several methods for solving the sets of equations described in the previous section with respect to the unknowns are known. In Thermo-Calc, as well as in BINGSS/BINFKT and TERGSS/TERFKT, the Newton–Raphson iteration method is used, so this method alone shall be described here.

It is a generalization of Newton's method for finding the abscissa, where the value of a function is zero (Fig. 2.2).

At a starting point x_0 the function $f(x)$ and its derivative $\mathrm{d}f/\mathrm{d}x$ are calculated. The intersection between the tangent to the function and the x-axis is taken as the next iteration point and the procedure is repeated until $f(x)$ is below a preselected limit ϵ. The $(i+1)$th iteration step, x_{i+1}, is calculated from the previous one, x_i, by

$$\left(\frac{\mathrm{d}f}{\mathrm{d}x}\right)_{x=x_i} \cdot \Delta x_i = -f(x_i); \qquad x_{i+1} = x_i + \Delta x_i \tag{2.35}$$

Figure 2.2 gives the impression that Newton's method converges very rapidly. Usually that is the case, but there exist also cases in which the method diverges. This is explained in Fig. 2.3 with the function $f_1(x) = \tanh(x)$. Starting at x_1 yields a step far to the right and the following step would be much further to the left. Starting at x_2, however, would lead to convergence.

The function $f_2(x)$ in Fig. 2.3 has different solutions for x for the condition $f_2(x) = 0$. Which of them is found depends on the starting value of x. Starting with x_3 will result

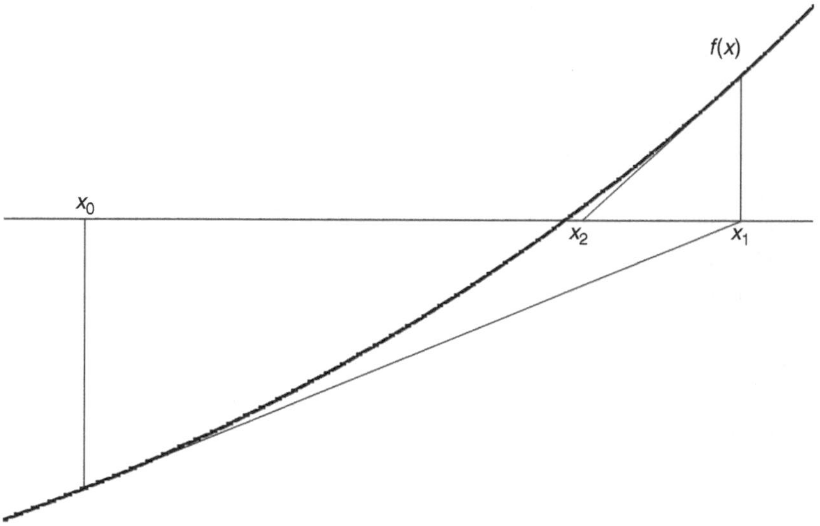

Figure 2.2 Newton's method.

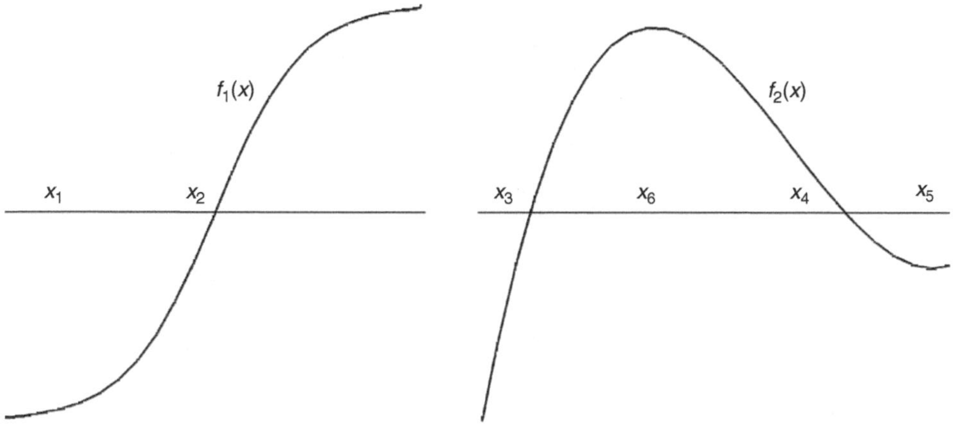

Figure 2.3 Divergence, or different solutions, of Newton's method.

in the solution on the left and starting with x_4 will give the solution on the right. x_5 will result in a first step far to the left and will finally result in the solution on the left. x_6 may give a first step to the vicinity of x_5 and thus also will result in the solution on the left. This illustrates that the influence of the starting value on the result may be complicated.

The Newton–Raphson method is the extension of Newton's method to more than one variable. If, for example, the values of the variables x and y are to be determined, where two functions, $f_1(x, y)$ and $f_2(x, y)$, are zero, Eq. (2.35) is replaced by the following two equations:

$$\left(\frac{\partial f_1}{\partial x}\right)_{x=x_i,y=y_i} \cdot \Delta x_i + \left(\frac{\partial f_1}{\partial y}\right)_{x=x_i,y=y_i} \cdot \Delta y_i = -f_1(x_i, y_i)$$

$$\left(\frac{\partial f_2}{\partial x}\right)_{x=x_i,y=y_i} \cdot \Delta x_i + \left(\frac{\partial f_2}{\partial y}\right)_{x=x_i,y=y_i} \cdot \Delta y_i = -f_2(x_i, y_i) \tag{2.36}$$

$$x_{i+1} = x_i + \Delta x_i; \qquad y_{i+1} = y_i + \Delta y_i$$

The method is extended similarly to a set of n equations with n unknowns.

2.3.4 *Global minimization of the Gibbs energy*

All iterative techniques like the Newton–Raphson one need an initial constitution for each phase in order to find the minimum of the Gibbs energy surface for the given conditions. If a phase has a miscibility gap and its initial constitution is on the "wrong" side of the miscibility gap, the phase may be metastable when it should be stable or may become stable with the wrong composition; see Fig. 2.4.

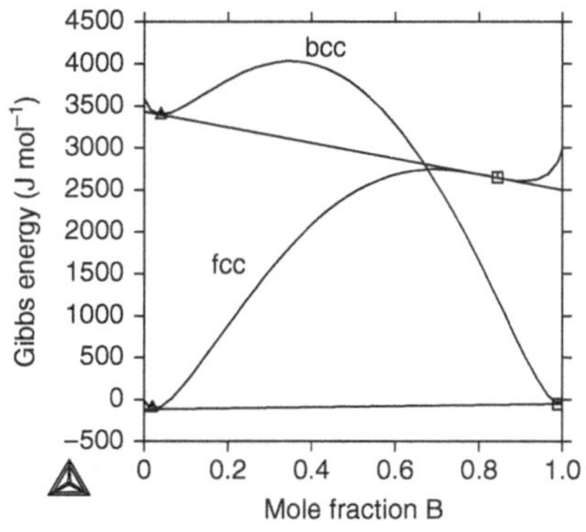

Figure 2.4 The Gibbs-energy curves for two phases with miscibility gaps. If the initial constitution of the phases is on the "wrong" side of the miscibility gap when starting the search for the equilibrium, one may find the metastable equilibrium represented by the upper tangent, whereas the stable equilibrium is represented by the lower tangent.

Miscibility gaps can occur only in solution phases. For a system in which all phases are compounds with fixed compositions, the equilibrium set of phases is given by the tangent "hyperplane" defined by the Gibbs energies of a set of compounds constrained by the given overall composition and with no compound with a Gibbs energy below this hyperplane. There must be exactly as many compounds in this set as there are components in the system and this equilibrium is "global."

There are several minimization techniques that can be used to find the global equilibrium (Connolly and Kerrick 1987, Chen *et al.* 1993) from a set of compounds; one such has been implemented in Thermo-Calc, also for systems with solution phases. In this case the Gibbs energy surface of all solution phases is approximated with a large number of "compounds," each of which has the same Gibbs energy as the solution phase at the composition of the compound. The compositions of the compounds are selected to cover the whole surface in a reasonably dense grid, often more than 10^4 compounds are used for each phase. In a multicomponent system with many solution phases, this may generate more than 10^6 such compounds. A search for the hyperplane representing the equilibrium for the compounds is then carried out. When this has been found, the compounds in this equilibrium set must be identified with regard to which solution phase they belong to; many may be inside the same phase and Thermo-Calc may then automatically create new composition sets. Each compound in the equilibrium set gives an initial constitution of the solution phase, which can be used in a Newton–Raphson calculation to find the equilibrium for the solution phases. This technique of approximating solution phases with compounds can thus be considered as a method by which to obtain good initial constitutions for the Newton–Raphson method and avoid the risk of starting from a constitution of a phase that is on the "wrong" side of a miscibility gap.

The main limitation of the "global" method is that the conditions for the equilibrium must be such that T, p, and the overall composition are known. If other conditions are used, for example the value of the activity of a component, that a phase should be stable, or that a composition of a phase is known rather than the overall composition, then it is not possible to calculate the global equilibrium directly. However, after an equilibrium calculation according to a Newton–Raphson technique, the overall composition can be calculated and used in a global minimization to check whether the result is correct; if it is not, one can use the global minimum as a starting point for a new equilibrium calculation.

2.3.5 Conditions for a single equilibrium

The equilibrium conditions described above as sets of equations contain fewer equations than unknowns. The difference is the number of degrees of freedom f, so f extra equations (conditions) must be added in order to select definitely a single equilibrium. This corresponds, for example, to the selection of a single tie-line from a monovariant binary two-phase equilibrium. In the optimization procedure the "calculated value" corresponding to a measured one is selected in this way.

One means of selection is by fixing some of the unknowns to given values. This may be counted as diminishing the number of unknowns, but formally it can also be counted as adding an equation of the type "unknown state variable" = "constant value," for example

$$T = 1273; \qquad p = 101\,325; \qquad x_i = 0.1; \qquad \mu_i = -40\,000 \tag{2.37}$$

With the constraints Eqs. (2.23)–(2.27) the simplest additional conditions are to fix temperature T, pressure p, and the amounts of all components N_i. For each calculation step, however, it must be selected which and how many phases are present, since Gibbs-energy descriptions exist only for phases. Calculation steps with different sets of phases may be compared. The phase set with the lowest Gibbs energy describes the stable equilibrium.

If the equilibrium is calculated with a given set of phases, the equilibrium μ-values describe a simplex (straight line in binary, plane in ternary, hyperplanes in higher-order systems), as shown in the next section. Whenever another phase has a positive "driving force," see section 2.3.6, there exists an equilibrium containing this phase, which is more stable than that calculated for the selected set of phases. In the next calculation step this phase must be added. If the calculation finds a negative amount for one of the selected phases, this phase must be removed from the selected set. Phases with miscibility gaps may have more than one driving force at different compositions and this test must be performed separately for each of these compositions.

In Thermo-Calc this comparison of equilibria of different sets of phases is done automatically. As initial values an arbitrary set of phases is selected, either by explicitly setting starting values or by using "automatic starting values." In BINGSS and BINFKT the set of phases is selected by the user. Experimental data with undefined phases (for example, a liquidus temperature with undefined primarily crystallizing solid phase) thus cannot be used in BINGSS. For the calculated value to be compared with an experimental one, the phases must thus be determined. A check is necessary only if, in the finally assessed dataset, some phases appear to be stable where they were experimentally found not to be stable.

In the set of equations (2.22) the phase amounts and amounts of components are eliminated. The same is true for the set (2.24)–(2.27) without (2.23). The degree of freedom is that defined by Gibbs' phase rule and thus depends on the number of stable phases. The elimination of the amounts of components includes the elimination of an overall composition from the equations.

Instead of just fixing single unknowns, other conditions may give relationships involving several unknowns. By the use of such conditions an overall composition may be re-introduced into the set of equations (2.22).

A point of the boundary between the α field and the $\alpha+\beta$ field in the isopleth through a ternary system may be selected by adding one of the following two sets of two equations to the six equations of the form of Eq. (2.22) for finding the eight values T, $\mu_{\mathrm{A}}^{\mathrm{Eq}}$, $\mu_{\mathrm{B}}^{\mathrm{Eq}}$, $\mu_{\mathrm{C}}^{\mathrm{Eq}}$, x_{B}^{α}, x_{C}^{α}, x_{B}^{β}, and x_{C}^{β} describing the isobaric $\alpha+\beta$ equilibrium,

$$x_{\mathrm{B}}^{\alpha} = x_{\mathrm{B}}^{0}; \qquad x_{\mathrm{C}}^{\alpha} = x_{\mathrm{C}}^{0} \tag{2.38}$$

or

$$T = T^0; \qquad \begin{vmatrix} x_{\mathrm{A}}^{\alpha} & x_{\mathrm{B}}^{\alpha} & x_{\mathrm{C}}^{\alpha} \\ x_{\mathrm{A}}^{(1)} & x_{\mathrm{B}}^{(1)} & x_{\mathrm{C}}^{(1)} \\ x_{\mathrm{A}}^{(2)} & x_{\mathrm{B}}^{(2)} & x_{\mathrm{C}}^{(2)} \end{vmatrix} = 0 \tag{2.39}$$

Equations (2.38) fix the composition of α to x_{B}^{0}, x_{C}^{0}, which, of course, must be on the isopleth, whereas Eqs. (2.39) fix the temperature and give the condition that the composition of α is on the straight line defined by the two points (1) and (2), which is the abscissa of the isopleth. Thus the abscissa is given by Eqs. (2.38) and the ordinate calculated; with Eqs. (2.39) the ordinate is given and the abscissa calculated. Therefore Eqs. (2.38) are valid for the calculation of a steep boundary, whereas Eqs. (2.39) are well suited to calculate flat boundaries. In BINGSS and BINFKT this can be selected by the user. Thermo-Calc does that automatically.

The condition of three points being on a straight line can be used in BINGSS and BINFKT to find the single point where phase α of a ternary three-phase equilibrium passes through the isopleth defined by points (1) and (2),

$$\begin{vmatrix} x_{\mathrm{A}}^{\alpha} & x_{\mathrm{B}}^{\alpha} & x_{\mathrm{C}}^{\alpha} \\ x_{\mathrm{A}}^{(1)} & x_{\mathrm{B}}^{(1)} & x_{\mathrm{C}}^{(1)} \\ x_{\mathrm{A}}^{(2)} & x_{\mathrm{B}}^{(2)} & x_{\mathrm{C}}^{(2)} \end{vmatrix} = 0 \tag{2.40}$$

or to find an azeotropic maximum or minimum of a three-phase equilibrium $\alpha+\beta+\gamma$,

$$\begin{vmatrix} x_{\mathrm{A}}^{\alpha} & x_{\mathrm{B}}^{\alpha} & x_{\mathrm{C}}^{\alpha} \\ x_{\mathrm{A}}^{\beta} & x_{\mathrm{B}}^{\beta} & x_{\mathrm{C}}^{\beta} \\ x_{\mathrm{A}}^{\gamma} & x_{\mathrm{B}}^{\gamma} & x_{\mathrm{C}}^{\gamma} \end{vmatrix} = 0 \tag{2.41}$$

or to find in a ternary isothermal section the $\alpha+\beta$ tie-line passing through the point x_{B}^{0}, x_{C}^{0}:

$$\begin{vmatrix} x_{\mathrm{A}}^{\alpha} & x_{\mathrm{B}}^{\alpha} & x_{\mathrm{C}}^{\alpha} \\ x_{\mathrm{A}}^{\beta} & x_{\mathrm{B}}^{\beta} & x_{\mathrm{C}}^{\beta} \\ x_{\mathrm{A}}^{0} & x_{\mathrm{B}}^{0} & x_{\mathrm{C}}^{0} \end{vmatrix} = 0 \tag{2.42}$$

Thermo-Calc does not support these nonlinear conditions between mole fractions. Nevertheless, the same equilibria can be calculated. The first one is found during "mapping" of the isopleth and can be picked out from tabulated output of this mapping. The second is obtained by "stepping" one mole fraction of one phase of the three-phase equilibrium across this azeotropic extremum and selecting the equilibrium with maximal or minimal temperature, respectively, in the tabulated output. The third one is just the equilibrium at the fixed overall composition (x_{A}^{0}, x_{B}^{0}, x_{C}^{0}).

Generally, boundaries in an isopleth can be found as "zero-phase-fraction" lines. The boundary between the $\alpha+\beta+\gamma$ and $\beta+\gamma$ fields is found by searching for the $\alpha+\beta+\gamma$ three-phase equilibrium using Eqs. (2.23)–(2.27) with the condition $m^{\alpha}=0$. The N_j in Eq. (2.23) must be selected in such a way as to define a point on the abscissa of the isopleth; then T, the ordinate of the isopleth, can be calculated or vice versa.

Maxima and minima of binary two-phase fields (azeotropic points) are found by setting conditions between the mole fractions of the phases equal. The single condition

$$x^{\alpha}-x^{\beta}=0 \tag{2.43}$$

selects an azeotropic extremum of the binary $\alpha+\beta$ two-phase field. The two conditions

$$x_{\mathrm{B}}^{\alpha}-x_{\mathrm{B}}^{\beta}=0; \qquad x_{\mathrm{C}}^{\alpha}-x_{\mathrm{C}}^{\beta}=0 \tag{2.44}$$

select azeotropic extrema of a ternary two-phase field.

If no azeotropic extrema exist, the corresponding sets of equations must, of course, have no solution, except purely mathematical solutions outside the range $0 \leq x_j^i \leq 1$ ($i=\alpha, \beta, \gamma$; $j=\mathrm{A, B, C}$).

In quaternary systems conditions similar to those used for isopleths are valid as a means to find boundaries in plane sections through an isothermal-composition tetrahedron. All the other conditions in quaternary and higher systems are constructed in accord with the same ideas as Eqs. (2.37)–(2.44).

2.3.6 *The driving force for a phase*

At equilibrium the partial Gibbs energies for each component have the same value in all stable phases, i.e. the Gibbs-energy curves of the stable phases touch the same tangent (hyper)plane of chemical potentials. The Gibbs energy surfaces of all metastable phases are above this plane. For each metastable phase one can determine the distance in terms of Gibbs energy between the stable tangent plane and a tangent plane parallel to the

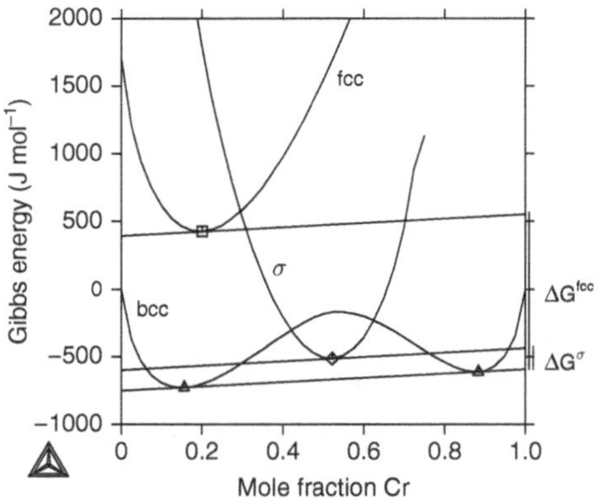

Figure 2.5 The Gibbs-energy curves for some stable and metastable phases in the Cr–Fe system at 500 K. In the middle of the system the stable phases are the two bcc phases on either side of the miscibility gap which are connected with a common tangent. The endpoints of this tangent represent the chemical potentials of the components, μ_i. The triangular symbols on the tangent represent the equilibrium compositions of the two bcc phases. The other two phases, fcc and σ are metastable since their Gibbs energies are always above the stable tangent.

metastable phase. This distance is regarded as the "driving force" and is illustrated in Fig. 2.5. Tangents to the Gibbs-energy curves for fcc and σ phases that are parallel to the stability tangent have been drawn and the compositions of the fcc and σ phases at the tangents are marked by a square and a diamond, respectively; fcc or σ phases with these compositions are closest to being stable. The differences in chemical potentials between these tangents and the stability tangent are called the driving forces of these two phases and denoted ΔG^{σ} and ΔG^{fcc}. Most often the driving force is divided by RT and thus rendered dimensionless.

The driving force is an important quantity in the theory of nucleation of a phase and, as mentioned in the previous subsection, it can be used during minimization of the Gibbs energy to find whether there is another phase that is more stable than the calculated equilibrium set of phases.

2.3.7 *Phase diagrams*

2.3.7.1 Definition and types

If two or three of the conditions mentioned in section 2.3.5 are systematically changed within certain intervals, the resulting two- or three-dimensional manifold of equilibria can be visualized as a phase diagram. Also four-dimensional and "higher-order" phase diagrams are possible, but usually these are represented by two- (or three-)dimensional sections. The largest dimension of a phase diagram is the number of free variables under the equilibrium conditions, for example T, p, and N_j in Eq. (2.20), which is two more than the number of components. This maximal number is diminished by one, since extensive variables are represented as relative values by dividing them by one of them, usually N. For example, the amounts of components N_i are replaced by the overall compositions

$x_i = N_i/N$, the total enthalpy of the system, H, by the molar enthalpy of the system H/N, etc. The variable used as divisor (the total amount N in the above examples) is of interest only as a measure of the size of the system. By keeping one of those relative variables or an intensive variable constant, the dimension of the corresponding phase diagram is again diminished by one. An isobaric binary system thus has two-dimensional phase diagrams.

A phase diagram differs from a "normal" diagram (called a property diagram here), in which one quantity is expressed as a function of another quantity. In a phase diagram the coordinate axes all represent independent variables and the coordinate space shows the "state of the system," i.e. how many and which phases are in equilibrium at a selected coordinate point. In a phase diagram also the points between the lines are meaningful; in a two-dimensional phase diagram they describe the existence of a divariant equilibrium. In a property diagram, in contrast, only points on the curve have meaning.

Lines in a phase diagram nevertheless can be interpreted as functional dependences. Under the condition of monovariant equilibrium of a set of specified phases, one coordinate of the phase diagram is an implicitly defined function of the other one. This will be explained in more detail in the next subsection.

The coordinates as well as the state variables kept constant in a phase diagram must satisfy some restrictions. Hillert *et al.* (1997a) elaborated the rules regarding possible combinations of coordinates and fixed state variables. From each pair of conjugate variables of a set, Eq. (2.5) or (2.6), one variable must be selected and it can either be used as an axis or kept constant.

Schmalzried and Pelton (1973) classified plane phase diagrams into three types in terms of the type of coordinate axes: a first type with two intensive variables, a second type with an intensive variable and a quotient of extensive variables, and a third type with two quotients of extensive variables. Examples of these types are given in Fig. 2.6.

The two-dimensional phase diagrams in Fig. 2.6 belong to a unary system, a binary system at a constant value of one intensive variable (p), or a ternary system with constant values of two intensive variables (p, T). Similar plane phase diagrams exist for n-component systems with $n - 1$ intensive variables kept constant.

Lines in phase diagrams represent monovariant equilibria, which in all the diagrams of Fig. 2.6 are two-phase equilibria. The intensive variables are the same for all phases in equilibrium, but the extensive variables differ. Therefore, in diagrams of the first type (Figs. 2.6(a) and (b)), from Fernández Guillermet and Gustafson (1984) and Coughanowr *et al.* (1991), respectively, the image of each monovariant equilibrium is a single line. Coordinates, being quotients of extensive variables, represent overall state variables of the whole system as well as individual state variables of the phases. In diagrams of the second and third type (Figs. 2.6(c)–(f)), those of the latter type from Liang *et al.* (1998), the variables of the individual phases are represented by a pair of lines belonging to the two coexisting phases. Corresponding points on these pairs of lines are connected by straight lines called tie-lines. In diagrams of the second type (Figs. 2.6(c) and (d)) the intensive coordinate is constant along the tie-lines, so they are parallel to the axis of the quotient of extensive variables, whereas in diagrams of the third type (Figs. 2.6(e) and (f)) the directions of the tie-lines are an important part of the information which can be

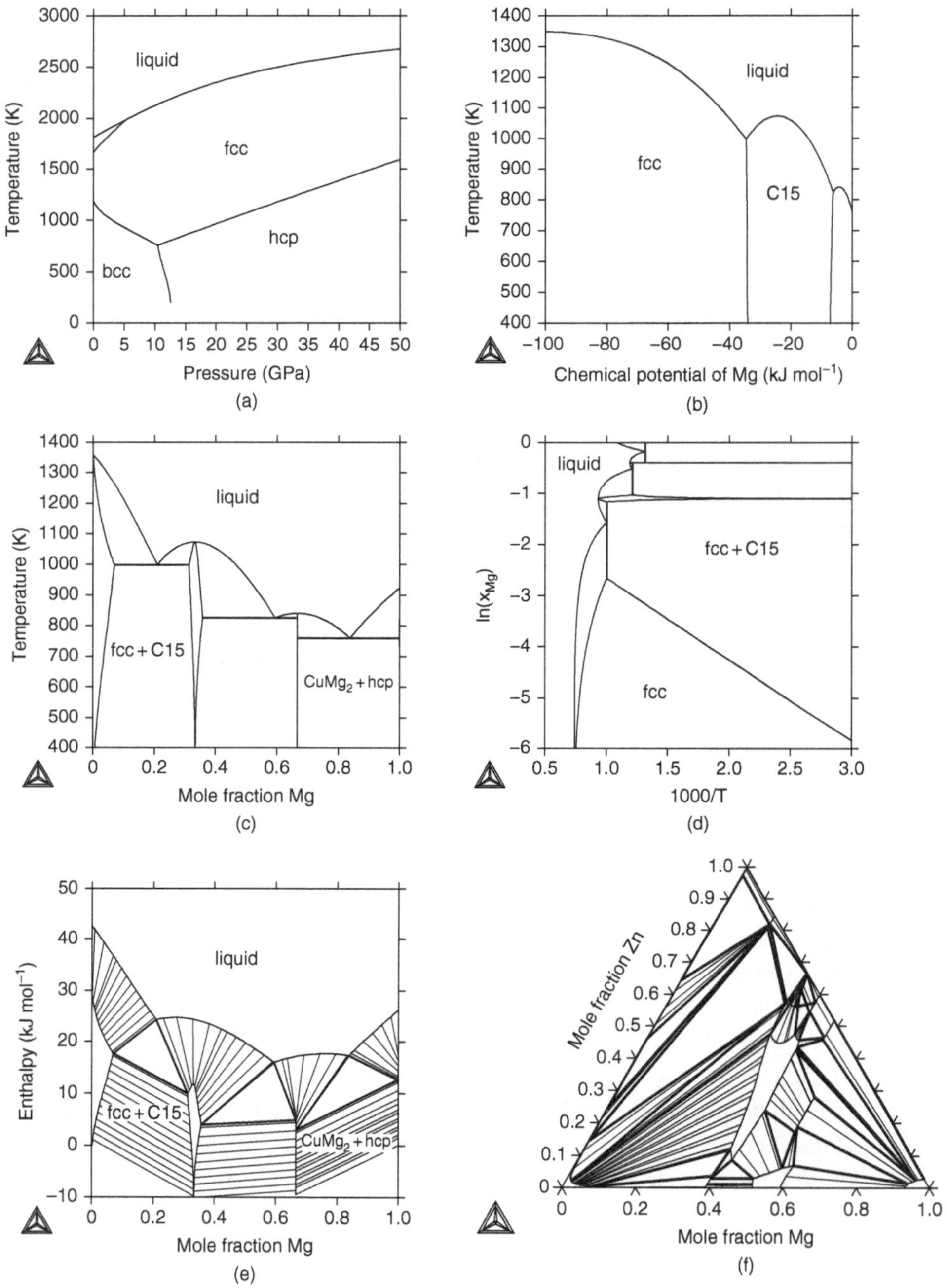

Figure 2.6 Types of phase diagrams: (a) is the $T-p$ diagram of pure Fe, (b) is the T versus μ_{Mg} diagram of Cu–Mg, (c) is the normal binary phase diagram of Cu–Mg, (d) is the $\ln(x_{\mathrm{Mg}})$ versus $1000/T$ diagram of Cu–Mg, (e) is the H/N versus x_{Mg} diagram of Cu–Mg, and (f) is the isothermal section of Al–Mg–Zn at 608 K.

given by the phase diagram. At the points on the tie-lines in the areas between these pairs of lines the quotients of extensive variables represent overall variables of the system only.

The monovariant equilibria meet in invariant equilibria, where the variables of the phases are represented by points, a single point in diagrams of the first type and separate points for each of the (three) phases in the other diagrams. In diagrams of the second type the three points are on a straight line at the same value of the intensive coordinate, whereas in diagrams of the third type they form a triangle (a tie-triangle). The coordinates of points on the tie-lines or within the tie-triangles represent only overall variables of the system.

The difference between a phase diagram and a functional dependence may be explained by an example. The functional dependence of the molar volume on temperature can be plotted in the coordinates abscissa = temperature and ordinate = molar volume for a single phase with given composition and at a given constant pressure (for example $V_m = (V_0/T_0)T$ for an ideal gas). A phase diagram of a unary system can be given in the same coordinates. Here, however, instead of selecting a single phase and taking V_m as a function of T, both T and V_m are taken as independent variables. Thus the pressure as the intensive variable conjugate to volume cannot be freely selected, but depends on the two coordinates temperature and molar volume. The diagram is a typical phase diagram of the second type, showing how many and which phases are in equilibrium at selected coordinate values.

The coordinates pressure and molar volume cannot be used simultaneously as coordinates of a phase diagram, since they are conjugate and thus can replace each other, but they cannot be chosen independently. A functional dependence, however, can be shown for these coordinates, namely the dependence of molar volume on pressure at constant temperature.

Two-dimensional phase diagrams not covered by the three types of Schmalzried and Pelton are sections through higher-dimensional phase diagrams, cut by a condition relating extensive variables. The most common examples of this type are "vertical sections" of ternary systems (isopleths). In these diagrams the tie-lines usually are not in the plane of section, which means that the extensive values of the phases are not all inside the coordinate space. Owing to that, this type of diagram is often considered not to be a phase diagram, but merely a section through a higher-dimensional phase diagram. This distinction may be of minor importance, since any phase diagram shows only part of the state variables, which are defined by the equilibrium conditions. For example, Fig. 2.6(b) does not show the composition of the phases and Fig. 2.6(c) does not show μ-values. Also in isoplethal diagrams the quotients of extensive variables represent overall state variables of the system, which can be interpreted as individual state variables of a single phase only in single-phase areas including their boundaries simultaneously.

The lines in this type of diagram do not necessarily belong to monovariant equilibria, but have another common feature: they show "zero-phase-fraction" equilibria. All the lines show boundaries between an n- and an $(n-1)$-phase field. They can be calculated from the equilibrium conditions for the n-phase equilibrium with the additional condition that the amount of one phase is zero, although still present in equilibrium. This one phase is the phase missing from the adjacent $(n-1)$-phase field. Points where lines meet in this diagram can be calculated by setting the amounts of two phases to zero.

2.3.7.2 Mapping a phase diagram

Several strategies for the calculation of phase diagrams are in use. One of them, which is called "mapping" in Thermo-Calc, shall be outlined here. Two or three variables of the conditions are selected as axis variables. For each axis a lower and upper limit as well as a maximal step are given. All additional conditions are kept constant throughout the whole diagram. If any of these additional conditions set constraints on mole or mass fractions (or, generally, on quotients of extensive variables), the diagram will be an "isoplethal" section of a full phase diagram. The selected axis variables in the plot are not necessarily those of the calculation of the phase diagram, because, besides the axis variables themselves, all other variables not fixed by the additional conditions are calculated and stored in the raw output. The conditions kept constant, however, restrict the output and thus also the diagrams to be drawn.

If the fixed conditions constrain only intensive variables, the mapping procedure searches for all equilibria that are monovariant or invariant under these conditions, since only monovariant equilibria appear as lines in complete phase diagrams. In isoplethal diagrams it searches for all zero-phase-fraction equilibria.

The mapping procedure starts with a single "initial equilibrium," whereby arbitrary values for the axis variables are fixed inside the selected ranges of the axes. Starting values regarding a selected set of phases and estimated values of the unknowns to be calculated may be given by the user. Otherwise, Thermo-Calc creates automatic starting values. Now the Newton–Raphson calculation is started and, if necessary, the set of phases is changed using the driving forces of all phases considered (those having the status "entered") until the most stable set of phases and the equilibrium values of all variables are found.

Now one of the axis variables is changed in a stepwise manner and the corresponding equilibria are calculated until, in the case of non-isoplethal phase diagrams, the resulting set of stable phases corresponds to a monovariant equilibrium. In isoplethal sections the first change of the set of stable phases defines a point of a zero-phase-fraction line.

In non-isoplethal phase diagrams this monovariant equilibrium is traced, keeping all except one of the axis variables constant and incrementing the condition of the only variable axis in steps until an invariant equilibrium is found or one of the selected limits of the axis is exceeded. The magnitude of the steps of incrementation is controlled by the curvature of the line. At the invariant equilibrium there are initially $c+1$ monovariant equilibria, where c is the number of components. One of them is now traced similarly; the other ones are stored for tracing later. This procedure is continued until all monovariant equilibria have ended at axis limits or at "known invariant equilibria" (meaning those stored already).

In isoplethal diagrams the zero-phase-fraction line is traced by setting the appropriate conditions: the set of stable phases is constituted by all phases appearing in both areas adjacent to the line plus the phase appearing only in one area; however, this phase is assigned the fixed amount of zero. This tracing is continued until an axis limit is reached or the set of stable phases changes again, i.e., a phase appears on acquiring a positive driving force or a phase disappears on receiving a negative amount. At such a point always four lines meet, separating the fields of stability of Θ, $\Theta+\alpha$, $\Theta+\alpha+\beta$ and $\Theta+\beta$, where Θ stands for a set of phases numbering between one and $c+2$ in a c-component system. Now the other lines starting at this point are traced similarly to the monovariant equilibria in the non-isoplethal mapping procedure.

The mapping procedure sometimes ends with the message "sorry, can not continue." This indicates that the Newton–Raphson iteration does not converge. As pointed out in section 2.3.3, this may happen due to bad starting values. Since here the starting values of all unknowns are their equilibrium values in the previous step, this seldom occurs, but there is no way to avoid it absolutely.

If in a phase diagram a set of lines is completely separated from another set, the mapping procedure must be continued with an additional starting equilibrium leading to the other set. This is likely to happen if a large homogeneity range of a single solid solution extends throughout the whole phase diagram. An additional starting equilibrium may also help to solve a problem with non-converging Newton–Raphson iteration.

In the Thermo-Calc software, equilibrium values of all variables are stored for each calculated equilibrium during the mapping. This enables one to plot any state variable or function of state variables after the mapping. For example, Figs. 2.6(b)–(e) could all be plotted from a single mapping.

BINFKT and TERFKT have no automatic mapping procedure. Monovariant equilibria of a selected set of phases are calculated in stepwise fashion along a range of an axis variable, most often the temperature, but also the chemical potential or mole fraction of one of the participating phases may be used as the axis variable. Starting from a boundary system with one phase fewer is supported. In this boundary system (a pure component in BINFKT, a binary system in TERFKT) the corresponding equilibrium is invariant, but the derivatives of the monovariant lines into the system itself can be calculated (Lukas *et al.* 1982). Invariant equilibria involving a selected set of phases can be calculated easily. Possibly the set of starting values of the unknowns will lead to divergence of the iteration and a different set must be tested.

These two programs were constructed primarily to calculate diagrams for comparison of calculated with experimental data after a run of an optimization procedure. Thus, for the purpose of calculating phase diagrams from a database, they may be somewhat less user-friendly than Thermo-Calc. Nevertheless, they enable also calculation of complete phase diagrams and isopleths through ternary systems. The distinction between stable and metastable equilibria is not made in each calculation step, but there are some final checks allowing one to determine whether the accepted equilibria are the most stable ones.

A good strategy for BINFKT is to calculate first the two-phase fields starting from the transformation points of the pure elements. Azeotropic maxima or minima are automatically found. Then some invariant three-phase equilibria are tentatively calculated. The monovariant two-phase equilibria starting from them can now be calculated. Where two of the monovariant equilibria intersect, an invariant equilibrium must be present, for which now the unknowns are approximately known and can be used as a good set of starting values. Both in Thermo-Calc and also in BINFKT and TERFKT various plots can be generated from the same calculation. In a binary μ–T plot a crossing of lines indicates that a calculated equilibrium becomes metastable there. Thus μ–T plots enable one to check whether the most stable equilibrium has been selected everywhere.

2.3.7.3 Implicitly defined functions and their derivatives

Although the lines in a phase diagram are not intended to show a functional dependence, they can be interpreted as such a dependence for the condition of a monovariant equilibrium. For example, in Fig. 2.6(b) for equilibrium between liquid and the C15-$MgCu_2$ phase, the corresponding line shows a functional dependence of the temperature T on the chemical potential of Mg, μ_{Mg}. Mathematically, this functional dependence is given as an implicitly defined function by the equilibrium conditions, expressed for the two phases selected.

Since this two-phase equilibrium is monovariant, one additional equation of the type (2.37), namely $\mu_{Mg} = \xi$, completely defines equilibrium. Now all state variables can be calculated, solving the equilibrium conditions by the Newton–Raphson method. Since the result depends on the value of ξ, all the state variables, including T, are implicitly defined functions of ξ and the calculation yields the function values for the chosen value of ξ. Also the inverse function, $\mu_{Mg}(T)$, can be calculated by selecting a temperature instead of a Mg chemical potential as the independent variable ξ. In the present case this inverse function has no value above 1074 K and two values below this temperature. Thus, to plot the two-phase equilibrium liquid + C15-$MgCu_2$ on the phase diagram, it is better to calculate T as a function of μ_{Mg} rather than μ_{Mg} as a function of T. For other two-phase equilibria the contrary may be true.

Also in a diagram of the second type, a possible point of view is to interpret the lines as representations of implicitly defined functions. In Fig. 2.6(c) the boundary of the two-phase field C15-$MgCu_2$ + Mg_2Cu against C15-$MgCu_2$ shows T as an implicitly defined function of $x_{Mg}^{MgCu_2}$ as well as the inverse function $x_{Mg}^{C15\text{-}MgCu_2}(T)$. Here it is preferable to select T and calculate the value for the inverse function, $x_{Mg}^{C15\text{-}MgCu_2}$, but, for example, to calculate the liquidus line of the liquid + C15-$MgCu_2$ two-phase equilibrium, it is better to select x_{Mg}^{liquid} as the independent variable.

The concept of implicitly defined functions also defines the derivatives of these functions. If the Newton–Raphson method is used for the calculation of function values, it is easy to get also values of the derivatives. To show that, one can assume a set of two conditions $f_1(x, y, t) = 0$ and $f_2(x, y, t) = 0$ relating the three variables x, y, and t. These conditions implicitly define the two functions $x(t)$ and $y(t)$. For example, x and y may represent the mole fractions of the two phases of a selected two-phase equilibrium and t the temperature. The conditions $f_1 = 0$ and $f_2 = 0$ must be satisfied throughout the whole range of variables of the implicitly defined functions. Therefore also their total derivatives with respect to t must be zero:

$$\begin{aligned} \frac{\mathrm{d}f_1}{\mathrm{d}t} &= \frac{\partial f_1}{\partial t} + \left(\frac{\partial f_1}{\partial x}\right)\frac{\mathrm{d}x}{\mathrm{d}t} + \left(\frac{\partial f_1}{\partial y}\right)\frac{\mathrm{d}y}{\mathrm{d}t} = 0 \\ \frac{\mathrm{d}f_2}{\mathrm{d}t} &= \frac{\partial f_2}{\partial t} + \left(\frac{\partial f_2}{\partial x}\right)\frac{\mathrm{d}x}{\mathrm{d}t} + \left(\frac{\partial f_2}{\partial y}\right)\frac{\mathrm{d}y}{\mathrm{d}t} = 0 \end{aligned} \tag{2.45}$$

or

$$\left(\frac{\partial f_1}{\partial x}\right)\frac{\mathrm{d}x}{\mathrm{d}t}+\left(\frac{\partial f_1}{\partial y}\right)\frac{\mathrm{d}y}{\mathrm{d}t}=-\frac{\partial f_1}{\partial t}$$
$$\left(\frac{\partial f_2}{\partial x}\right)\frac{\mathrm{d}x}{\mathrm{d}t}+\left(\frac{\partial f_2}{\partial y}\right)\frac{\mathrm{d}y}{\mathrm{d}t}=-\frac{\partial f_2}{\partial t} \tag{2.46}$$

The derivatives $\mathrm{d}x/\mathrm{d}t$ and $\mathrm{d}y/\mathrm{d}t$ have to be taken over the equilibrium states and thus are the sought derivatives of the implicitly defined functions. The values of these derivatives corresponding to the selected value of the independent variable t can be found by solving Eq. (2.46) as a set of two linear equations for the two unknowns $\mathrm{d}x/\mathrm{d}t$ and $\mathrm{d}y/\mathrm{d}t$.

Comparing Eq. (2.46) with (2.36) shows that the matrix on the left-hand side is the same as that used to find the corrections to the variables in the last iteration step of the Newton–Raphson method and thus has already been calculated. The right-hand side of Eq. (2.46) consists of the partial derivatives of the conditions f_1 and f_2 with respect to the variable t, the independent variable of the implicitly defined functions.

In this way any of the unknowns of the equilibrium conditions can be treated as an implicitly defined function of another one and values of the function itself as well as of its derivatives can be calculated by the Newton–Raphson method.

As an example of derivatives, we may derive the famous Gibbs–Konovalov rule (Goodman *et al.* 1981) from Eq. (2.46). We formulate the equilibrium conditions for two phases in a binary isobaric system by modifying Eq. (2.21):

$$F_1 \equiv G_1^\alpha - G_1^\beta = 0$$
$$F_2 \equiv G_2^\alpha - G_2^\beta = 0 \tag{2.47}$$

Using Eq. (2.15) and expressing the derivatives of the partial Gibbs energies with respect to T as partial entropies, Eq. (2.46) becomes

$$-x^\alpha\left(\frac{\partial^2 G_\mathrm{m}^\alpha}{\partial x^{\alpha 2}}\right)\frac{\mathrm{d}x^\alpha}{\mathrm{d}T}+x^\beta\left(\frac{\partial^2 G_\mathrm{m}^\beta}{\partial x^{\beta 2}}\right)\frac{\mathrm{d}x^\beta}{\mathrm{d}T}=S_1^\alpha - S_1^\beta$$
$$(1-x^\alpha)\left(\frac{\partial^2 G_\mathrm{m}^\alpha}{\partial x^{\alpha 2}}\right)\frac{\mathrm{d}x^\alpha}{\mathrm{d}T}-(1-x^\beta)\left(\frac{\partial^2 G_\mathrm{m}^\beta}{\partial x^{\beta 2}}\right)\frac{\mathrm{d}x^\beta}{\mathrm{d}T}=S_2^\alpha - S_2^\beta \tag{2.48}$$

On solving this set of two equations for the first unknown $\mathrm{d}x^\alpha/\mathrm{d}T$, we get

$$\frac{\mathrm{d}x^\alpha}{\mathrm{d}T}=\frac{(S_1^\alpha - S_1^\beta)(1-x^\beta)+(S_2^\alpha - S_2^\beta)x^\beta}{(x^\alpha - x^\beta)\partial^2 G_\mathrm{m}^\alpha/\partial x^{\alpha 2}} \tag{2.49}$$

The numerator is the change in entropy associated with reversible dissolution of β in α. It can be replaced by the corresponding change in enthalpy divided by T. Then we get the Gibbs–Konovalov rule in the form referenced by Goodman *et al.* (1981) as the derivative $\mathrm{d}T/\mathrm{d}x^\alpha$ of the inverse function

$$\left(\frac{\mathrm{d}T}{\mathrm{d}x^\alpha}\right)_\mathrm{coex}=\frac{(x^\alpha - x^\beta)T\,\partial^2 G_\mathrm{m}^\alpha/\partial x^{\alpha 2}}{(H_1^\alpha - H_1^\beta)(1-x^\beta)+(H_2^\alpha - H_2^\beta)x^\beta} \tag{2.50}$$

where the denominator is the molar enthalpy of reversible dissolution of an infinitesimal amount of β in α, denoted $-\Delta H^{\beta}_{\mathrm{FP}}$ by Goodman *et al.* (1981) (numerator and denominator here have opposite signs to those in the formulation of Goodman *et al.*).

2.4 Optimization methods

2.4.1 *The principle of the least-squares method*

In this section the least-squares method shall be briefly described as it was first introduced by Gauss, as is done in many textbooks. Mainly the connection with our problem shall be outlined.

The general problem is as follows: a set of n measurable values W_i depends on a set of m unknown coefficients C_j via functions F_i with values of independent variables x_{ki}:

$$W_i = F_i(C_j, x_{ki}) \qquad i = 1, \ldots, n, \quad j = 1, \ldots, m \tag{2.51}$$

The index k distinguishes the various independent variables (temperature, concentrations, ...) belonging to measurement number i.

If n is greater than m, it is usually not possible to get a set of coefficients C_j for which the W_i calculated using Eq. (2.51) are equal to the corresponding measured values L_i. Here the criterion for the "best" set C_j is that the sum of squares of the "errors" must be minimal, where the "errors" v_i are defined as the differences between calculated, F_i, and measured, L_i, values times a weighting factor p_i:

$$(F_i(C_j, x_{ki}) - L_i) \cdot p_i = v_i \tag{2.52}$$

In many textbooks the square of the quantity p_i of Eq. (2.52) rather than p_i itself is called the weighting factor. Equation (2.52) is called the error equation.

The condition for the best values C_j is taken as

$$\sum_{i=1}^{n} v_i^2 = \mathrm{Min} \qquad \text{(with respect to the } C_j\text{)} \tag{2.53}$$

From this condition the m following equations relating the m unknown coefficients C_j can be derived:

$$\sum_{i=1}^{n} v_i \cdot \frac{\partial v_i}{\partial C_j} = 0 \qquad j = 1, \ldots, m \tag{2.54}$$

To solve these equations, Gauss expanded the v_i into a Taylor series and truncated it after the linear terms:

$$v_i(C_j, x_{ki}) \approx v_i^0(C_j{}^0, x_{ki}) + \sum_{l=1}^{m} \frac{\partial v_i}{\partial C_l} \cdot \Delta C_l \tag{2.55}$$

where the ΔC_l are corrections to the coefficients C_l. If the $F_i(C_j, x_{ki})$ and therefore also the v_i are not linear in the coefficients C_j, this is an approximation, which might not be acceptable if the values of the C_j^0 are too far from the final values of the coefficients.

To calculate the corrections ΔC_l, Eq. (2.55) is inserted into Eq. (2.54) and, after rearranging, that yields

$$\sum_{l=1}^{m}\left(\sum_{i=1}^{n}\frac{\partial v_i}{\partial C_j}\cdot\frac{\partial v_i}{\partial C_l}\right)\Delta C_l = -\sum_{i=1}^{n} v_i^0 \cdot \frac{\partial v_i}{\partial C_j} \qquad j = 1, \ldots, m \tag{2.56}$$

This is a set of m linear equations for the m unknowns ΔC_i, called "Gaussian normal equations," which in matrix notation are written

$$\left(\left(\sum_{i=1}^{n}\frac{\partial v_i}{\partial C_j}\cdot\frac{\partial v_i}{\partial C_l}\right)\right)_{m,m} \cdot ((\Delta C_l))_{m,1} = \left(\left(-\sum_{i=1}^{n} v_i^0 \cdot \frac{\partial v_i}{\partial C_l}\right)\right)_{m,1} \tag{2.57}$$

These equations are set up, using an initial set C_l^0 of the coefficients, and solved for the corrections ΔC_l. After adding the corrections to the initial set, the calculation may be repeated until the corrections are below a given limit.

As a measure of the fit between the resulting coefficients and the measured values, the mean square error can be used. It is defined by

$$\text{mean square error} = \sum_{i=1}^{n}\frac{v_i^2}{n-m} \tag{2.58}$$

The accuracy of the calculated coefficients is proportional to the square root of the mean squared error. The factor of proportionality of each coefficient is the square root of the corresponding diagonal element of the reciprocal of the left $m \times m$ matrix of Eq. (2.57).

2.4.2 The weighting factor

In the BINGSS software, described in section 7.2, special care is taken to deal with the fact that, when different types of measurements are used, the errors v_i cannot be compared directly. Quantities of different dimensions cannot be added, therefore the squares of errors in Eq. (2.53) should have the same dimension. The weighting factor p_i in Eq. (2.52) can be used to make the errors v_i dimensionless, if p_i is taken as the reciprocal of the estimated accuracy ΔL_i of the measured values (Lukas *et al.* 1977):

$$p_i = (\Delta L_i)^{-1} \tag{2.59}$$

If also the independent variables, x_{ki} (composition, temperature,... of the sample investigated), have limited accuracies, Δx_{ki}, that may also be taken into account (Lukas *et al.* 1977). In this case the weighting factor p_i is defined as

$$p_i = \left(\Delta L_i^2 + \sum_k((\partial F_i/\partial x_{ki})\cdot \Delta x_{ki})^2\right)^{-\frac{1}{2}} \tag{2.60}$$

This choice may be interpreted as the v_i being dimensionless relative errors defined as fractions of the mean errors of the corresponding measurements. Therefore, the errors of different kinds of measured values can easily be compared.

The ΔL_i may be estimated from different points of view varying from one series of measurements to another. During the optimization in BINGSS, this can be corrected: for measurements marked by a common label the weight given by Eq. (2.59) can be multiplied by a dimensionless constant factor. This includes the factor "0" to discard a doubtful series of measurements in the following calculations.

In the PARROT software, see section 7.3, the term added to the sum of squares of errors for each experimental value i is equal to

$$\left(weight_i \cdot \frac{(experimental\ value)_i - (calculated\ value)_i}{(estimated\ uncertainty)_i} \right)^2 \tag{2.61}$$

This is equivalent to a combination of Eqs. (2.52) and (2.59) without considering the uncertainties of the independent variables (Eq. (2.60)). The additional dimensionless factor "$weight_i$" is left to the responsibility of the user to assign weights that reflect the relative importance of the data. The default settings of all weights are equal to 1.

2.4.3 *Marquardt's algorithm*

If all the equations of error are linear in the coefficients, Eq. (2.57) is correct and the final solution should be found in one step. A second or third step may be useful merely due to errors produced by rounding. However, if the equations of error are nonlinear in the coefficients (either due to the analytical description of thermodynamics or due to the error equation) the step after Eq. (2.57) may fail and the mean square of error increase.

To solve this problem D. W. Marquardt (1963) combined the Newton–Raphson method with the steepest-descent method. The identity matrix multiplied by a factor called the Marquardt parameter is added to the normalized matrix of Eq. (2.57). If the Marquardt parameter is large, this term is dominant and the corrections correspond to a steepest-descent step, in which the length of the vector is the reciprocal of the Marquardt parameter. If the Marquardt parameter is small, the iteration step is nearly the pure Newton–Raphson step. If the mean square error increases, the last correction is discarded and the Marquardt parameter enlarged by a factor (for example 10) and new corrections are calculated using the matrix of Eq. (2.57) of the previous step. If two consecutive steps were successful, the Marquardt parameter is diminished by the same factor for the next calculation.

There are many other methods of improving the Gauss algorithm for nonlinear equations of error described in the literature. Most of them are mainly intended to diminish the calculation time compared with that of Marquardt's algorithm or to avoid the analytical calculation of derivatives. Since in our special case the calculation time depends mostly on the calculation of the sum of squares of error, rather than on solving the system of linear equations (2.57), the calculation time cannot significantly be reduced from that of Marquardt's algorithm. Also the analytical calculation of derivatives is no problem in our case. Therefore no other algorithm is used in BINGSS.

2.5 Final remarks

In this chapter, diverse topics of science such as thermodynamics, crystallography, numerical methods, and phase diagrams have been mentioned. As stated at the beginning, the aim of CT is to make use of experimental and theoretical information from all these topics to formulate a thermodynamic description using for each phase in a multicomponent system a single Gibbs energy function that can describe the equilibrium state of the system. Such a description must, of course, be very approximate and can be improved with new data and better models, but it has already proved to be of great value in understanding, predicting, and simulating chemical processes and properties of materials.

The thermodynamic description using a single Gibbs energy for each phase means that all thermodynamic properties such as enthalpy, heat capacity, and chemical potentials can be consistently calculated from this. The equilibrium state for a system at given temperature, pressure, and amounts of components, the most common external conditions, can be calculated by finding the minimum of the Gibbs energy, including all possible phases. From the calculated equilibrium all thermodynamic quantities can be obtained, as can, for example, amounts of the stable phases and the partitioning of the components.

The phase diagram describes the equilibrium state of a system as a function of two or more state variables. For multicomponent systems there are many different types of phase diagrams that can be useful. Any such diagram can be calculated from the Gibbs energy functions and they will all be consistent. The thermodynamic functions can be used for much more than just calculating equilibrium phase diagrams and properties of the stable states. There are no drastic changes in these functions when they are extrapolated outside the stability ranges of a phase. A simple kind of extrapolation is to suspend a phase and calculate a metastable phase diagram from the functions. For example, the metastable Fe–C phase diagram with cementite can be calculated if graphite is suspended.

By extending this idea further, one can calculate any equilibrium for a phase or an assemblage of phases, irrespective of whether these phases represent the stable state of the system. Of course, the calculated metastable equilibrium is only approximate and the further away from the stable state the less confidence can be attributed to the calculated values of the state variables. However, often the modeled Gibbs energies are the only thermodynamic information available for a metastable state. This use of the Gibbs energy functions requires that the assessment of the stable states be done with great care. The selection of models and model parameters should be made with the intention that the Gibbs energy functions and their derivatives will behave reasonably outside as well as inside the stability ranges of the phases.

The same rules apply to these metastable equilibria as to the stable one. The only difference is that the total Gibbs energy (for a system at constant T and p) is higher than that for the stable equilibrium. However, it is still a minimum for the phases included in the calculation and the chemical potentials for all components are the same in all phases.

3 First principles and thermodynamic properties

Viewing thermodynamics with an insight into statistical mechanics freed scientists from the empiricism of the nineteenth century (Callen 1985), and extending this insight to quantum-mechanical calculations of electronic structures, which has been enormously enhanced in the last 20 years, has made possible the contemporary integrated view of materials' properties.

In chapter 2 the thermodynamic state functions and crystallography were said to be related. If one considers the electronic structure, a straightforward link is made, since the crystal structure at $T = 0$, for a given set of atoms at constant pressure p, is the one for which the enthalpy H, $H = U + PV$, has a minimum.

It is appealing to unify knowledge, and one can start with unities, $1\mu\text{Ry/atom} = 1.32\,\text{kJ}\,\text{mol}^{-1}$, $1\,\text{eV/atom} = 96\,\text{kJ}\,\text{mol}^{-1}$ and the internal energy U at $T = 0$ is usually called the total energy, E_{total}, by the electronic-structure community. Learning about these quantum-mechanical calculations and their development should help assessors to envisage how thermodynamic knowledge can be enhanced by using state-of-the-art first-principles (*ab initio*) results.

Suppose that one could calculate quantities like the energy U (Table 2.1) per cell of volume $V = (V_{\text{cell}})$ for a given crystal, just using a purely quantum-mechanical approach. One could use equations like (2.12) for the U calculated from first principles and obtain two other fundamental quantities:

$$\begin{aligned} &\text{energy} && U = U(V) \\ &\text{pressure} && p = -\left(\frac{\partial U}{\partial V}\right) \\ &\text{bulk modulus} && \kappa = -V\left(\frac{\partial^2 U}{\partial V^2}\right) \end{aligned} \tag{3.1}$$

These would be impressive and useful since the volume and bulk modulus could then be predicted for any material. But are these calculations possible? These quantities can nowadays truly be calculated for a crystal frozen at $T = 0$, specifying only Z_i (atomic numbers) and r_i (atomic coordinates like the ones in Table 2.3), without any adjustable parameters (Hafner *et al.* 2006).

This is made possible by using the density-functional theory (DFT) (for which the 1998 Nobel Prize was awarded) together with the local spin-density approximation (LSDA) formulated by Kohn and co-workers in the mid 1960s (Kohn and Sham 1965). (For a tutorial approach, see the books by Martin (2004), Pettifor (1996), and Pisani (2000), and also the report from the 5th Ringberg workshop (Turchi *et al.* 2007).)

But are these calculated values predictive? To test whether they are, one can calculate, for example, the equilibrium volume, V^0, where U is minimum or $p = 0$ or κ at zero pressure for the crystal structure of interest. Since these quantities are experimentally observable, one can extrapolate the measured values to $T = 0$ and compare the results with the values predicted by the theory. In many cases these predictions obtained using DFT are surprisingly good; surprisingly, because the theoretical approach used is, in principle, very simple, as one can see in the next sections.

3.1 The density-functional theory (DFT) and its approximations

To treat a set of atoms with quantum mechanics requires one to solve the Schrödinger equation for many atoms, Eq. (3.2), which means calculating the kinetic and potential energies (Hamiltonian) for all of them, and they are many (of the order of the Avogadro number). Simplifications can be made in order to solve the problem, for example, treating the dynamics of the electrons decoupled from the dynamics of the nuclei, as the adiabatic approximation does; but the many-body problem still remains. We have

$$\left(-\frac{\hbar^2}{2m_e}\sum_i \nabla_i^2 + V(x_1, x_2, \ldots, x_{N_i}, R_1, R_2, \ldots, R_{N_I}) - \sum_I \frac{\hbar^2}{2M_I}\nabla_I^2 + \frac{1}{2}\sum_{I \neq J}\frac{Z_I Z_J}{|\mathbf{R}_I - \mathbf{R}_J|}\right)\Psi = E\Psi \tag{3.2}$$

where $\Psi \equiv \Psi(x_1, x_2, \ldots, x_{N_i}, R_1, R_2, \ldots, R_{N_I})$.

3.1.1 *The Hohenberg–Kohn theorems*

It was the ideas of the creators of DFT and subsequent approximations that made the situation treatable because they reformulated the problem.

The first step was to state (and prove) that "for any system of electrons in an external potential $V_{ext}(r)$ (from the nuclei), that potential is determined uniquely, except for a constant, by the ground-state density $n(r)$ (r is a point in the space)."

Another statement was "a universal functional for the energy $E(n)$ of the density $n(r)$ can be defined for all electron systems. The exact ground-state energy is the global minimum for a given $V_{ext}(r)$, and the density $n(r)$ which minimizes this functional is the exact ground-state density."

With this, one concludes that the functional $E(n)$ alone is sufficient to determine the ground-state energy and the density. However, this exact functional is unknown! Furthermore, there was no hint regarding how to obtain it. Later there appeared the Kohn–Sham *Ansatz* that showed how to get it.

3.1.2 *The Kohn–Sham* Ansatz

The Kohn–Sham (K–S) approach simplifies the problem in a very clever way: instead of having to work with the complicated Hamiltonian, they defined an effective simpler one, for non-interacting electrons, which were assumed to have the same density as that in the real system:

$$\left(-\frac{\hbar^2}{2m_e}\nabla^2 + V_{KS}(x_1, R_1, R_2, \ldots, R_{N_I})\right)\phi_i(x_1, R_1, R_2, \ldots, R_{N_I}) = \epsilon_i\phi_i(x_1, R_1, R_2, \ldots, R_{N_I}). \tag{3.3}$$

The K–S equations are Schrödinger-like equations, and they are much simpler to solve, because the effective potential V depends on the coordinates of only one electron. According to the first Hohenberg–Kohn theorem (Hohenberg and Kohn 1964), it is a unique function of the electron density because neither the potential nor the density is known in advance. The K–S equations are solved iteratively, starting with some reasonable guess for the charge density (which is often taken as a superposition of atomic charge densities), see Fig. 3.1. The second Hohenberg–Kohn theorem ensures that the converged solution corresponds to the charge density in the ground state. The total energy of the system is a unique functional of the density:

$$E_{\mathrm{tot}}[n] = T_s[n] \int \mathrm{d}^3r\, V_{\mathrm{ext}}(\mathbf{r})n(\mathbf{r}) + \iint \mathrm{d}^3r\,\mathrm{d}^3r'\, \frac{n(\mathbf{r})n(\mathbf{r}')}{|\mathbf{r}-\mathbf{r}'|} + E_{II} + E_{\mathrm{xc}}[n]. \tag{3.4}$$

There are various ways to solve these equations numerically and there are several methods to handle the problem. There are specific computational techniques and they are implemented in various software codes. Each different way to solve the problem has its particular advantages and disadvantages, but the important fact is that one needs to know what each code is calculating and to have complete control of the results and knowledge

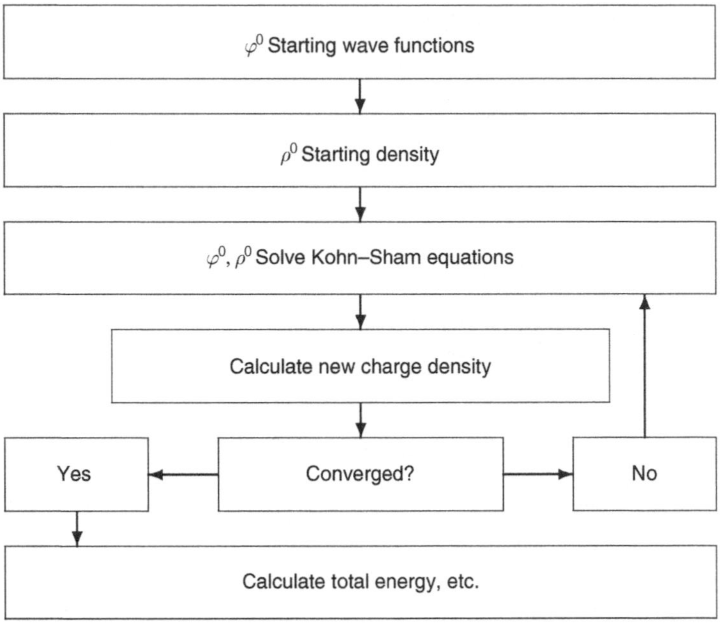

Figure 3.1 The DFT method, showing the Kohn–Sham routine for solving the Schrödinger equation for one effective electron.

of how to interpret them (Turchi *et al.* 2007). Bearing this in mind, one should be careful when analyzing DFT results.

3.2 The DFT results at 0 K

There are cases for which the theory gives wrong predictions. For example, within the DFT-L(S)DA (local (spin-) density approximation) bcc Fe is unstable, and even with the best LDA, germanium is a metal. For the Fe case, use of the so-called generalized gradient approximation (GGA) (Martin 2004) cures the problem. But it seems that there is no help to be obtained from the LDA/GGA for solving the band-gap problem in semiconductors and in this case an approach that goes beyond the DFT is being developed, see Fig. 3.2.

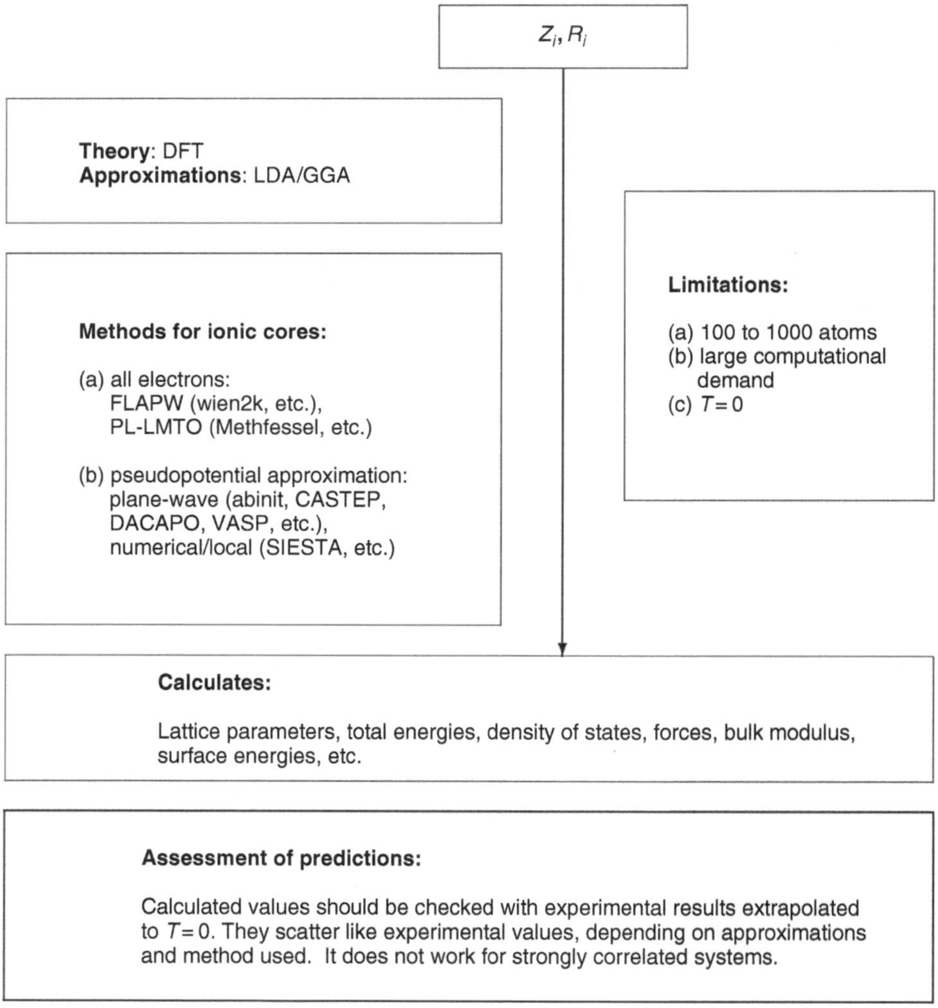

Figure 3.2 A schematical view of the first-principles (or *ab initio*) approach without fitting parameters: theory, approximation methods, results, and limitations. The items in parentheses are some of the codes related to the methods.

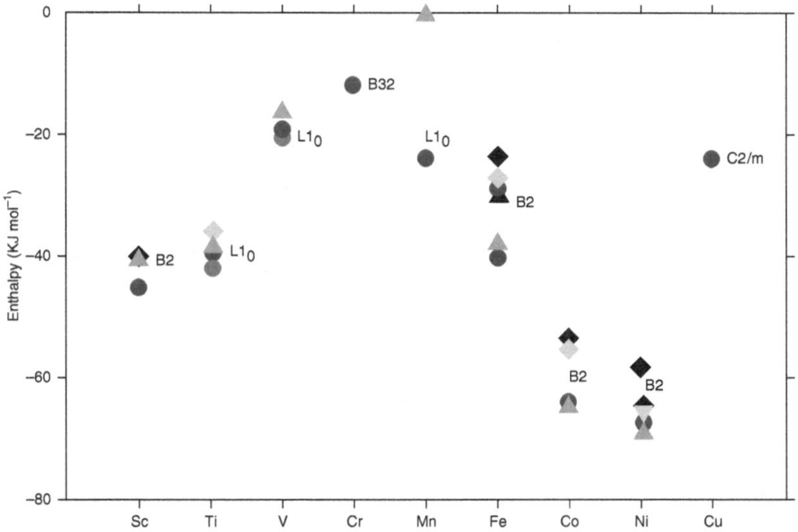

Figure 3.3 Total energies for a number of ordered aluminides calculated from first principles by various authors using various methods. Courtesy of Catherine Colinet.

The good results obtained with the GGA for Fe cannot be used as an argument that the GGA is better than the LDA in all cases. For example, the GGA is known to provide better calculated lattice parameters for solids than does the LDA, and as a consequence it gives better descriptions of the bulk properties. However, the predictions of the GGA for the estimation of the surface properties are still under discussion. With much experience having been accumulated by scientists after many calculations using DFT in different approximations in recent years, some insight has been obtained into the question of for which quantities each approximation works best.

In Fig. 3.3, total energies for a number of ordered aluminides calculated by various authors using different methods are plotted in order to give a feeling for the magnitude of the scatter of the calculated values. One can easily conclude that first-principles results scatter in a very similar way to experimental data. It is, then, of fundamental importance to evaluate those results. The best way to do this is to have an expert in these techniques to help in the evaluation.

Excellent combinations of results from experiments and DFT calculations are given in papers by Ghosh and Asta (2005) and Ghosh *et al.* (2006). In the former they investigate the phase stability, phase transformations, and elastic properties of a phase, Cu_6Sn_5, relevant for lead-free solders and in the latter they study ordering in Ti–Zn alloys. The way they present the theoretical results can be used as standard for publications of the same type. All values are listed in tables, LDA and GGA results for formation energies and lattice constants are compared with each other and with experimental results in a very detailed way; for instance, reported atomic positions are compared with the experimental ones obtained from X-ray diffractometry. Their paper demonstrates that the DFT results can be quickly validated by a few experiments using the techniques described in the next chapter.

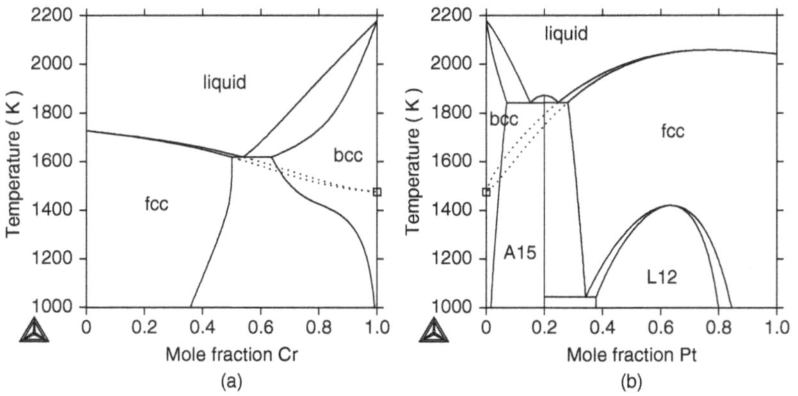

Figure 3.4 Extrapolations of the liquid–fcc solubility curves for Cr–Ni and Cr–Pt (dashed lines) must end at the same melting temperature of metastable Cr. This point is marked with □ on the temperature axis.

3.2.1 *Calphad enthalpy estimates at 0 K*

The stabilities of elements in the various types of lattices are needed in the Calphad technique. In the book by Kaufman and Bernstein (1970) a combination of first-principles data and extrapolation techniques is used to estimate these "lattice stabilities." In most cases these were obtained from estimates of the entropy of melting and of the melting point for the metastable phase. The latter could sometimes be estimated from phase diagrams, for example, for fcc Cr, from phase diagrams with high solubility of fcc Cr. In Fig. 3.4 this extrapolation is shown by dashed lines in the phase diagrams calculated for Cr–Ni by Lee (1992) and for Cr–Pt by P. J. Spencer (unpublished work, 1988).

This gives the following Gibbs energy for the fcc and bcc phases for pure Cr relative to the liquid:

$$G_{\rm Cr}^{\rm Liquid} - G_{\rm Cr}^{\rm bcc} = \Delta S_{\rm f}^{\rm L/b}(T_{\rm f}^{\rm bcc} - T) \tag{3.5}$$

$$G_{\rm Cr}^{\rm Liquid} - G_{\rm Cr}^{\rm fcc} = \Delta S_{\rm f}^{\rm L/f}(T_{\rm f}^{\rm fcc} - T) \tag{3.6}$$

where $\Delta S_{\rm f}$ is the entropy of melting and $T_{\rm f}$ is the melting temperature. Subtracting Eq. (3.6) from Eq. (3.5) gives the Gibbs-energy difference between fcc and bcc and, assuming that the heat capacity is the same for both phases, an estimate of the enthalpy difference between the two phases at 0 K, ΔH, can be obtained:

$$G_{\rm Cr}^{\rm fcc} - G_{\rm Cr}^{\rm bcc} = (\Delta S_{\rm f}^{\rm L/b}\ T_{\rm f}^{\rm bcc} - \Delta S_{\rm f}^{\rm L/f}\ T_{\rm f}^{\rm fcc}) + (\Delta S_{\rm f}^{\rm L/f} - \Delta S_{\rm b}^{\rm L/b})T \tag{3.7}$$

$$\Delta H_{\rm Cr}^{\rm f/b} = \Delta S_{\rm f}^{\rm L/b}\ T_{\rm f}^{\rm bcc} - \Delta S_{\rm f}^{\rm L/f}\ T_{\rm f}^{\rm fcc} \tag{3.8}$$

A calculation of lattice stabilities for transition-metal elements done by Skriver (1985) gave very different values of the enthalpies at 0 K for some of the elements and this led to severe criticism of the Calphad lattice stabilities. Later it was shown that the enthalpy of fcc Cr is in fact impossible to calculate by *ab initio* methods because the lattice is dynamically unstable and transforms to a bcc lattice by the well-known Bain transformation (Bain 1924). The Calphad lattice stabilities have recently been reviewed by Pettifor (2003) and Sluiter (2006) and the Calphad values for fcc Cr are now generally accepted, as long as the fcc phase is dynamically stable (Turchi *et al.* 2007).

Similar problems with dynamically unstable lattices still occur in *ab initio* calculations of more complicated lattices in binary and higher-order systems. The relaxed lattice must be identified carefully.

3.3 Going to higher temperatures, adding the statistics

3.3.1 Quantum molecular dynamics and Monte Carlo simulations

The theory and results discussed thus far are for $T = 0$. In order to have predictions using DFT for the effects of the temperature, the elegant quantum molecular dynamics (QMD) approach of Car and Parrinello (1985) is being developed. This technique allows one to calculate melting temperatures. It has been applied for some elements, for example C, for which the pressure versus temperature phase diagram was obtained (Grumbach and Martin 1996).

Monte Carlo simulations can also be coupled to first-principles methods, but they are more applicable to gases or isolated atoms. However, some progress has been achieved also for solids.

3.3.2 The cluster-expansion method, plus a CVM or plus Monte Carlo: the so-called first-principles phase diagrams

Another way to get the statistics for obtaining properties at higher temperatures is to use the technique called cluster expansion (Connolly and Williams 1983, Sanchez *et al.* 1984, Zunger *et al.* 1990). In this method, starting from the total energies, the configuration and interactions are decoupled and configuration-dependent quantities are expanded in clusters. Normally the cluster energies decrease rapidly with size and only a few are needed, but sometimes very large clusters are needed (Zarkevich and Johnsson 2004). After this step, two routes can be used to obtain the statistics: a cluster-variation method (CVM), whereby the configuration entropy is obtained as a function of composition and temperature; or a Monte Carlo simulation whereby it can also be calculated. From these quantities the phase diagram can be calculated. Figure 3.5 shows schematically the three steps for obtaining a first-principles phase diagram.

An example of this method applied to a ternary system is given by Lechermann *et al.* (2005). The mapping of bondings for the pure elements Al, Fe, and Ni calculated by the same authors is shown in Fig. 3.6.

Cluster expansion coupled to a CVM has been reviewed by Colinet (2002). An example of its use with Monte Carlo simulation is given in Tepesch *et al.* (1998).

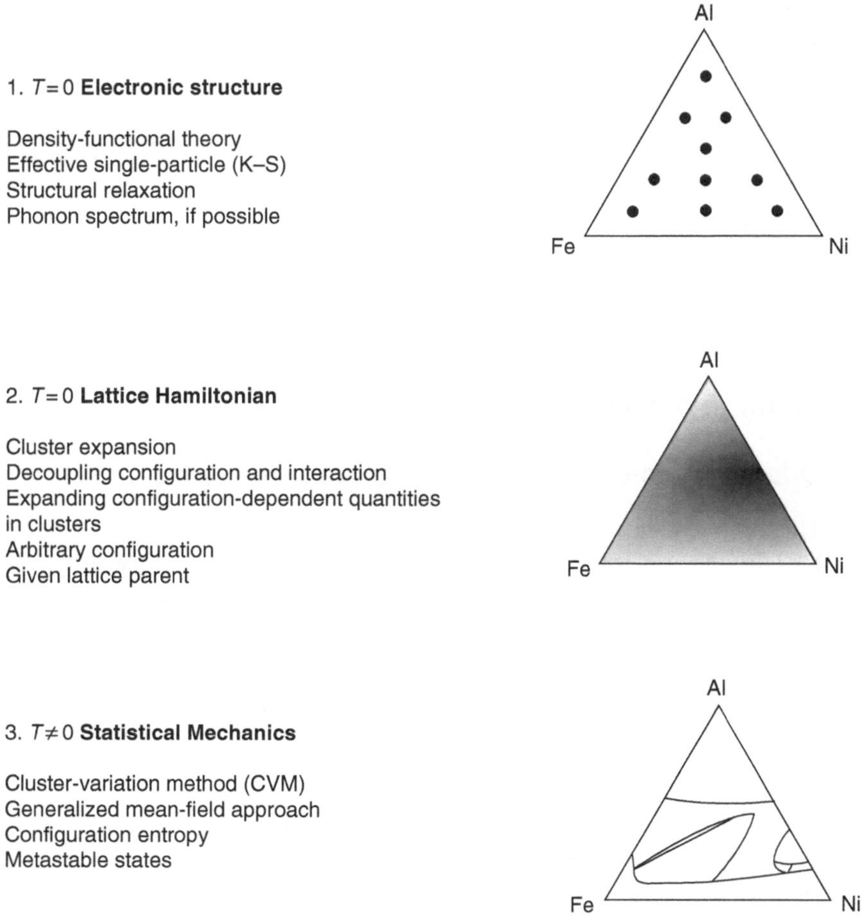

Figure 3.5 A schematic view of a method by which to obtain a first-principles phase diagram. Courtesy of Frank Lechermann.

3.3.3 *First-principles-enhanced Calphad extrapolations of fcc and bcc with ordering for the Al–Fe–Ni ternary system*

Calphad extrapolations from binary subsystems to ternary systems require descriptions of metastable phases, which, in the case of the binary subsystems of the Al–Fe–Ni system (for Al–Fe by Seiersten and Tibballs (1993), Fig. 6.5, for Al–Ni by Ansara *et al.* (1997b), Fig. 9.16(a), and for Fe–Ni by A. T. Dinsdale and T. G. Chart (unpublished work, 1986) and I. Ansara (unpublished work, 1995), in Fig. 5.4(b) later) were not particularly well described in the initial Calphad assessments.

Later the metastable extension of the fcc phase in Al–Ni was assessed, see Fig. 9.16(b), as well as the metastable extension of the fcc ordering in Al–Fe and the metastable bcc ordering in Fe–Ni. In the assessment of the metastable phases *ab initio* calculations of energies of the ordered phases were used and the reciprocal parameter according to Eq. (5.151) was included to describe the short-range order.

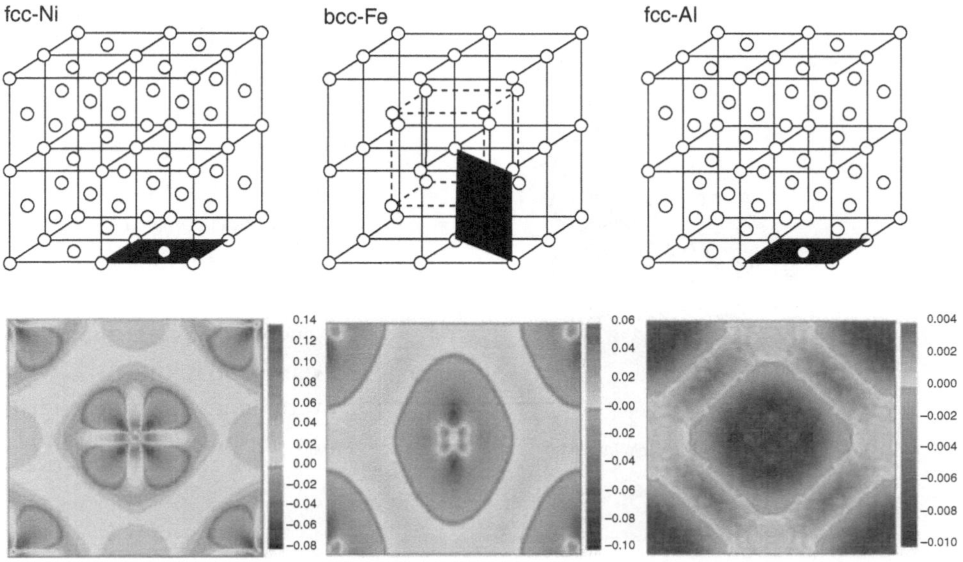

Figure 3.6 The bondings for Al, Fe, and Ni calculated in the plane indicated in the cells. Courtesy of Frank Lechermann.

From the metastable extensions of the binary phases one can extrapolate more reliably into the ternary system, using the four-sublattice CEF model as explained in section 5.8.4.4 for the ordered fcc phase and a two-sublattice model for the ordered bcc as described in section 5.8.2.4. The Gibbs energies of the ternary tetrahedra in the fcc model representing the ordered compounds Al_2FeNi, $AlFe_2Ni$, and $AlFeNi_2$ can initially be estimated from the binary bond energies u_{AlFe}, u_{AlNi}, and u_{FeNi} and the numbers of bonds of each type as

$$\begin{aligned} {}^{\circ}G_{\text{Al:Al:Fe:Ni}} &= 2u_{\text{AlNi}} + 2u_{\text{AlFe}} + u_{\text{FeNi}} \\ {}^{\circ}G_{\text{Al:Fe:Fe:Ni}} &= u_{\text{AlNi}} + 2u_{\text{AlFe}} + 2u_{\text{FeNi}} \\ {}^{\circ}G_{\text{Al:Fe:Ni:Ni}} &= 2u_{\text{AlNi}} + u_{\text{AlFe}} + 2u_{\text{FeNi}} \end{aligned} \tag{3.9}$$

The notation for the parameters is explained in chapter 5. Altogether this gives a dataset, from which the isothermal ternary phase diagram at 1023 K shown in Fig. 3.7(a) was calculated by B. Sundman (unpublished work, 2003). It can be compared with the phase diagram in Fig. 3.7(b), which was constructed from experimental data by Bradley (1951).

The Al–Fe–Ni phase diagram calculated at 1250 K by Lechermann *et al.* (2005) using only *ab initio* results as described in section 3.3.2 is shown in Fig. 3.7(c).

These calculations illustrate that skilled use of first-principles results for modeling metastable regions can make Calphad extrapolations more realistic and that these extrapolations can, inversely, give feedback to the first-principles diagrams when cluster expansion is used because a Calphad assessment can provide extrapolations into metastable regions whereas experimental data are usually scattered and limited to the stable range of the phases as in Fig. 3.7(b).

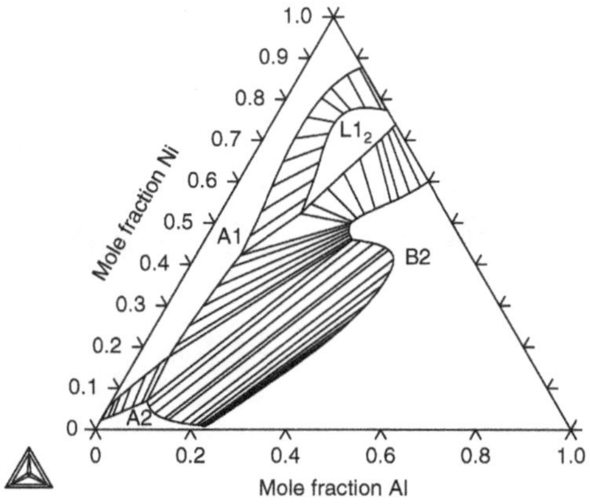

(a)

Calculated extrapolation from the assessed binary Al–Fe, Al–Ni, and Fe–Ni systems into the ternary system at 1023 K. The miscibility gap between the A2 and B2 phases and the direction of the L1$_2$ phase are correctly predicted in comparison with the phase diagram in (b) constructed from experimental data, even though the stability of L1$_2$ is overestimated. The high-Al corner has not been calculated.

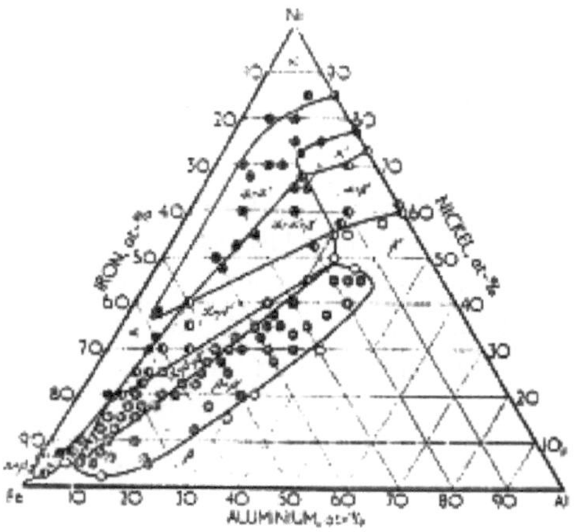

(b)

The Al–Fe–Ni phase diagram at 1023 K constructed from experimental data by Bradley (1951). Reprinted with permission from the Institute of Materials, Minerals and Mining.

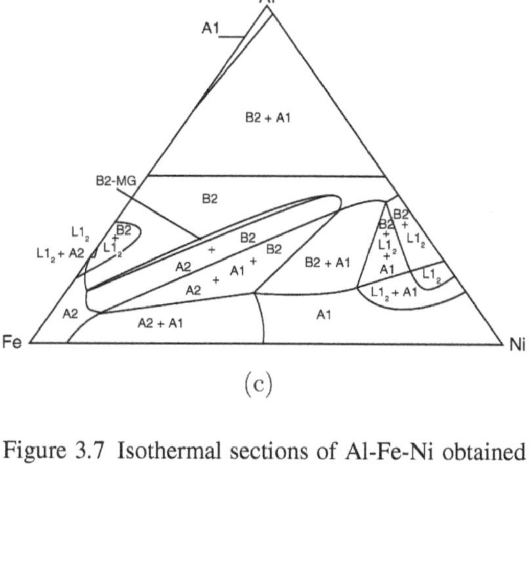

(c)

An *ab initio* prediction of the Al–Fe–Ni phase diagram at 1250 K. Reprinted from Lecherman *et al.* (2005). Copyright 2005, with permission from Elsevier.

Figure 3.7 Isothermal sections of Al-Fe-Ni obtained with different techniques.

3.3.4 *Electronic-structure calculations for ordered and disordered systems based on the CPA approach*

In contrast to the methods quoted above, the method based on perturbation theory performs configurational averaging analytically. One of the most well-known approximations in alloy theory is the coherent-potential approximation (CPA), which was recently reviewed in the context of Calphad by Turchi *et al.* (2007).

3.4 Final remarks

In the next chapters the connections between Calphad and the DFT results will be emphasized as much as possible. It is clear that further development will bring the theoretical results closer and closer to reality. Combining the calculated results with experimental data using empirical methods like Calphad will give important feedback and point out the good and bad trends of the theoretical predictions, at the same time as providing better predictions for thermodynamic applications.

Thermodynamic functions and phase diagrams calculated from first-principles data usually show correct trends and topology when compared with experimental data, but the calculated values are often far from the accuracy needed by scientific or industrial applications. The Calphad method provides the means to combine the first-principles calculations with experiments using models with adjustable parameters to reach the required accuracy for applications. At the same time the thermodynamic models and their extrapolations to metastable states are improved, as shown in section 3.3.3.

When publishing a paper including first-principles calculations, one should always compare the results with results from calculations done by other scientists and with experimental data.

Some years ago first-principles results were described as a kind of caricature of reality. Nowadays, even though the methods being used are much better, one has the impression that the picture the theoreticians are trying to paint has no contact with reality. If all the known experimental data for a system are compiled and checked for consistency according to Calphad methods, this is a clear image of reality. What one can then expect the theoreticians to achieve is that, when painting that image, the picture will be able to talk.

4 Experimental data used for the optimization

For numerical use the parametric functions described in chapter 5 must be assessed using experimental data. To get a maximum of information, all types of measurements that are quantitatively related to any thermodynamic function of state must be considered. From this dataset quantitative numerical data for the adjustable parameters of the Gibbs energy functions are obtained using the methodology described in chapter 6.

In order to evaluate the reliability and accuracy of the experimental data, it is of great help to know about the various experimental techniques used. Therefore, the main experimental methods in thermodynamic and phase-diagram investigations shall be described here. Nevertheless, this cannot be done here as deeply as in textbooks teaching experimental techniques, for example Kubaschewski *et al.* (1993).

Here the main emphasis is on how to use various types of data for the optimization and how to connect typical as well as more-exotic measured values with the thermodynamic functions of state.

Since experiments are expensive and time-consuming, all data available in the literature should be sought and their validity checked before one's own experiments are planned. An optimization using only literature data may be a good start, to give an overview, and may reveal, where the knowledge is poor, which further experiments are best suited to fill these gaps. Careful planning of one's own experiments taking this overview into account can very effectively keep the effort involved to a minimum and results in a very significant improvement of the optimization.

The measurements can be classified into a few principal types. In the following part, it will be explained how data may be measured and reported differently, although they are equivalent, using examples from the literature. As the highest level of classification, "thermodynamic data" and "phase-diagram data" shall be distinguished.

4.1 Thermodynamic data

4.1.1 *Calorimetric data*

Calorimeters measure heat transfer from a sample to its surroundings or vice versa. Isobaric heat transfer is identical with the change of enthalpy of the sample, if no other energy is simultaneously transferred. To be useful for the optimization, the sample must represent a "system" that is in a well-defined state before the measurement as well as there-after. Before the measurement the sample may be separated into several parts, such that each is in a well-defined equilibrium state, but these parts are not in mutual equilibrium and react on combination.

4.1.1.1 Mixing and reaction calorimetry

The principle of this method is that two (or more) well-defined different samples are combined in the calorimeter and react to give a single sample, which again must be well defined. Well defined means that all the variables upon which the enthalpy depends must be known: each sample must be in internal equilibrium; its temperature and composition must be known. (The pressure is assumed not to vary significantly and must be the same for all three samples, since the heat change can be identified with the enthalpy change only for isobaric changes.) The initial states, denoted $'''$ and $''$, may be in metastable equilibrium, as long as they are well characterized:

$$\text{sample}''' + \text{sample}'' \Longrightarrow \text{sample}'$$

The measured heat loss ΔQ has to be compared with the corresponding calculated enthalpy difference:

$$\begin{aligned} \Delta Q &= \text{enthalpy of reaction product} - \text{enthalpy of reactants} \\ &= \text{enthalpy of sample}' - \text{enthalpy of } [\text{sample}'' + \text{sample}'''] \end{aligned} \tag{4.1}$$

In most cases the two samples before and the single sample after the measurement are single phases, but samples in heterogeneous equilibrium may in principle also be used. For single-phase samples the superscripts denoting the samples ($'''$, $''$, and $'$) can be identified with phase indices. The enthalpy difference of Eq. (4.1) can be calculated by summing up the molar enthalpies (H^i) of the samples multiplied by their amounts (m^i), where the amount of the reaction product is the sum of amounts of the reactants. Each molar enthalpy is calculated with the appropriate expression from chapter 5 using the independent variables temperature (T^i) and composition (x^i), which may be different for all three samples. They must be referred to the same reference state, for example H^{SER} (SER denotes the stable-element reference state at 298.15 K and 1 bar):

$$\Delta Q = (m'' + m''') \cdot H'(T', x') - [m'' \cdot H''(T'', x'') + m''' \cdot H'''(T''', x''')] \tag{4.2}$$

The independent variables m'', m''', x'', x''', T', T'', and T''' of the samples have to be evaluated from the reported data; x' is calculated from m'', m''', x'', and x'''.

Equation (4.2) may be normalized with respect to amount 1 mol for the reaction product; then for simplification m''' is called m and m'' is called $1 - m$:

$$\Delta Q^{\text{norm}} = H'(T', x') - [(1 - m) \cdot H''(T'', x'') + m \cdot H'''(T''', x''')] \tag{4.3}$$

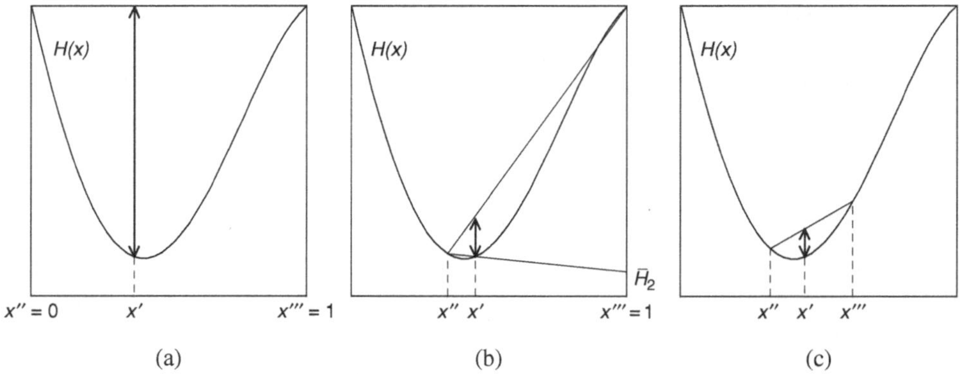

Figure 4.1 (a) integral enthalpy; (b) partial enthalpy; (c) curvature of $H(x)$.

Measurements of this type appear in many variations in the literature, depending on what is mixed in the calorimeter, as shown in Figs. 4.1(a)–(c)

1. **Enthalpies of mixing of binary liquid alloys** (Fig. 4.1a). The two liquid pure elements (amounts $1-m$ and m, $x''=0$, $x'''=1$) are separated inside the calorimeter. After the calorimeter has been heated and is ready for measurement, the two liquids are mixed, for example by breaking a glass crucible or by opening a stopper. In this case the three temperatures T', T'' and T''' in Eq. (4.2) are equal and identical to the calorimeter temperature:

$$\begin{aligned} \Delta Q &= H(x') - (1-m)\cdot H(x'') - m\cdot H(x''') \\ x' &= (1-m)\cdot x'' + m\cdot x''' \end{aligned}$$

2. **Series of mixing-enthalpy measurements** (Fig. 4.2). Known amounts of one of the pure elements are dropped consecutively from room temperature T''' into the liquid alloy inside the calorimeter, which is the result of the previous measurement. Here $T''=T'$ is the calorimeter temperature. The enthalpy difference $H_i^{\text{liq}}(T') - H_i^{\text{SER}}(T''')$ of the pure element which has been dropped in is often subtracted from the measured value by the authors and only this difference is reported as the result. Furthermore, if the sample inside the calorimeter is the other pure element in the first measurement, each measurement of the series is reported as the sum of all previous measurements, giving directly the integral molar enthalpy of the liquid. For use in the optimization it is strongly recommended that one recalculate from the reported values the heat effect of each single measurement by subtracting the effects of the previous measurements. In measurements with a solid element, the original temperature T''' may also be important, because the value of $H_i^{\text{liq}}(T') - H_i^{\text{SER}}(T''')$ included in the accepted parameters for G of the pure element i may differ from the value adopted by the authors of the experiments.
3. **Partial enthalpies** (Fig. 4.1b). If the amount $m'''=m$ of the dropped pure element in Eq. (4.2) is small relative to the amount $m''=1-m$ of the liquid in the calorimeter, the measurements yield partial enthalpies, approximating $\partial H/\partial N_i$

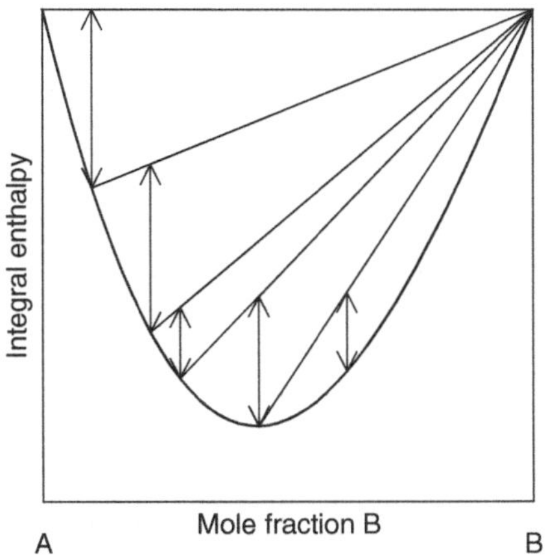

Figure 4.2 Measuring a series of enthalpies of mixing of a liquid by dropping successively small portions of the second element (B) into the calorimeter, starting with the pure first element (A) there.

by $\Delta Q/m$. This approximation is exact for one composition between x' and x'', but this composition is not known exactly. If x' and x'' are not very different, the mean $(x'+x'')/2$ is a good approximation for this mole fraction, where $\Delta Q/m$ is exactly the partial enthalpy. Unfortunately, some papers report only the resulting partial enthalpies. In others, however, these partial enthalpies are summed up with all the previous heat effects of the series and reported as integral enthalpies of mixing as described in the previous paragraph. The partial enthalpies are then available only by reconstructing the originally measured values from the reported ones.

4. **Direct-reaction calorimetry**. Inside a calorimeter a mixture of the powders of the pure elements (amounts m'' and m''') is quickly heated in a small furnace until it reacts to form a compound. The power input of the furnace is subtracted from the heat measured. The measurement must include all the heat released until the sample has cooled down to the starting temperature. Then $\Delta Q/(m''+m''')$ directly represents the enthalpy of formation of the compound at the starting temperature of the calorimeter. The completeness of the reaction has to be checked.
5. **Solution and combustion calorimetry**. The terms of Eq. (4.2) may be measured separately, once by dissolving the compound or solid solution (single-prime variables) and once by dissolving a mechanical mixture of the two pure elements (double- and triple-prime variables) in the same solvent inside a calorimeter. If the final solutions in both cases have the same state (T, x), the difference of the two heats of solution is equivalent to ΔQ of Eq. (4.2). The solvent may be an aqueous acid at room temperature or a liquid metal (Sn, Al,...) in a high-temperature calorimeter.

   Combustion in a bomb calorimeter is treated similarly. The heats of combustion of the pure components are subtracted from the heat of combustion of the compound.

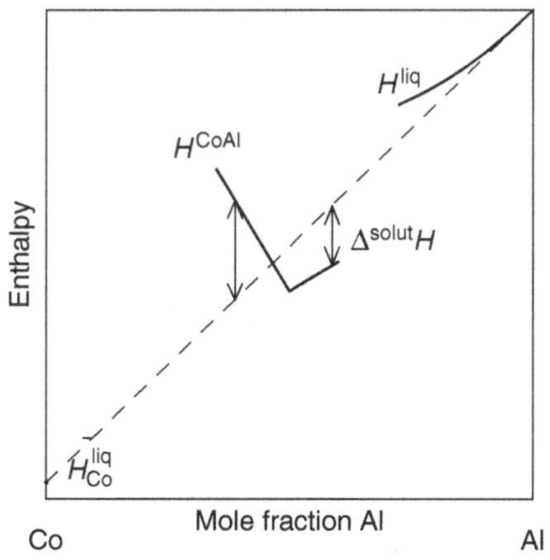

Figure 4.3 Solution calorimetry of an intermetallic compound (CoAl) in liquid of one of the pure components (Al). The heat of solution of the compound is relatively small and the shape of the $H(x)$ curve can be measured more accurately than can the value itself.

An advantageous special case of solution calorimetry is when the solvent liquid metal is one of the elements of the binary system itself. As an example the determination of the enthalpy of mixing of the $CoAl_{1\pm x}$ phase by solution calorimetry in liquid Al (Henig *et al.* 1980) is given in Fig. 4.3. The enthalpies of solution of the CoAl phase, $\Delta^{\text{solut}}H$, are relatively small and measured accurately to within about 5%, which contributes less than ± 1 kJ mol^{-1} to the enthalpy of formation. The heat of solution of the pure components is the negative of the partial enthalpy of Co in liquid, $\Delta H_{\text{Co}}^{\text{liq}}$, times the mole fraction of Co in the compound, because the partial enthalpy of Al in dilute molten Al is virtually zero. To get the enthalpy of formation from the heat of solution of the phase, one has to subtract the partial enthalpy of pure Co in nearly pure molten Al. This heat of solution of pure Co, corresponding to the amount of Co in the phase, is given in Fig. 4.3 by the dashed straight line. The heats of solution of the phase have to be added to the values given by this straight line. This is visualized by double arrows for two phases with different Co contents.

The partial enthalpy of Co is much larger than the heat of solution of the phase and is also measured accurately to within about 5%, contributing about ± 7 kJ mol^{-1} to the enthalpy of formation of CoAl. This contribution, however, is the same for all CoAl samples. The difference of the accuracies of the samples is thus less than 1 kJ mol^{-1}. The shape of the $H(x)$ curve is therefore determined to within ± 1 kJ mol^{-1}, although the values themselves may have uncertainties of ± 7 kJ mol^{-1}.

6. **The shape of the $H(x)$ curve** (Fig. 4.1c). If samples 2 and 3 are liquid solutions with mole fractions that are not very different, the heat effect on mixing mainly reflects the curvature of the plot of integral enthalpy of mixing versus mole fraction, $H(x)$. Experiments of this type are seldom reported in the literature because the results cannot be presented directly in terms of conventional thermodynamic functions, but, as experiments done especially for the optimization, they may be very useful.

Further modifications of these techniques may be found in the literature. The treatment of the experimental results for the optimization, however, should give no additional problems.

4.1.1.2 Drop and scanning calorimetry

The principle of these two methods is that a single, well-defined sample in internal equilibrium is brought from one temperature to another temperature, at which it again reaches internal equilibrium, and the heat loss or gain ΔQ is measured. ΔQ is identified with the enthalpy difference between the two equilibrium states:

$$\Delta Q = H'(T', x) - H''(T'', x) \tag{4.4}$$

In drop calorimetry the temperature change is usually large, whereas in scanning calorimetry it is usually small. The sample in any of the two states may be either single-phase or in heterogeneous equilibrium. The overall composition x is the same for both states. The phases in the two states may, however, be different.

In scanning calorimetry, if both states contain the same single phase, then, instead of using Eq. (4.4), the measurement may be identified with the heat capacity C_p:

$$C_p(T, x) \approx \frac{\Delta Q}{\Delta T} \tag{4.5}$$

In continuous scanning calorimetry even $\partial Q/\partial T$ may be measured directly. With scanning calorimetry across a melting or transformation temperature, the enthalpy of isothermal melting or transformation may be measured by extrapolating to $\Delta T = 0$. For more details see Hoehne *et al.* (1996).

4.1.2 *Chemical-potential data*

These measurements are classified into evaluations of the emf of reversible galvanic cells, of vapor pressure, of equilibria with a well-defined gas mixture, and of the solubility in a nearly inert solvent (treated as a Henrian solution). In these cases two different equilibrium states are considered, between which just one element can be transferred by a process that, in a *Gedankenexperiment*, can be carried out reversibly. One of the states is usually the pure element itself.

The data may be reported either as differences of chemical potentials (partial Gibbs energies) $\Delta\mu$ or as activities a, which are connected by the formula $\Delta\mu = RT\ln(a)$. In the least-squares method it is important to have nearly equal probabilities for positive and negative "errors," so the function μ or $\ln(a) = \mu/(RT)$ is in most cases more suitable than the activity a.

4.1.2.1 Galvanic cells

In a reversible galvanic cell the more electronegative element can be transferred from the pure element to the alloy (single-phase solution or two-phase equilibrium state) or vice versa by an electric current. If the current is interrupted, in a reversible cell the exchange is totally inhibited. The electric energy per mole transferred is identical to the difference in partial Gibbs energy,

$$\mu_i^{\text{alloy}} - {}^\circ G_i = U \cdot Z \cdot F \tag{4.6}$$

where U is the voltage of the cell, Z the charge of the ion, and F the Faraday constant. $Z \cdot F$ is the electric charge transported with one mole of the element and thus $U \cdot Z \cdot F$ is the electric energy which can be gained by diluting one mole of the pure element in the alloy or which must be provided to retrieve it from the alloy.

There are various experimental realizations of galvanic cells, but they always contain two electrodes separated by an electrolyte. The electrolyte may be a molten salt (one that is often used is the eutectic mixture of KCl and LiCl with some chloride of the metal, the ions of which are to be transferred), a solid ionic conductor (Y-doped ZrO_2, CaF_2), or an aqueous solution. The electrode may be of the "first kind," i.e., the metal is in equilibrium with its own ion (Zn/Zn^{2+}), or of the "second kind," i.e., the electrode is surrounded by an insoluble salt and discharging of cations releases anions into the electrolyte or vice versa ($Ag/AgCl/Cl^-$, $Ni/NiO/O^{2-}$). An electrode of the second kind may be measured against a reference electrode based on an anion, for example

Cu (in an Au–Cu alloy) + Cu_2O/ZrO_2 (doped with Y_2O_3)/O_2(Pt)

or the reference electrode may be an electrode of the second kind with another cation, for example

(Mo)/Cu (in an Au–Cu alloy) + Cu_2O/ZrO_2 (doped with Y_2O_3)/Ni + NiO/(Mo)

The condition of reversibility of the cell is that the conductivity of the electrolyte must be purely ionic and that only one element must be discharged or ionized at each electrode and the valence of ionization must be well defined. The reversibility is proved by reaching the same stable equilibrium voltage after passing a current in both directions.

Usually the results are reported as the voltages U of the cells, sometimes only the resulting values of $\Delta\mu$ ($U \cdot Z \cdot F$ in Eq. (4.6)) or $\ln(a)$ ($U \cdot Z \cdot F/(RT)$ in Eq. (4.6)) are reported.

Since it is possible to make many measurements with the same cell by changing the temperature, the measured values are often reported not directly but rather as linear functions smoothing the measured values of each cell, $U = A + B \cdot T$ or $\mu = A + B \cdot T$. More exactly, that should be a very slightly curved function with a three-term formula $U = A + B \cdot T + C \cdot T \cdot \ln(T)$, but usually the values are not accurate enough for one to determine C, which represents the difference ΔC_p between ${}^\circ C_p$ of the pure element i and the partial $C_{p,i}$ of the alloy. When using such information for an optimization, the values for one sample may be represented by two or three values from this formula, taken from the lowest and highest quarter of the temperature range investigated and, if C is given, also from the center of the range.

$\Delta\mu_A$ values in a two-phase field between an intermediate phase Φ and the nearly pure element B are usually reported as Gibbs energies of formation of the intermediate phase Φ. This is possible because generally $G^\Phi = \sum x_i \cdot \mu_i$ and, in a two-phase field with pure element B, $\mu_B - {}^\circ G_B = 0$ and therefore $G^\Phi - \sum(x_i \cdot {}^\circ G_i) = x_A \cdot (\mu_A - {}^\circ G_A)$, i.e., the Gibbs energy of formation of phase Φ and $\Delta\mu_A$ in the two-phase field between Φ and the pure element B are proportional. If the formula for Φ is A_pB_1, the Gibbs energy of one mole of Φ (i.e., of $p+1$ moles of atoms of Φ) is identical to $\Delta\mu_A$ since $x = 1/(p+1)$.

If the galvanic cell contains a reference electrode other than the pure element, the voltage of a cell with this reference electrode and the pure element must be subtracted from the measured voltage in order to get the voltage U of Eq. (4.6). If this was not measured in the same investigation, it must be calculated from a thermodynamic database considering the Gibbs energy change of the reaction connected with the cell containing the reference electrode and the pure element; and then this value of G must be divided by $Z \cdot F$.

4.1.2.2 Vapor pressure

If one of the elements of an alloy is much more volatile than the other one(s), the vapor phase above this alloy contains virtually this element only. Its partial pressure can be compared with that above the pure element at the same temperature. In a *Gedanken-experiment* the element can be transferred reversibly between the two vapor phases by an ideal-gas engine. The work connected with the transfer of 1 mol of vapor of element i is equal to the change in chemical potential of element i. It is calculated by integrating $V\,\mathrm{d}p$ between the two pressures, assuming the volume V to be expressed by the ideal-gas equation of state, $V = R \cdot T/p$:

$$\mu_i^{\text{alloy}} - {}^\circ G_i = R \cdot T \cdot \int_{{}^\circ p_i}^{p_i^{\text{alloy}}} \frac{\mathrm{d}p}{p} = R \cdot T \cdot \ln(p_i^{\text{alloy}}/{}^\circ p_i) \tag{4.7}$$

where ${}^\circ p_i$ is the vapor pressure of the pure component i at the same temperature.

This is true for monatomic vapor only. For diatomic vapor molecules Eq. (4.7) gives half the μ-value and for n-atomic vapor molecules one nth of the μ-value. If the vapor phase is a mixture of monomer atoms and polymeric molecules, the partial pressure of a single species should be measured rather than the total vapor pressure above the alloy as well as above the pure element.

There are several experimental methods for vapor-pressure measurements.

4.1.2.2.1 Direct pressure measurement Direct measurement of the vapor pressure by a manometer can seldom be done at high temperature. Firstly the material of the manometer must not react with the sample and secondly the manometer must not be affected by high temperature. The vapor pressure of Te was measured in this way: the sample was included in a sealed silica-glass vessel with a deformable thin-walled part enclosed in an atmosphere of nitrogen with adjustable pressure. By use of a lever with a mirror a deviation of the position of the thin-walled part could be recognized. The pressure of the outer atmosphere was always controlled to keep the mirror at the same position and thus was equal to the vapor pressure of the sample.

4.1.2.2.2 Optical spectroscopy of the vapor phase If the volatile component does not attack glass, the sample may be sealed into a glass vessel with two parallel windows, which is put into a furnace. By measuring the absorption at a characteristic wavelength of the gas, the partial pressure is measured. For calibration, the pure volatile element may be measured similarly.

4.1.2.2.3 The gas-transport method The substance is in a boat inside a furnace, where an inert gas (usually argon) flows a measured flow rate. The volatile component evaporates, is transported with the gas, and, before the end of the furnace, is condensed on a water-cooled finger of glass or ceramic. The amount of condensed material is determined by measuring the change in weight of the finger or by chemical analysis of the dissolved deposit, and divided by the total volume of gas that has flowed through the furnace. The measurement is usually done with different flow rates of the gas and the measured concentration of the volatile component in the gas is extrapolated to zero flow rate. This is assumed to be the concentration of the saturated gas and is transformed into a partial pressure.

4.1.2.2.4 Isopiestic and dew-point methods The sample is sealed in an evacuated tube, within which at some distance a sample with known partial pressure of the same volatile component is situated. The tube is kept in a furnace, where the temperatures of sample and reference sample can be controlled independently. The volatile component may now move from the reference sample to the sample or vice versa until both have the same vapor pressure. Then the tube is rapidly cooled and the sample analyzed.

The partial pressure of the reference sample should not vary rapidly with composition; preferably it is a binary two-phase sample or the pure volatile element itself at a lower temperature. It has to be checked that no part of the tube has a temperature lower than the boiling point or sublimation temperature of the pure component.

A modification of this method is the measuring of the dew point: the pure element is present not as a sample, but at the end of the tube, which is usually of glass. The furnace has a small window and the temperature there is slowly decreased until a dew of liquid or solid pure element condenses there. The temperature is raised and decreased again by small values, to see whether condensation and re-evaporation can be done reversibly and to get the equilibrium temperature as accurately as possible. The vapor pressure of the pure element must be known as a function of the temperature:

$$\ln(p) = \frac{A}{T} + B + C \cdot \ln(T) \tag{4.8}$$

which reflects the ΔG between condensed and gaseous pure element; then the logarithm of the activity $\ln(a) = \ln(p/p_0)$ can be directly calculated from the temperature difference ΔT between sample and dew point:

$$\begin{aligned} \ln(p/p_0) &\approx -\frac{\mathrm{d}\ln(p)}{\mathrm{d}T} \cdot \Delta T \\ &= \left(-\frac{A}{T^2} + \frac{C}{T}\right) \cdot \Delta T \end{aligned} \tag{4.9}$$

The temperature T in this formula is approximated by taking the mean of the sample temperature and the dew-point temperature.

4.1.2.2.5 The Knudsen cell Low vapor pressures, below 10^{-1} Pa, are measured by a Knudsen cell. The sample is enclosed in a container with a small orifice. The volume inside the container virtually reaches the equilibrium vapor pressure and the molecular beam emanating from the orifice is proportional to this vapor pressure. To measure the intensity of this molecular beam several methods have been used.

1. It may be collected on a target during a measured time and then analyzed.
2. The Knudsen cell is suspended on a very sensitive micro-balance and the weight loss is recorded continuously.
3. The molecular beam is bombarded with low-energy (≈10-eV) electrons and the resulting ions are analyzed by mass spectrometry. This is the most-often-used technique nowadays.

In using the Knudsen cell together with mass spectrometry, the main problem is deviations of the calibration after changing the sample, which makes it difficult to get comparable values for p_i^{alloy} and $^\circ p_i$ in Eq. (4.7). The main sources of variation are the size of the orifice, the orientation of the Knudsen cell with respect to the ionization chamber, and the effectivity of the ionization, which is the factor correlating the density of the molecular beam with that of the ion beam and, finally, with the recorded signal.

Various methods to diminish the influence of these effects have been reported. Most of them cannot be discussed here, but for each use of reported experimental Knudsen-cell data in an optimization the description of the experimental procedure must be carefully studied with respect to this problem in order to judge the quality of the data.

Neckel and Wagner (1969) proposed that, if the vapor pressures of the two components are of the same order of magnitude in a binary system, one could measure the ratio of the two vapor pressures over the whole composition range. Their only assumption is that the factor of proportionality between the vapor pressure and the recorded signal changes equally for both components in different experiments. This eliminates the influence of the size of the orifice, orientation of the Knudsen cell, and electron-beam intensity in the ionization chamber (which is not the mean energy of an electron).

4.1.2.3 Equilibria with gases of known activity

A gas mixture containing H_2 and H_2O molecules at high temperatures provides a well-defined chemical potential of oxygen. Depending on the H_2/H_2O ratio, between 0.01 and 100 a range of $\Delta\mu_O = RT\ \ln(10\,000)$ is covered. A sample in contact with such a gas mixture will take up or release oxygen until μ_O reaches the equilibrium value. By analyzing the final oxygen content of the gas the chemical potential of oxygen can be determined.

4.1.2.4 Solubilities in Henrian solutions

In dilute solutions, according to Henry's law the activity of a solute is proportional to its mole fraction. Therefore the mole fraction of an element in a dilute solution may be used as a measure for the activity. A sample may be immersed in a liquid that does not dissolve in it and itself dissolves only small amounts of the sample at a particular temperature, at which equilibrium is reached within a reasonable time. The equilibrium mole fraction(s) of one or several components in the liquid are compared with those of samples with known activities of these components, which have been treated in the same way.

4.2 Binary phase-diagram data

The quantities measured in binary phase diagrams are either (i) temperatures of invariant (three-phase) equilibria (points (a) in Fig. 4.4), or points (x', T) on the boundaries of two-phase fields. The latter points can be measured either (ii) for samples of known composition x' by determining the temperature T (points (c) in Fig. 4.4), or (iii) for a series of samples of different composition x, annealed to equilibrium at the same temperature T, determined to be single-phase or two-phase (points (b) in Fig. 4.4). The corresponding values calculated with the least-squares method are (i) the calculated temperature, where besides the amounts of the three phases (arbitrary, but >0) only the pressure is given; (ii) the calculated temperature of the two-phase equilibrium, where besides the pressure, the amounts of the two phases and the composition of one phase are given; or (iii) the calculated composition x' of one phase in the two-phase equilibrium at given temperature and pressure. (Note that in cases (ii) and (iii) the two phases of the equilibrium are treated differently in the least-squares calculation.)

There are several experimental methods to measure phase-diagram data.

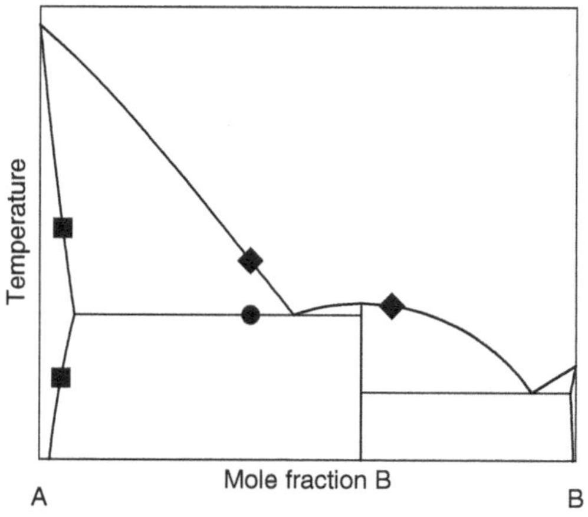

Figure 4.4 Types of measurements in a binary phase diagram: (a) ● the temperature of three-phase equilibrium, (b) ■ the mole fraction x' measured at a given temperature T, and (c) ♦ the temperature T measured at a given mole fraction x'.

4.2.1 *Thermal analysis*

A sample is heated or cooled and its temperature recorded with time. When the sample is going from a single-phase equilibrium state into a two-phase field, some heat of precipitation is released. This leads to a kink in the temperature versus time curve. The sensitivity of the method can be very much enlarged by the use of differential thermal analysis (DTA). The sample and an inert reference sample are symmetrically located in a block. Besides the temperature of the sample, the temperature difference between sample and reference sample is recorded (Fig. 4.5). As long as there is no reaction in the sample, the temperature difference follows the furnace temperature in approximately the same manner as does the temperature of the reference sample and the $\Delta T(t)$ curve remains near zero. If now the sample crosses the boundary to a two-phase field, the released heat delays the temperature change of the sample and $\Delta T(t)$ shows a kink. The temperature difference versus time curve ($\Delta T(t)$) can be much more amplified than the $T(t)$ curve of a normal thermal analysis.

In a three-phase (invariant) equilibrium the reaction evolving the heat takes place at constant temperature. In the thermal analysis this shows up as a horizontal part in the $T(t)$ curve. The length of the horizontal part is roughly proportional to the amount of matter reacting in the three-phase equilibrium. Plotting this length versus mole fraction for a eutectic reaction gives a triangle, called a Tammann triangle, whose vertex indicates the mole fraction of the eutectic liquid. If the DTA is performed in a scanning calorimeter, the result of the Tammann triangle is quantitatively correct. For peritectic reactions, however, due to usually severe segregation, the Tammann triangle must be used with caution.

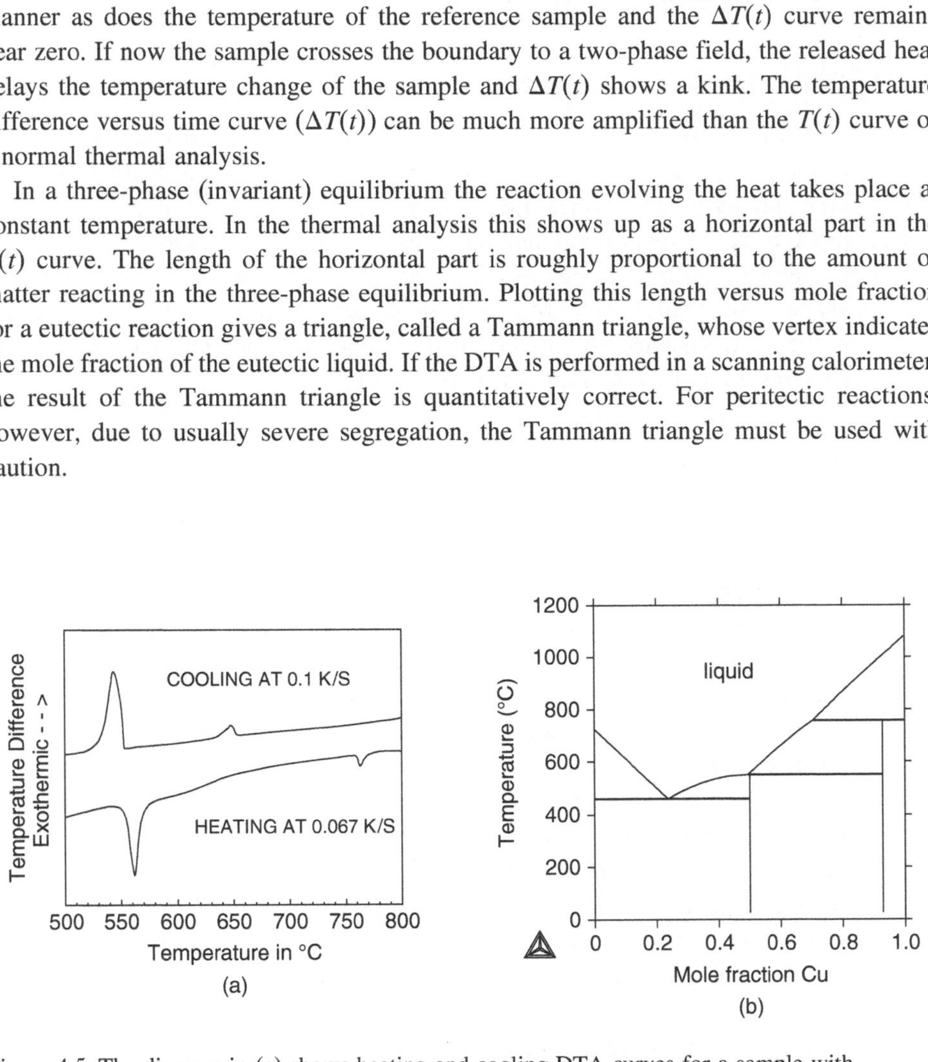

Figure 4.5 The diagram in (a) shows heating and cooling DTA curves for a sample with 81 mol% Cu in Ba–Cu, courtesy of R. Schmid-Fetzer. The phase diagram assessed by Konetzki *et al.* (1993) is shown in (b). The curve illustrates the large undercooling for the high-temperature peritectic and very small undercooling for the medium-temperature three-phase equilibrium.

For a two-phase field boundary the sensitivity of the thermal analysis is much larger for flat boundaries than for steep boundaries. First the amount of matter reacting in a given temperature interval, for example 1 K, can be derived by the lever rule from a tie-line just 1 K below the boundary and is the larger, the flatter the boundary. Secondly the Gibbs–Konovalov rule (Goodman *et al.* 1981) Eqs. (2.50) and (2.50), shows that also the "enthalpy of reversible precipitation" is larger for flatter boundaries. The enthalpy of reversible precipitation in principle is identical to the heat released in the DTA experiment. An overview on the field of thermal analysis is given by Shull and Joshi (1992). A very detailed paper about DTA measurements and interpretation of the corresponding results is that by Boettinger and Kattner (2002).

4.2.2 *Properties versus temperature*

In thermal analysis a singularity of the enthalpy versus temperature curve $H(T)$ is used to identify the boundary between different fields of a phase diagram. In principle any other property can be measured versus T instead of the enthalpy H. A kink of the property versus temperature plot indicates a boundary in the phase diagram. Examples of properties used for this purpose are dilatometric measurements (the length of a sample versus temperature) in the Mg–Zn system, used by Grube (1929); electric conductivity in the Zr–O systems, used by Gebhardt *et al.* (1961); and magnetic susceptibility in the Fe–Nb system, used by Ferrier and Wachtel (1964).

4.2.3 *Properties versus composition*

Similarly to how a property can be plotted versus temperature, it can be plotted versus composition (mole fraction). The main difference is that the temperature of a single sample may be changed continuously and the property recorded. For the composition dependence, however, many samples with different compositions are annealed to equilibrium at the same temperature and the property is measured for these different samples. In principle the same properties can be plotted versus temperature or composition, but several properties are much more usefully plotted versus temperature whereas others are better plotted versus composition.

A property preferably plotted versus composition is the lattice parameter measured by X-ray diffraction. It is constant in a two-phase field and varies in a single-phase field. If the decomposition of supersaturated single-phase samples can be prevented by quenching, this facilitates its use. Several samples of a phase ϕ are equilibrated at the temperature of maximum solubility and quenched and a lattice parameter versus mole fraction curve $a^{\phi}(x)$ of ϕ is generated. A single two-phase sample containing ϕ is now equilibrated at different temperatures and quenched, and the lattice parameter of ϕ is measured. Using the $a^{\phi}(x)$ curve as the inverse function, $x(a^{\phi})$, the corresponding mole fraction of ϕ is read. For non-cubic phases this procedure should be carried out with all the different lattice parameters of the phase.

The electric conductivity ρ is a property that can be used as a function of either temperature or mole fraction. An example is in the paper of Grube (1929), which was used in the assessment of the Mg–Zn system (Agarwal *et al.* 1992). In a $\rho(x)$ plot as well as in a $\rho(T)$ plot, there is just a kink of two not necessarily linear curves at the boundary between a single-phase and a two-phase field.

4.2.4 Metallography

A very useful tool in phase-diagram determination is micrography, which, for metals, is often called metallography. Much of the information from metallography is qualitative and, although very useful in general, cannot directly be used for the thermodynamic optimization.

Boundaries in phase diagrams are often deduced from series of samples of different compositions equilibrated at several selected temperatures. The results for a "single phase" or "two phases" are plotted in a temperature versus mole fraction diagram (a conventional phase diagram). The boundaries between single-phase and two-phase fields are now mapped in such a way as to satisfy these results. Quantitatively, a result may be expressed as a point on the boundary (represented by x' and T), centered between two consecutive samples found to be respectively "single-phase" and "two-phase" $\pm$ half the distance ($\Delta x'/2$ or $\Delta T/2$).

A large advantage of micrography is that phases decomposing on quenching can usually still be identified from their shapes. For example, droplets of liquid present in the sample at the annealing temperature, although they become solidified during quenching, are usually well detected as rounded areas of fine-grained material between solid crystallites of much larger grain size.

Electron micrography and scanning micrography may in principle be used for the determination of boundaries in phase diagrams in the same manner as optical micrography.

4.2.5 Quantitative metallography

In the micrograph of a two-phase sample the ratio of the areas covered by the images of the two phases can be measured and it is approximately identical to the volume ratio of the two phases. If the molar volumes of the two phases are known, the molar ratio can be calculated from the volume ratio. This molar ratio may, for example, be expressed as moles of atoms of phase 1 in a total amount of one mole of atoms, m'. This value appears also as the calculated value in the description of an equilibrium in Eq. (2.23) and thus can be used in the least-squares calculation.

Unfortunately, in the literature this type of measurement is seldom used for the construction of a phase diagram. In the above-mentioned use of micrography the "two-phase" samples have the potential to give more information: even a very rough estimate of the amount of the second phase (here usually a very small amount) quantitatively gives the distance of the boundary from the composition of this sample.

For the quantitative use of this type of measurement, a source of systematic errors must be taken into account, namely that during grinding and polishing one of the phases may be more easily removed than the other. Then, compared with the ideal geometry of the micrograph surface, this other phase presents a larger surface area than it should do according to the above statistical consideration.

4.2.6 Microprobe measurements

In the electron microprobe, areas of the order of 1 μm^2 can be chemically analyzed by X-ray spectroscopy. If a two-phase sample is annealed for long enough for it to have grain sizes of several diameter micrometers and to have both phases in equilibrium, the equilibrium composition of both phases can directly be analyzed. The literature contains several methods used to take care of the X-ray absorption in the sample by deducing amounts of elements from measured X-ray intensities of their characteristic wavelengths. Transmission electron microscopy (TEM) instruments can also do this, with a very much higher resolution. A good general reference for TEM is the book by Williams and Carter (1996). With the technique called ALCHEMI (Spence and Tafto 1983) the state of ordering can be determined and point defects identified (Jones 2002).

4.3 Ternary phase-diagram data

The methods used to localize boundaries in ternary phase diagrams are in principle the same as for binary ones, but there are now two independent variables describing the composition and consequently there may be two different types of measurements of mole fractions for the same phase with two different interpretations.

4.3.1 Thermal analysis in ternary systems

The temperature of the beginning of primary crystallization (on cooling a single-phase liquid until the first precipitation occurs) and the temperature of an invariant equilibrium are found in the same manner as in a binary system. The Tammann triangle, constructed from the duration of isothermal solidification, now becomes a pyramid constructed over two independent composition variables, e.g., mole fractions. Between primary crystallization and invariant equilibrium, however, there may be the beginning of a secondary crystallization with an additional kink in the thermal-analysis line (section 4.3.4).

4.3.2 Two-phase tie-lines

At the end of a two-phase tie-line the two independent variables defining the composition may be interpreted as two vectors, which can be combined to give one in the direction of the tie-line and another one parallel to the boundary of the two-phase field against a single-phase field. Changing the latter vector, without looking also to the second phase, simply means selecting another tie-line. The first vector, however, describes the length of the tie-line and thus the position of the boundary, for example a ternary solubility. By the same methods as in binary systems, two different types of measurements may be performed: either the temperature may be measured, where this boundary is at a given composition; or the composition may be measured, where this boundary is found at a given temperature. At a given composition the temperature may be measured by thermal analysis, whereas the composition can be determined by metallography, X-ray lattice parameter measurements, microprobe measurements of quenched samples, or any other of the methods used for binary systems. The lattice parameters, however, in contrast to those of a binary system, may vary also within the two-phase field, unless the samples are situated along the same tie-line.

If temperature is the measured quantity, the "calculated value" F_i of Eq. (2.52) is the temperature of the tie-line, calculated for the composition of phase 1 fixed at the experimentally determined composition using Eq. (2.38). Both phases must be identified for the calculation.

If the composition of α is the measured quantity at a given temperature, the tie-line passing through this experimentally determined composition is calculated for the given temperature by using Eq. (2.42), see Fig. 4.6. The difference between "calculated" and "measured" compositions must be taken parallel to the tie-line and may be related to the total length of the tie-line. This is equivalent to calculation of the phase amount of β referred to the measured composition of α as the overall composition (symbols ■ in Fig. 4.6).

$$m^{\beta} = (x_i^{\alpha}(\text{calc}) - x_i^{\alpha}(\text{meas}))/(x_i^{\alpha}(\text{calc}) - x_i^{\beta}(\text{calc})) \qquad (i = \text{B, C}) \qquad (4.10)$$

This "amount" can directly be interpreted as the "error" $F_i - L_i$ according to Eq. (2.52). It is zero if the calculation exactly reproduces the experiment. If the measured composition is outside the calculated two-phase field ($\alpha + \beta$ in Fig. 4.6), then this "amount" formally becomes negative.

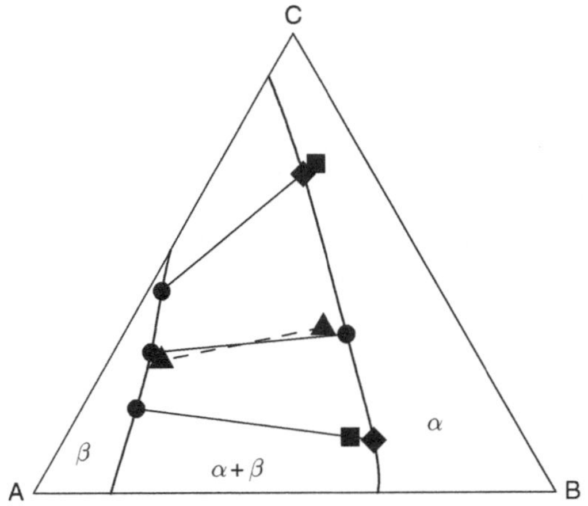

Figure 4.6 Tie-lines of a ternary two-phase field at a given temperature: ■ measured composition of phase α, ♦ calculated composition of α in the tie-line passing through ■, ▲ two points experimentally determined to be on the same tie-line (not necessarily the endpoints of the tie-line), and ● calculated endpoints of tie-lines not directly related to measured points.

4.3.3 *Directions of two-phase tie-lines*

The composition vector parallel to the boundary between two-phase and single-phase fields becomes meaningful when it is compared with the other end of the same tie-line. Measuring both ends of the same tie-line means measuring the "direction of the tie-line." An often-used method for that is the simultaneous determination of the compositions of both phases by microprobe measurements. Another common procedure is measuring the lattice parameters of one or both phases for two series of samples along two lines in the two-phase field that are approximately parallel and near the two boundaries. Samples with the same lattice parameters but in the two different series lie on the same tie-line (Fig. 4.7).

As a measure of the "error" $F_i - L_i$ in Eq. (2.52), the vector product of calculated and measured tie-lines may be used, both tie-lines being interpreted as vectors in the x_B–x_C plane. The calculated tie-line may be selected using the center of the measured tie-line as the overall composition in Eq. (2.42),

$$\begin{vmatrix} {}^{c}x_B^{\alpha} - {}^{c}x_B^{\beta} & {}^{c}x_C^{\alpha} - {}^{c}x_C^{\beta} \\ {}^{m}x_B^{\alpha} - {}^{m}x_B^{\beta} & {}^{m}x_C^{\alpha} - {}^{m}x_C^{\beta} \end{vmatrix} = \text{error} \tag{4.11}$$

where ${}^{c}x_i^{\Phi}$ and ${}^{m}x_i^{\Phi}$ denote the calculated and measured mole fractions of component i in phase Φ, respectively.

A tie-line connects composition points where the chemical potentials μ_i are the same in both phases. This point of view is often helpful if the two phases are nearly binary in two different subsystems of the ternary. An experimental tie-line of this kind indicates equality of the chemical potential of the element common to both binary subsystems. The

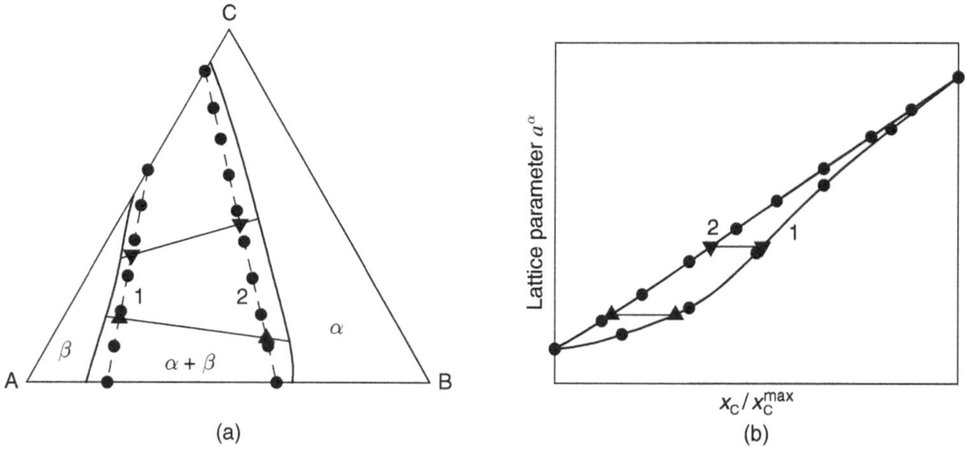

Figure 4.7 Determination of the direction of tie-lines by lattice-parameter measurements. With two-phase $\alpha+\beta$ samples of compositions along the two dashed lines 1 and 2 in (a) (•) the lattice parameter of the α phase, a^α, is measured and plotted against a composition variable in (b) (here x_C/x_C^{max}, i.e. 0–1 along dashed lines 1 and 2). Points with the same lattice parameter a^α are on the same tie-line. Two pairs of points, (▲) and (▼), are selected in (b) and re-translated to (a).

calculated values of these chemical potentials, however, are given by the descriptions of the previously optimized binary systems. Such a measurement is therefore a check of the compatibility of the two binary descriptions.

4.3.4 *Ternary three-phase equilibria*

The composition of one phase of a three-phase equilibrium is described by two independent mole fractions. Similarly to the case of a ternary two-phase field, two different measurements of the composition of one of the equilibrium phases can be defined for a given temperature. They may be described as two different linear combinations of the mole fractions, perpendicular or parallel to the line connecting the compositions of the other two phases. The measurement perpendicular to this line corresponds to the length of a two-phase tie-line, whereas the measurement parallel to it corresponds to the direction of a tie-line. The latter, like the directions of two-phase tie-lines, may help as a check of the binary descriptions. Besides the two composition variables of one of the phases of a three-phase equilibrium, also those of a point on the boundary between the three-phase equilibrium and one of the adjacent two-phase equilibria can be measured.

The experimental methods are the same as for isothermal two-phase equilibria. Lattice parameters are constant within a three-phase field.

Instead of measuring the composition at a given temperature, the temperature can be measured for a given composition, for example by thermal analysis. The secondary effect of a thermal analysis corresponds to the temperature at which the boundary between the three-phase field and an adjacent two-phase field crosses the overall composition of the sample. If the phase of primary crystallization has a range of homogeneity, its composition usually changes during solidification and the measured temperature, due to this segregation, might not be the equilibrium temperature. Therefore this kind of measurement should be used for the optimization only, if the primary phase is stoichiometric.

In principle, for temperature measurements at a given composition the same cases should be distinguished as for isothermal composition measurements. Fixing both independent mole fractions of an overall composition is, however, usually incompatible with getting a single phase of a three-phase equilibrium.

Besides measurements of boundaries of a phase diagram, phase amounts in equilibrium can be measured by various methods, for example by quantitative metallography, for three-phase as well as for two-phase equilibria at a given overall composition. The corresponding calculated value is determined by applying the lever rule, if it is not taken directly from the solution of the equilibrium conditions Eqs. (2.23)–(2.27).

4.4 Multicomponent and other types of experimental data

Experimental data from multicomponent systems are usually not used directly in an assessment, but, if the extrapolation from the lower-order systems gives wrong results, one may use them to modify parameters describing the lower-order systems. For example, phases that are not stable in a binary system may have parameters assessed from a higher-order system such as the fcc phase in the Cr–Mo system. Great care must be taken to ensure that such modifications do not change the descriptions of the lower-order systems, for example by letting a metastable phase appear to be stable.

In PARROT any measured quantity of a state variable, or a combination of state variables, can be used as experimental data, provided that there is at least one extra quantity measured in addition to those needed to be set as conditions for the equilibrium.

4.5 X-ray and neutron diffraction

These techniques are especially important for the determination of crystal structures and require a single-crystalline sample. Lattice parameters and site occupancies as functions of composition and temperature can be obtained. In cases of crystals with elements having similar X-ray-scattering factors, these techniques can be used complementarily since the atoms can be distinguished on the basis of their different nuclear scattering factors (Grytsiv *et al.* 2005). Ordering can be observed as shown in Fig. 4.8, where superstructure lines related to carbon-vacancy ordering are observed only by neutron diffraction (Grytsiv *et al.* 2003).

4.5.1 Rietveld refinement

The Rietveld method allows the determination of site-occupancy parameters by analyzing intensity ratios of X-ray- or neutron-diffraction spectra of polycrystalline samples. An overview of this technique is given by Joubert (2002).

4.6 Mössbauer spectroscopy and perturbed angular-correlation measurements

These techniques can measure local states of ordering and site occupancy as well. The former, described by Lee *et al.* (2005), is very appropriate for systems containing Fe. The latter requires the incorporation of a radioactive probe into the sample. Binczycka *et al.* (2005) reported a combination of these two methods and compared their results with neutron-diffraction results. Both techniques can measure properties as functions of temperature, composition, and pressure.

4.7 Final remarks

There are certain experimental data that cannot be used with the selection of models described in the next chapter; they are, however, important and should also be scanned in order to help provide an educated guess about the models to be used. For example, even if lattice vibrations are not modeled explicitly, the literature data about phonon spectra

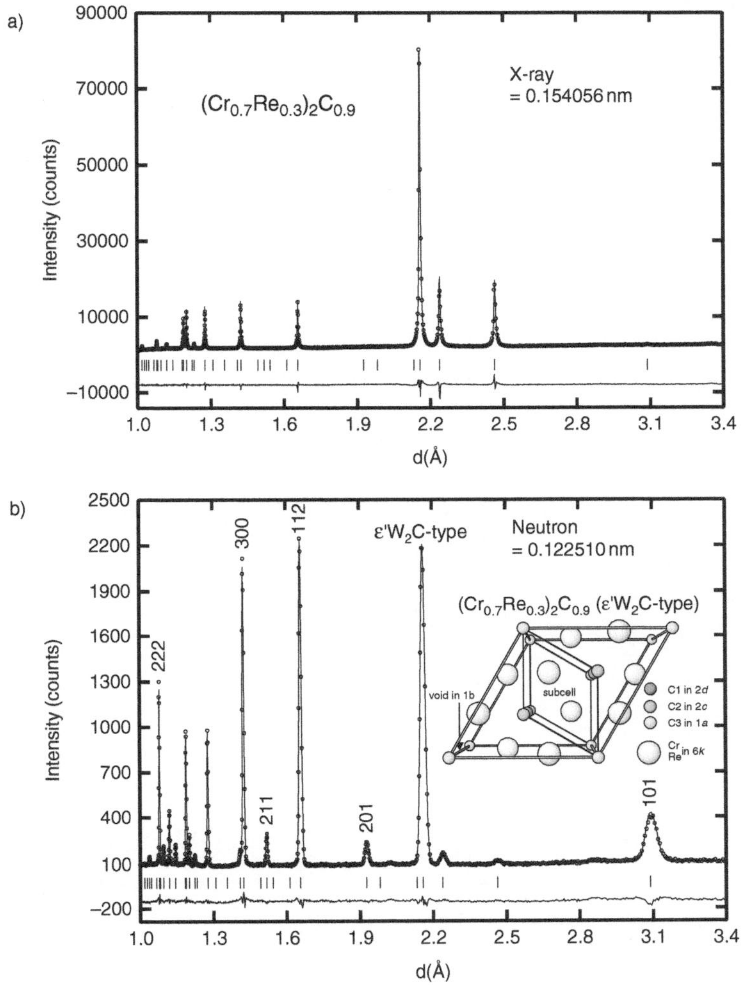

Figure 4.8 X-ray and neutron data showing superstructure formation due to carbon-vacancy ordering. Courtesy of Peter Rogl.

should be scanned. Information about the order of magnitude of these effects can give hints about the size and sign of the excess terms for the Gibbs energies. Techniques measuring the elastic constant, bulk modulus, and thermal expansion are important and give results that should also be taken into account since they are related to derivatives of the Gibbs energy.

5 Models for the Gibbs energy

In this chapter a number of models for the thermodynamic properties of various phases will be described. The integral Gibbs energy will be used as the modeled thermodynamic property. The reason to model the Gibbs energy rather than any other thermodynamic function is that most experiments are done at constant temperature and pressure. From the Gibbs energy all other important quantities can be obtained according to Eqs. (2.12). Using the Gibbs energy means that the modeling is limited to a "mean-field" approximation. Thermodynamic calculations using "Monte Carlo" methods or "molecular dynamics" are outside the scope of this presentation, but these techniques can provide important information about the type of mean-field model to be selected.

A phase may sometimes have a particular physical or chemical feature that requires a special model in order for it to be described accurately. It is not uncommon that the mathematical expression for such a model may be identical to the expression derived to describe another physical feature. That simply means that the mathematical expression is more general than the physical model. Whenever such a generalized expression can be obtained, it will be called a *formalism*. A general formalism should be able to handle cases when various constituents added to a phase behave differently, for example some may dissolve interstitially or cause chemical ordering. Most of the models used in this book are special cases of the compound-energy formalism (CEF). The physical meanings of the CEF parameters in various models differ and this will be discussed in detail.

The models for the phases in a thermodynamic system can be selected independently, except for the case when two phases are members of a structure family. Phases with chemical ordering form structure families and they may be modeled with the same Gibbs energy function, but usually only rather simple structural relations can be described with a single Gibbs energy function, for example A2/B2 and A1/$L1_2$.

The Gibbs energy for a phase can be attributed to two main factors, the bonding between the constituents and their configuration. The configurational part always enhances the mixing of unlike atoms. The bond energy between two unlike atoms may be either more negative than the bond energy between two equal atoms (creating a tendency for compound formation or ordering) or more positive (creating a tendency for there to be a miscibility gap).

The selection of the model for a phase must be based on the physical and chemical properties of the phase, for example crystallography, type of bonding, order–disorder transitions, and magnetic properties. Most of the models for crystalline phases described here will take into account the crystal structure of the phases by dividing it into sublattices characterized by different crystallographic symmetries and numbers of nearest neighbors. Sometimes a sublattice has just a single constituent, but usually several constituents can enter a sublattice. By selecting one constituent in each sublattice, one has a stoichiometric compound with fixed composition. The Gibbs energy of formation of that compound contains the most important part of the bond energies. In the simplest case the compound can be a pure element if the phase has a single set of sites and the elements as constituents. Any solution phase has at least two such compounds and they are then called the "**end members**" of the solution phase. The end members define the limit of solubility, but the model may have end members with a composition inside the composition range of the phase. Examples of this are a gas phase with molecules such that each molecule is an end member and an ordered phase in which some end members represent the possible ordered states. See also section 5.2.3.

In order to describe the measured phase equilibria and the thermodynamic properties of a system, it is necessary to adjust a number of parameters in the Gibbs-energy model of the phases. The aim of this book is to teach the reader how to make a good choice of models and obtain the best possible set of model parameters that describe the experimental information and are useful for extrapolations to multicomponent systems. The term **parameter** will be used for a quantity that is part of a model, like *excess parameter*. Some parameters can be a function of temperature, pressure, or even composition, and thus can be split into several other parameters. Each parameter may consist of several coefficients and a **coefficient** is always just a single numerical value. In the PARROT software, see section 7.3, the term **variable** is used with the same meaning as coefficient.

The selection of models described here has been based on those currently implemented in either of the two software packages that are described in more detail in this book. New models are continuously being developed and added to the software, so a future user can expect to have an even larger choice of models than that described here.

5.1 The general form of the Gibbs-energy model

The total Gibbs energy of a phase θ is expressed as

$$G_{\mathrm{m}}^{\theta} = {}^{\mathrm{srf}}G_{\mathrm{m}}^{\theta} + {}^{\mathrm{phys}}G_{\mathrm{m}}^{\theta} - T \cdot {}^{\mathrm{cnf}}S_{\mathrm{m}}^{\theta} + {}^{\mathrm{E}}G_{\mathrm{m}}^{\theta} \tag{5.1}$$

The superscript θ denoting the phase will normally be omitted in the rest of this chapter since all expressions are valid for a particular phase. The pre-superscript "srf" stands for "surface of reference" and represents the Gibbs energy of an *unreacted* mixture of the constituents of the phase. Not included in this is the quantity ${}^{\mathrm{phys}}G_{\mathrm{m}}$ which represents the contribution to the Gibbs energy due to a physical model such as magnetic transitions, as described in section 5.4. These models may be composition-dependent through particular physical quantities like the Curie temperature and the Bohr magneton number.

The pre-superscript "cnf" stands for the configurational entropy of the phase and is based on the number of possible arrangements of the constituents in the phase given by Eq. (2.9). This can easily be extended to include random arrangements on several sublattices. More elaborate methods to predict the entropy of configuration, like the cluster-variation method (CVM), can be derived for specific crystalline structures and will be discussed in section 5.7.2.2.

The term with pre-superscript "E" stands for the excess Gibbs energy and describes the remaining part of the real Gibbs energy of the phase when the first three terms have been subtracted from the real Gibbs energy. This partitioning means that there is no attempt to model the physical origins of the Gibbs energy, except those included in ${}^{\mathrm{phys}}G_{\mathrm{m}}$. The terms ${}^{\mathrm{srf}}G_{\mathrm{m}}$ and ${}^{\mathrm{E}}G_{\mathrm{m}}$ will thus include configurational as well as vibrational, electronic, and other contributions.

In this chapter a number of models for the various parts will be described. The temperature and pressure dependences of the Gibbs energy for a phase with fixed composition will be described first. Then the magnetic model will be described as an example of a physical property that is modeled separately. The reason for a separate description of this contribution to the Gibbs energy is that the magnetic properties of the phase depend on the critical temperature for magnetic ordering and the Bohr magneton number, and the composition dependences of these quantities must be described separately. The description of the models for composition dependence will start with the simplest cases and gradually introduce more complex features of real phases.

5.2 Phases with fixed composition

A phase with fixed composition can be a pure element, a stoichiometric compound, or a solution phase whose composition is kept constant externally. The Gibbs energy of such a phase can depend on temperature and pressure only. An important class of such phases is constituted by the "end members" of solutions, a term introduced at the beginning of this chapter.

5.2.1 *Temperature dependence*

Except for the ferromagnetic transition, described in section 5.4.2, the temperature dependence of the molar Gibbs energy, G_{m}, of a stable end member of a phase θ is often described by a power series in temperature like

$$G_{\mathrm{m}}^{\theta} - \sum_i b_i H_i^{\mathrm{SER}} = a_0 + a_1 T + a_2 T \ln(T) + a_3 T^2 + a_4 T^{-1} + a_5 T^3 + \cdots, \quad T_1 < T < T_2 \tag{5.2}$$

where b_i is the stoichiometric factor of element i in θ and $\sum_i b_i H_i^{\mathrm{SER}}$ represents the sum of the enthalpies of the elements in their reference states, usually the stable state at 298.15 K and 1 bar, denoted SER. This term is needed because there is no absolute value of the enthalpy of a system and one must select some reference state. Such a power series can be valid for a limited temperature region only, here delimited by T_1 and T_2.

The coefficients in Eq. (5.2) expressing the temperature dependence of the end-member parameter cannot be related directly to any other thermodynamic quantity, but, using the relations in Eqs. (2.12), one obtains

$$H_{\rm m}^{\theta} - \sum_i b_i H_i^{\rm SER} = a_0 - a_2 T - a_3 T^2 + 2a_4 T^{-1} - 2a_5 T^3 \cdots \tag{5.3}$$

$$S_{\rm m}^{\theta} = -a_1 - a_2(1+\ln(T)) - 2a_3 T + a_4 T^{-2} - 3a_5 T^2 \cdots \tag{5.4}$$

$$C_p^{\theta} = -a_2 - 2a_3 T - 2a_4 T^{-2} - 6a_5 T^2 \cdots \tag{5.5}$$

From the expression for the heat capacity, C_p, it can be seen that the coefficient for the $T\ln(T)$ term in Eq. (5.2) originates from the temperature-independent heat-capacity coefficient. These quantities for pure Cu, from Dinsdale (1991), are plotted in Fig. 5.1.

The expression above is suitable for expressing the Gibbs energy for a limited temperature range and above the Debye temperature. If the coefficients are fitted to experimental data and the expression is then used far outside the known temperature range, or down to 0 K, severe problems with using this power series will occur. Thus there is an interest in developing and using temperature-dependent models based on the physical properties of the phase, namely lattice vibrations, thermal vacancies etc. At present, however, physical models are rarely used for modeling the heat capacity except for a model for ferromagnetic transitions described below. The lower temperature limit is usually 298.15 K, which is sufficiently low to allow calculations of the equilibrium in most heterogeneous systems that require diffusion in order to reach the equilibrium state. At lower temperatures diffusion is usually not possible and thus there is little practical interest in extending the model to lower temperatures.

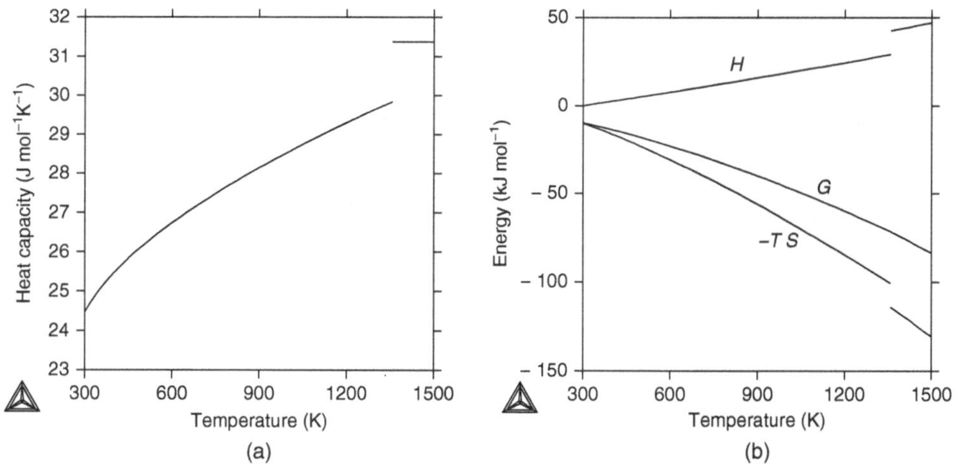

Figure 5.1 The thermodynamic properties of pure Cu: in (a) the heat capacity and in (b) the Gibbs energy (G), enthalpy (H) and entropy (multiplied by T and with changed sign). Note that G is continuous at the melting point but H and S have jumps.

Equation (5.2) may require many coefficients if the temperature range is large. A method by which to decrease the number of coefficients is to use several temperature ranges and different expressions in each. This method is also needed in order to have backward compatibility with the time when thermodynamic data were usually described by giving the heat of formation and entropy at 298.15 K and 1 bar together with a four-coefficient heat-capacity expression. In such cases there were usually at least two and often up to four or five temperature regions with different coefficients. Note that it is mandatory that the first and second derivatives of the Gibbs energy must be continuous through such a breakpoint, otherwise it would behave like a phase transition.

5.2.2 *Pressure dependence*

The pressure-dependent properties such as volume and thermal expansivity are often ignored in thermodynamic models. For condensed phases they are important for the equilibrium only at very high pressures, but the molar volume can be important in phase transformations and knowledge of it is necessary in order to obtain the volume fraction of a phase. For the gas phase, except close to the critical point or the boiling point, it is sufficient to describe the pressure dependence by one term, $RT\ln(p/p_0)$. For the condensed phases a model suggested by Murnaghan (1944) is useful for limited pressure ranges.

5.2.2.1 The Murnaghan pressure model

In the Murnaghan model one assumes that the bulk modulus can be expressed by a linear pressure dependence. The following expression is used for the compressibility, κ, which is the inverse of the bulk modulus:

$$\kappa(T,p) = \frac{K_0(T)}{1+nK_0(T)p} \tag{5.6}$$

where $K_0(T)$ is the compressibility at zero pressure and n is a constant independent of temperature and pressure. Experimentally n is found to be about 4 for many phases. The thermal expansivity can usually be described as a power series in temperature and, in order to have reasonable extrapolations from low temperatures, at which most measurements are made, to high temperatures, one should use an expression of the form

$$\alpha = \alpha_0 + \alpha_1 T + \alpha_2 T^{-2} + \cdots \tag{5.7}$$

and avoid higher powers than one. This model was used in an assessment of pure iron by Fernández Guillermet and Gustafson (1984) and Fig. 5.2(a) shows the phase diagram for iron for varying temperatures and pressures. In Fig. 5.2(b) the pressure axis has been changed to the molar volumes.

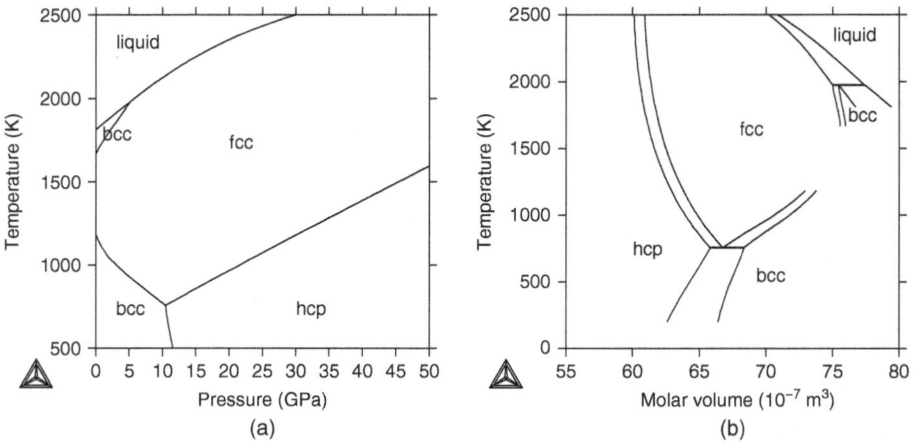

Figure 5.2 Two variants of the phase diagram for pure Fe: (a) the $p-T$ phase diagram and (b) the volume–temperature phase diagram. In (b) there is a gap in the fcc/bcc curves at high volume because the pressure is below zero there.

For solution-phase modeling the Murnaghan model must be integrated into a Gibbs energy and the expression is

$$G_m(T,p) = G_m(T,p=0) + \frac{V_0 e^{\int_{298}^{T} \alpha(T) dT}}{(n-1)K_0(T)} [(1 + nK_0(T)p)^{1-1/n} - 1] \tag{5.8}$$

For high pressure, such as at the center of the Earth, the Murnaghan model is not sufficient and one must use higher-order terms. In most high-pressure models the volume is then used as independent variable, but in such a case it is no longer a model for the Gibbs energy.

5.2.2.2 A new pressure model

A new pressure-dependent model was proposed by Lu *et al.* (2005) because the composition dependences of the parameters in the Murnaghan model, V_0, α, K_0, and n, make it very cumbersome to handle when calculating partial Gibbs energies. This model is based on the empirical relation proposed by Grover *et al.* (1973):

$$V(T,p) = V_0(T) - c(T) \ln\left(\frac{\kappa_0(T)}{\kappa(T,p)}\right) \tag{5.9}$$

where $V_0(T)$ and $\kappa_0(T)$ are the volume and compressibility at zero pressure and $c(T)$ is an adjustable function. This can be integrated to give a surprisingly simple Gibbs-energy expression:

$$G_m(T,p) = G_m(T,p=0) + \frac{c(T)}{\kappa_0(T)} \left[\exp\left(\frac{V(T,p) - V_0(T)}{c(T)}\right) - 1\right] \tag{5.10}$$

The expression for the volume as a function of temperature and pressure is more complicated and involves the exponential integral, Ei, defined as

$$\mathrm{Ei}(z) = \int_z^\infty \frac{\mathrm{e}^{-x}}{x}\,\mathrm{d}x \tag{5.11}$$

$$V(T,p) = c(T)\mathrm{Ei}^{-1}\left[\mathrm{Ei}\left(\frac{V_0(T)}{c(T)}\right) + p\kappa_0(T)\exp\left(\frac{V_0(T)}{c(T)}\right)\right] \tag{5.12}$$

where Ei^{-1} is the inverse of the exponential integral.

The advantage of this model is that one can include more easily the composition dependences of the parameters and it is thus better suited for modeling solution phases. Another advantage is that the model can be extended to higher pressure ranges than can the Murnaghan model.

5.2.3 Metastable states, lattice stabilities, and end members

When a new component can dissolve in a compound, it forms a solution phase and the stability range of the phase can be extended also in temperature. A consequence of this is that the Gibbs energy of the solution phase must have a value at the endpoint representing the compound also at temperatures outside its range of stability. Thus experimental data are not always enough to fit coefficients in the temperature models even for compounds. An example of such extrapolation is shown in Fig. 5.3, where the heat capacity and Gibbs energy of various forms of pure iron (Fernández Guillermet and Gustafson 1984) have been extrapolated from 298.15 to 2000 K.

In other cases the Gibbs energies of compounds that are never stable are needed. These may be inaccessible to measurement, but it may be possible to predict properties of such a

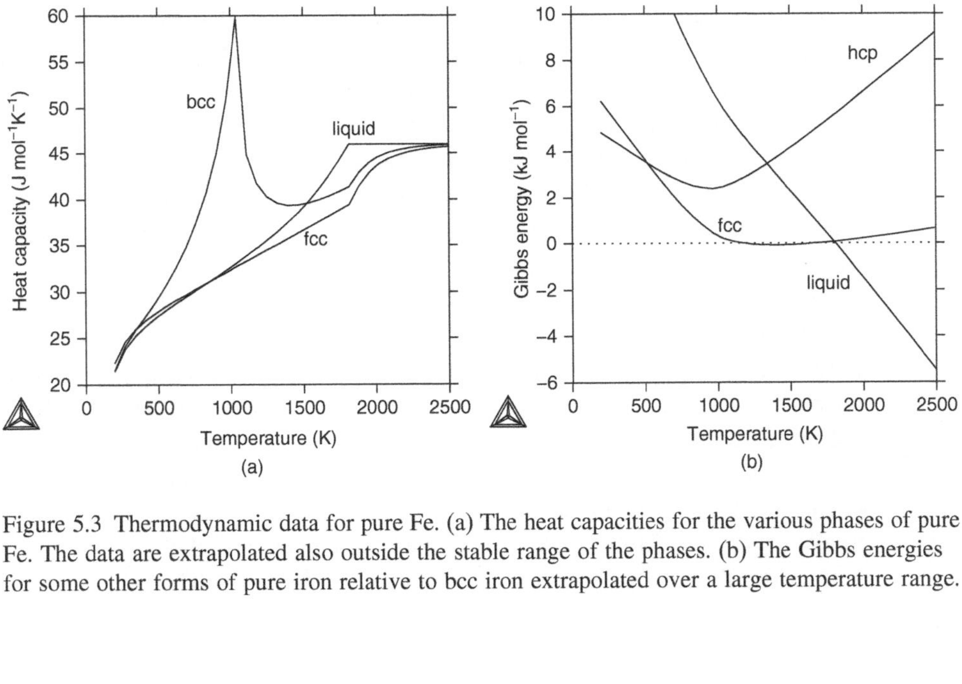

Figure 5.3 Thermodynamic data for pure Fe. (a) The heat capacities for the various phases of pure Fe. The data are extrapolated also outside the stable range of the phases. (b) The Gibbs energies for some other forms of pure iron relative to bcc iron extrapolated over a large temperature range.

compound by extrapolations or by *ab initio* calculations. With some care the results from total-energy calculations can be used as "experiments" together with real experimental data with the thermodynamic models described here.

Most elements have only one or two stable crystal structures, but they may have some solubility in many different phases. When the solubility is large, it is necessary to estimate a value of the relative stability of the metastable structure as an "end member" of the phase. These values for pure elements are called "lattice stabilities." Several estimates of such unary data have been made. Those currently most used are the comprehensive SGTE set published by Dinsdale (1991). The Ringberg meetings (Aldinger *et al.* 1995, 1997) may lead to a completely new set with improved models. It is important to agree on the values for these metastable states, even if they are very uncertain, because one cannot change the value of the Gibbs energy of formation of a stable or metastable end-member state without reassessing all parts of the thermodynamic database that depend on the previous value. This problem is not so big for structures stable in some range of T and p because their thermodynamic properties can be determined by measurements and thus one can always refer to experimental data, but for metastable states there are estimates that may differ by a factor of ten or more.

For compounds with no measured heat capacity it is usual to apply the Kopp–Neumann rule (Grimvall, 1999) that the heat capacity of the compound is equal to the stoichiometric average of the heat capacities of the pure elements in their SER. The Gibbs energy will then be

$$G^{\theta}_{\mathrm{m}} - \sum_i b_i G_i^{\mathrm{SER}} = a_0 + a_1 T \tag{5.13}$$

In comparison with Eq. (5.2), there are only two coefficients that have to be determined from experimental data because the heat capacity will be taken from the descriptions of the pure elements. If there are special reasons, for example the coordination number for a lattice site, one may refer an element to some phase other than the SER. The same equation is also applied to the end members used for modeling solution phases that have a composition within the composition range of the phase.

The main contribution to the heat capacity of a solution is due to the temperature dependences of the "end members" of solutions. Most other model parameters used to describe the properties of solutions are only linearly temperature-dependent; at most a $T\ln(T)$ term is used to describe a composition-dependent excess heat capacity, but any higher power of T is discouraged except for the "end members."

At the Ringberg meetings a new way to perform the extrapolation above and below the liquidus temperature for the pure elements has been proposed (Ågren *et al.* 1995); in particular, by including the glass transition. This will not be treated further here, since it has not yet been implemented in the software generally used for assessments.

Extrapolations of the Gibbs energy of the liquid below the melting temperature and of the Gibbs energy for the solid phases above the melting temperature are necessary when one considers solution phases. The SGTE group (Dinsdale 1991) has introduced breakpoints in the Gibbs energy at the melting temperature of the liquid and the stable solid phase. This usually works well for phase-equilibria calculations but sometimes gives strange results for calculated thermodynamic functions. Several other suggestions have been made for extrapolations, but most of them can be treated with the series expression above.

5.3 Variables for composition dependence

The Gibbs energy, G, of a phase depends on the amounts of the constituents in the phase. For modeling the composition dependence it is more convenient to use the molar Gibbs energy, G_m, defined in Eq. (2.14) and give the size of the system as the total amount of components, N, which is the sum of moles of all components N_i:

$$G = N \cdot G_m \tag{5.14}$$

$$N = \sum_i N_i \tag{5.15}$$

In this chapter "Gibbs energy" will be used to mean "molar Gibbs energy," except when clearly noted otherwise. The composition variables describing the composition dependence of G_m are in the simplest case the mole fractions of the components, x_i, defined as

$$x_i = \frac{N_i}{N} \tag{5.16}$$

The composition is sometimes expressed in mass fractions (often wrongly called weight fractions) and it is defined as

$$w_i = \frac{M_i}{M} \tag{5.17}$$

where M_i is the mass of component i and $M = \sum_i M_i$. The mole or mass percent is the respective fraction multiplied by 100.

For a closed system the amounts of the various components of a system are "external" variables that can be controlled from outside the system. In each phase in the system the components may form many different species or ions and enter as **constituents** various types of sites in a crystalline phase.

The "composition" of a phase in a system at equilibrium is the amounts or fractions of all components in the phase. This can be expressed using the mole fractions or the mass fractions defined above.

5.3.1 Components other than the elements

The number of components for a system is always equal to the number of elements, but one may prefer to use components other than the elements. For example, in an oxide system (Ca, Mg, O) one may prefer to use (CaO, MgO, O_2) as components if the composition is entirely inside the binary system (CaO, MgO). This case will be discussed in more detail in the section on quasibinary systems (section 6.2.4.5).

At equilibrium the chemical potential of a component j is related to the chemical potentials of the elements i by the reaction

$$\mu_i = \sum_j b_{ij}\mu_j \tag{5.18}$$

where b_{ij} is the stoichiometric factor of component i in constituent j and μ_j is the chemical potential of constituent j.

An example of the use of this relation is if one wants to know the chemical potential of H_2O in a gas with the components H_2 and O_2. The equation above gives

$$\mu_{H_2O} = \mu_{H_2} + 0.5\mu_{O_2} \tag{5.19}$$

This can be related to the classical "law of mass action" if one assumes that the gas is ideal, see section 5.5.4, because then one has for each molecule

$$\begin{aligned}\mu_{O_2} &= {}^\circ G_{O_2} + RT\ \ln(y_{O_2})\\ \mu_{H_2} &= {}^\circ G_{H_2} + RT\ \ln(y_{H_2})\\ \mu_{H_2O} &= {}^\circ G_{H_2O} + RT\ \ln(y_{H_2O})\end{aligned} \tag{5.20}$$

where y_i is the equilibrium constituent fraction of each molecule i in the gas. Inserting Eq. (5.20) into Eq. (5.19) and rearranging the terms gives

$${}^\circ G_{H_2O} - {}^\circ G_{H_2} - 0.5{}^\circ G_{O_2} = K = RT\ \ln\left(\frac{y_{O_2}y_{H_2}}{y_{H_2O}}\right) \tag{5.21}$$

where K is the reaction constant for the chemical reaction $H_2O = H_2 + 0.5O_2$.

The reference state for the chemical potential of H_2O is not well defined unless one compares it with the chemical potential of a gas consisting of pure H_2O, $\mu^\circ_{H_2O}$. In that case it can be related to the partial pressure of H_2O, p_{H_2O}, by

$$p_{H_2O} = \exp\left(\frac{\mu_{H_2O} - \mu^\circ_{H_2O}}{RT}\right) \tag{5.22}$$

Another example can be taken from the use of the sublattice model introduced in section 5.8, where, for example, carbon dissolves interstitially in fcc iron, which is modeled with two sublattices (Fe)(C, Va), where Va represents vacant interstitial sites. The chemical potential of C cannot be obtained directly from the model, but one has the following relations for the partial Gibbs energies of the end members for fcc iron:

$$\begin{aligned}G^{fcc}_{Fe:Va} &= G_{Fe}\\ G^{fcc}_{Fe:C} &= G_{Fe} + G_C\end{aligned} \tag{5.23}$$

On taking the difference between these partial Gibbs energies, one obtains for the chemical potential of C

$$\mu_C = G_C = G^{fcc}_{Fe:C} - G^{fcc}_{Fe:Va} \tag{5.24}$$

Some care should be taken when selecting components other than the elements, since it might not be possible to define a convenient reference state for the chemical potential or activity. As reference state one may use only a phase that can exist in pure form for the selected component. If solely the equilibrium state is interesting, one can ignore such problems, but one may find that the amount of the phase or the fraction of a component of a phase may be negative if the phase has a composition outside the composition range limited by the components defined by the user.

5.3.2 Internal degrees of freedom

In many cases a phase can have "internal" degrees of freedom that cannot be controlled externally. For example, the speciation of a gas phase at equilibrium is determined by the Gibbs energies of formation of the various gas species. A typical example is a gas formed by the elements H and O. In this gas one may at equilibrium expect to find the constituents H, H_2, H_2O, O, O_2, and O_3. For convenience one may also use H_2 and H_2O as components rather than the elements.

Sometimes a gas is formed by mixing various amounts of particular species, but, when the equilibrium has been established, new species may have formed and the equilibrium fractions of the initial species may be very different from the initial values. In a liquid or crystalline solid there may also be a formation of species and ions on different types of sites or different types of defects. These must be taken into account when modeling the phase and the fractions of these constituents can be obtained only from an equilibrium calculation.

5.3.3 The constituent fraction

The species that constitute the phase are called the "constituents." For a gas with several species the constituent fraction, y_i, of each species i in the gas describes the internal equilibrium in the gas. In a crystalline phase with several sublattices the constituent fraction is often called the "site fraction." The sum of the constituent fractions is unity (on each sublattice) and the mole fractions of the components can be calculated from the constituent fractions by using

$$x_i = \frac{\sum_j b_{ij} y_j}{\sum_k \sum_j b_{kj} y_j} \tag{5.25}$$

where b_{ij} are the same stoichiometric factors as in Eq. (5.18). If there are more constituents than components one cannot obtain the constituent fractions from the mole fractions without a minimization of the Gibbs energy of the phase.

For the gas phase the partial pressure is often used to describe its composition. For an ideal gas, Dalton's law can be applied and the partial pressure for a constituent i, p_i, and the constituent fraction, y_i, are related by the total pressure, p, by

$$p_i = p \cdot y_i \tag{5.26}$$

For phases with sublattices the constituent fractions, or site fractions, are defined as

$$y_i^{(s)} = \frac{N_i^{(s)}}{N^{(s)}} \tag{5.27}$$

where $N_i^{(s)}$ is the number of sites occupied by the constituent i on sublattice s and $N^{(s)}$ is the total number of sites on sublattice s. If some sites are empty, it is convenient to define a vacancy fraction

$$y_{\mathrm{Va}}^{(s)} = \frac{N^{(s)} - \sum_i N_i^{(s)}}{N^{(s)}} = 1 - \sum_{i \neq \mathrm{Va}} y_i \tag{5.28}$$

The vacancy, denoted Va, can be treated as a real component that always has its chemical potential equal to zero. This is a structural or constitutional vacancy, not a thermal vacancy. It is possible to have thermal vacancies in the models, too, and the fractions of such vacancies can be fitted by model parameters, but these vacancies will stabilize the phase and the lattice stability parameter for the crystalline phases has to be refitted to give the correct melting temperatures.

In this context it can be understood that it is convenient to define the molar Gibbs energy for a crystalline phase relative to the formula unit of the phase. In particular, if the phase is modeled with sublattices, which can be wholly or partially empty (i.e., containing vacancies), the molar Gibbs energy is most simply expressed per mole formula unit of the phase rather than per mole of real components.

The formula unit of a phase with sublattices is equal to the sum of the site ratios, $\sum_s a^{(s)}$, where the ratios $a^{(s)}$ usually are the set of smallest integers giving the correct ratio between the numbers of sites $N^{(s)}$ on each sublattice. It is recommended to use Cu_2Mg rather than $Cu_{0.6666667}Mg_{0.3333333}$ also in order to avoid rounding-off errors.

The mole fraction of the component i can be calculated from the site fractions of a crystalline phase as

$$x_i = \sum_s \frac{\sum_j b_{ij} y_j^{(s)}}{a^{(s)} \sum_k \sum_j b_{kj} y_j^{(s)}} \tag{5.29}$$

where b_{ij} has the same meaning as above. The vacancies must be excluded from the summations.

5.3.4 *Volume fractions of constituents*

If the constituents have very different sizes, some may represent chain molecules with thousands of atoms whereas others have maybe only a few atoms; this must be taken into account when considering the entropy of mixing. It is common to use volume fractions in such systems, but the volume is not a very good independent variable since it depends on temperature. Instead the volume of a constituent should be used directly in the model and the composition determined in terms of the constituent fraction of the species as defined above. An example is given by the Flory–Huggins model described in section 5.9.6.

5.3.5 *Other fraction variables*

For gases partial pressures have already been mentioned above. In aqueous solutions one frequently uses the molality as the fraction in the modeling. The molality of an aqueous species i is the number of moles of i per kilogram of solvent and it can be directly transformed to the constituent fraction of i.

5.4 Modeling particular physical phenomena

Several diverse physical phenomena may contribute to the thermodynamic properties of a phase. In some cases such contributions do not depend smoothly on the composition, but rather depend on some property that may itself vary with composition, for example the Curie temperature for ferromagnetic transitions. It is then advantageous to model these contributions separately even for the pure elements, and to model the composition dependence of the Curie temperature and similar properties as separate quantities. This introduces an "implicit" composition dependence into the Gibbs energy, which may make the equilibrium calculations more complicated but brings the modeling closer to reality.

5.4.1 *Lattice vibrations*

The contribution to the Gibbs energy due to lattice vibrations above 300 K is usually close to that given by the Dulong–Petit rule, $3RT$. As already mentioned, it is not the intention here to describe models that can be extrapolated to temperatures below 298 K, but it may in some cases be important to describe the Debye temperature of a phase because that is related to other properties. A possible method to apply in order to include this was proposed at the Ringberg workshop (Chase *et al.* 1995). This method has still not been implemented in most software and will not be discussed further here.

5.4.2 *A ferromagnetic transition model*

Some elements and compounds undergo *second-order* transitions, which can be due to magnetic ordering or other internal changes without a change in composition. The classification of a second-order transition is that there is a discontinuity in the second derivative of the Gibbs energy at the transition whereas all first derivatives, such as entropy and enthalpy, are continuous. A *first-order* transition means that there is a discontinuity in the first derivative. At the magnetic-transition temperature the heat capacity approaches infinity and it would be impossible to represent the heat capacity in the vicinity of the critical point using Eq. (5.2) without using many coefficients or possibly also temperature intervals. As an example of this, the heat capacity of pure iron was shown in Fig. 5.3(a). In such a case it is recommended that the contribution to the Gibbs energy due to the second-order transition in bcc be modeled separately. An additional reason is that the magnetic contribution to G_m^{bcc} in an alloy with iron is not proportional to the fraction of iron because the ferromagnetic transition is composition-dependent.

Equation (5.2) can, then, be considered to be valid for a hypothetical element or compound that does not undergo a second-order transition. For modeling a second-order transition it is necessary to start with its contribution to the heat capacity of the system. For the contribution to the heat capacity due to a magnetic transition the following empirical expressions have been proposed by Inden (1981):

$$C_p^{\mathrm{fm}} = K^{\mathrm{fm}} \ln\left(\frac{1+\tau^3}{1-\tau^3}\right) \qquad \tau < 1 \tag{5.30}$$

$$C_p^{\mathrm{pm}} = K^{\mathrm{pm}} \ln\left(\frac{\tau^5+1}{\tau^5-1}\right) \qquad \tau > 1 \tag{5.31}$$

where the superscript fm denotes the ferromagnetic state below the Curie temperature T_{C} and pm the paramagnetic state above T_{C}; τ is T/T_{C} and T is the absolute temperature. The two coefficients K are general functions for all magnetic transitions and can be determined by integrating the heat capacities as follows:

$$^{\mathrm{mag}}H = \int_0^1 C_p^{\mathrm{fm}}(\tau)\mathrm{d}\tau + \int_1^{\infty} C_p^{\mathrm{pm}}(\tau)\mathrm{d}\tau \tag{5.32}$$

The total magnetic contribution to the molar entropy is set equal to $-R\ln(1+\beta)$, where β is the mean magnetic moment measured in Bohr magnetons. This is equivalent to assuming that the magnetic entropy is due to the disordering of localized spins with average magnitude equal to β. The magnetic contribution to the Gibbs energy, $^{\mathrm{mag}}G = {}^{\mathrm{mag}}G - T \cdot {}^{\mathrm{mag}}S$, vanishes at the Curie temperature, giving $^{\mathrm{mag}}H = T_{\mathrm{C}} \cdot {}^{\mathrm{mag}}S$. The two contributions to $^{\mathrm{mag}}H$ below and above the Curie temperature may approximately be attributed to the contributions of long-range order (LRO) and short-range order (SRO), respectively. The ratio of these two contributions is introduced as an empirical constant, p, for which Inden found the value 0.28 for cobalt and nickel (fcc) and 0.4 for iron (bcc).

The heat capacities cannot be integrated to give a closed expression for the Gibbs energy unless it is expanded into a power series. Thus several simplifications of the model have been proposed and the simplified version of this model due to Hillert and Jarl (1978) has been used most widely. This gives

$$^{\mathrm{mag}}G_{\mathrm{m}} = nRTf(\tau)\ln(\beta+1) \tag{5.33}$$

where n is the number of atoms per formula unit that have the average magnetic moment β. The function $f(\tau)$ according to the simplified model is

$$f(\tau) = \begin{cases} 1 - \dfrac{1}{A}\left[\dfrac{79\tau^{-1}}{140p} + \dfrac{474}{497}\left(\dfrac{1}{p}-1\right)\left(\dfrac{\tau^3}{6}+\dfrac{\tau^9}{135}+\dfrac{\tau^{15}}{600}\right)\right] & \tau < 1 \\ -\dfrac{1}{A}\left(\dfrac{\tau^{-5}}{10}+\dfrac{\tau^{-15}}{315}+\dfrac{\tau^{-25}}{1500}\right) & \tau \geq 1 \end{cases}$$

with

$$A = \frac{518}{1125} + \frac{11\,692}{15\,975}\left(\frac{1}{p}-1\right)$$

As already mentioned, p depends on the structure. In a solution phase T_C and β vary with the composition. For maximum flexibility and simplicity, they are assumed to vary with composition in accord with the same models as are applied to the Gibbs energy itself, as described below. The total Gibbs energy for an element or end member with a magnetic transition is thus

$$G_{\mathrm{m}}^{\theta} = G_{\mathrm{m}}^{\theta,\mathrm{hyp}} + G_{\mathrm{m}}^{\theta,\mathrm{mag}} \tag{5.34}$$

where $G_{\mathrm{m}}^{\theta,\mathrm{hyp}}$ is described by Eq. (5.2) and $G_{\mathrm{m}}^{\theta,\mathrm{mag}}$ by Eq. (5.33). The large effect of the ferromagnetic transition in bcc Fe is clearly evident from Fig. 5.3(a). When these values are integrated to give a Gibbs energy, one can obtain the curves in Fig. 5.3(b), which show the value of the Gibbs energy relative to bcc Fe extrapolated also into the metastable ranges. These curves are much smoother than the heat capacities, but an effect of the magnetic transition is the increasing stability of bcc Fe at lower temperatures.

5.4.3 *Other physical contributions to the Gibbs energy*

The contributions to the Gibbs energy from other physical phenomena, such as the electronic heat capacity, size mismatch, and short-range order, are usually not modeled separately. The reason for this is that, although some of these contributions can be accurately modeled from atomistic theories, the contributions to the Gibbs energy are in most cases much smaller than other contributions that cannot be modeled accurately. In a theoretical approach one can treat some contributions accurately but may ignore others that are equally important. In the Calphad technique the total Gibbs energy is modeled using experimental data on the phase diagram and thermodynamics to fine-tune the models to reality.

5.4.4 *The composition dependence of physical models*

The physical models with composition-dependent physical quantities, such as the Curie temperature and Bohr magneton number in the model for ferromagnetic transitions, add to the phase a Gibbs-energy contribution that is "indirectly" composition-dependent. Usually the composition dependence of the physical quantity is described with the same composition variables as the Gibbs energy itself.

The magnetic model described in section 5.4.2 has two parameters, T_C, the critical temperature for magnetic ordering, and β, the Bohr magneton number. These can be composition-dependent in the same way as the parameters describing the chemical part of the Gibbs energy. Thus each "end member" of a solution can have a critical temperature for magnetic ordering and a Bohr magneton number. The composition dependences of these quantities are described using the same mathematical expression as that for the chemical excess Gibbs energy. Employment of particular models for other physical properties may be used more in the future.

An example of modeling the critical temperature for magnetic ordering is shown in Fig. 5.4. Its value for the bcc phase in the Fe–Cr system was assessed by Andersson

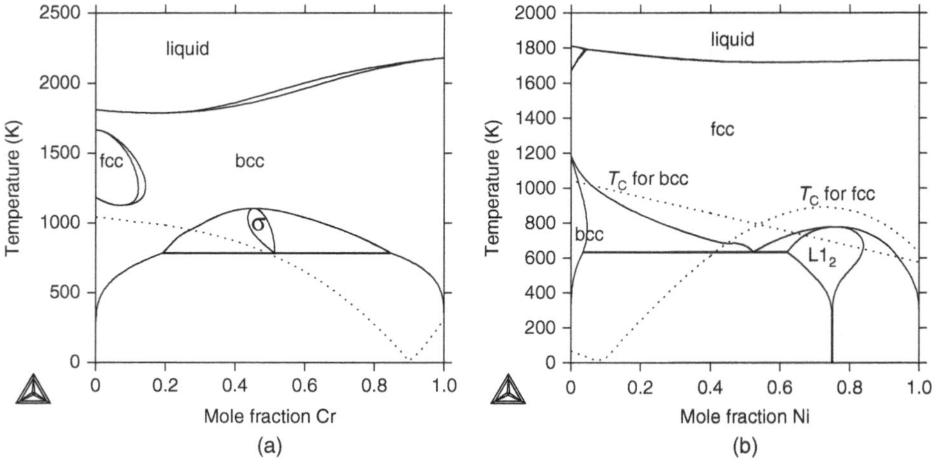

Figure 5.4 The phase diagrams for Fe–Cr and Fe–Ni with the Curie and Néel temperatures of the bcc and fcc phases plotted as dashed lines. Note that there is a small kink in the solubility curves where they are crossed by a magnetic-ordering curve.

and Sundman (1987) and is shown as a dashed line in Fig. 5.4(a). Since Cr is anti-ferromagnetic, this is modeled with a negative value of the critical temperature and the curve goes through zero at some Cr content and on the Cr-rich end it represents a Néel temperature. In Fig. 5.4(b) the critical temperatures for magnetic ordering for both fcc and bcc in Fe–Ni from an assessment by A. T. Dinsdale and T. G. Chart (unpublished work, 1986) are shown, with the ordered $L1_2$ phase added by I. Ansara (unpublished work, 1995). In the absence of any experimental and theoretical information when this system was assessed, the Curie temperature for bcc Ni was assumed to be the same as that for fcc Ni.

5.5 Models for the Gibbs energy of solutions

The term "solution" is in some contexts restricted to dilute solutions and "mixtures" used more generally, but in this book "solution" is used for all kinds of phases that can vary in composition.

All phases vary in composition to a greater or lesser extent, but for practical purposes it may be possible to ignore the compositional variations for some compounds. However, it is also important to state what the practical purpose is. For example, in electronic engineering the change in electronic properties of a semiconductor like PbS when its composition is varied over a small range is essential, but a metallurgist interested in extracting lead from a PbS ore may safely ignore this.

For solution-phase modeling it is necessary to describe the properties of the pure elements or compounds that form endpoints of the solutions. In many cases the solution-phase models will also require properties for compounds that cannot be measured separately.

For that one needs estimates and models that can predict, for example, the Gibbs energy of chromium in the fcc state. These problems were discussed in section 5.2.

In order to describe real solutions, it is necessary to model the terms in Eq. (5.1) and, together with the model parameters, this can be fitted to the experimental properties of the solution. One may place the emphasis on any one of ${}^{\mathrm{srf}}G_{\mathrm{m}}$, ${}^{\mathrm{cnf}}S_{\mathrm{m}}$, or ${}^{\mathrm{E}}G_{\mathrm{m}}$, but the main goal of the presentation in this book is to model the total Gibbs energy of a phase in such a way that it can be extrapolated to multicomponent systems.

5.5.1 *The compound-energy formalism*

There are many ways to derive the compound-energy formalism (CEF), but the original is based on the two-sublattice model formulated by Hillert and Staffanson (1970). This model was extended to an arbitrary number of sublattices and constituents on each sublattice by Sundman and Ågren (1981). In order to make the mathematics easier, they introduced the concept of a "constituent array." A constituent array specifies one or more constituents on each sublattice and is denoted I, while the individual constituents are denoted i, sometimes with a superscript (s) to denote the sublattice s. The constituent arrays can be of different orders and the zeroth order has just one constituent on each sublattice. Such a constituent array denotes a compound and this is the origin of the name compound-energy formalism.

The Gibbs-energy expression for the CEF is

$$ {}^{\mathrm{srf}}G_{\mathrm{m}} = \sum_{I_0} P_{I_0}(Y)\,{}^{\circ}G_{I_0} \tag{5.35} $$

$$ {}^{\mathrm{cnf}}S_{\mathrm{m}} = -R\sum_{s=1}^{n} a_s \sum_{i=1}^{n_s} y_i^{(s)} \ln(y_i^{(s)}) \tag{5.36} $$

$$ {}^{\mathrm{E}}G_{\mathrm{m}} = \sum_{I_1} P_{I_1}(Y)L_{I_1} + \sum_{I_2} P_{I_2}(Y)L_{I_2} + \cdots \tag{5.37} $$

where I_0 is a constituent array of zeroth order specifying one constituent in each sublattice and $P_{I_0}(Y)$ is the product of the constituent fractions specified by I_0. ${}^{\circ}G_{I_0}$ is the compound-energy parameter representing the Gibbs energy of formation of the compound I_0. In many applications of the CEF some constituent arrays I_0 do not represent any stable compounds, which means that those ${}^{\circ}G_{I_0}$ must be estimated in some way.

In the configurational entropy the factor a_s is the number of sites on each sublattice and $y_i^{(s)}$ is used to denote the constituent fraction of i on sublattice s. The first sum is over all sublattices and the second over all constituents on each sublattice.

The excess term, ${}^{\mathrm{E}}G_{\mathrm{m}}$, contains sums over the possible interaction parameters defined by component arrays of the first order, second order etc. A constituent array of first order has one extra constituent in one sublattice. For the second-order constituent array it is necessary to include both the case with three interacting constituents on one sublattice and that with two interacting constituents on two different sublattices, i.e. a reciprocal parameter. The use of constituent arrays of higher than second order is not advisable.

The partial Gibbs energy, Eq. (2.15), takes a special form for a phase with sublattices since it might not, in the general case, be possible to calculate the partial Gibbs energy for the components unless all constituents enter in all sublattices. Nonetheless, one can always calculate the partial Gibbs energy for a constituent array of zeroth order I_0 as follows:

$$G_{I_0} = G_\mathrm{m} + \sum_{s=1}^{n} \frac{\partial G_\mathrm{m}}{\partial y_i^{(s)}} - \sum_{s=1}^{n} \sum_j y_j^{(s)} \frac{\partial G_\mathrm{m}}{\partial y_j^{(s)}} \tag{5.38}$$

where the first sum is taken over the constituents i defined by the constituent array (one in each sublattice) and the second sum is over all constituents in all sublattices.

5.5.2 *The relation between the compound and cluster energies*

The CEF was developed for phases in which there are different types of atoms on the sublattices and different numbers of nearest neighbors in the sublattices, for example interstitial solutions, carbides, and intermetallic phases like the σ phase. More recently it has also been used to describe order–disorder transformations in phases with the structure type B2 (CuZn) and $L1_2$ (Ni_3Al) which are ordered forms of the A2 and A1 structure types, respectively. In such cases there are relations between the parameters in the CEF that must be respected in order to describe the physical properties of the phase.

The basic concept of the CEF is that each compound or "end member" has its own independent Gibbs energy of formation. This Gibbs energy is a function of temperature and pressure but independent of composition. One may compare this with a CVM treatment in which the "cluster energies" may appear to be similar to the compound energies, but they depend on the composition, or volume. However, one may expand the composition dependence of the cluster energy in the cluster fractions and obtain a separate end-member energy and an excess energy contribution. It is also possible to replace the configurational-entropy expression (5.36) by, for example, a quasi-chemical model or a tetrahedron CVM model, but, as long as the compound energies depend only on T and p, this should still be considered as a variant of the CEF.

It is not uncommon that there are no experimental data for many of the compounds or "end members." In some cases the end members are purely fictitious, in particular when the end members are charged, like in oxides. In those cases the concept of "bond energies" may be useful and the energy of such a compound may be estimated as the sum of the energies of bonds between the atoms in the compound.

Another technique to estimate the Gibbs energy of metastable end members in the CEF is to assume that an element in a sublattice with a coordination number of 12, or smaller, will contribute the same Gibbs energy as that element has in the fcc lattice. In a sublattice with coordination number 14 or higher it will contribute the same Gibbs energy as in a bcc lattice, since the second-nearest neighbors in a bcc lattice are not much further away than the nearest neighbors. This has frequently been used in the modeling of many phases, for example the σ phase. For example, a ternary end-member parameter can be estimated as

$$^\circ G^{\sigma}_{\mathrm{Fe:Mo:Ni}} = 10^\circ G^{\mathrm{fcc}}_{\mathrm{Fe}} + 4^\circ G^{\mathrm{bcc}}_{\mathrm{Mo}} + 16^\circ G^{\mathrm{bcc}}_{\mathrm{Ni}} \tag{5.39}$$

For models of ordered phases there are usually many possible ordered compounds, but only a few that are stable and can be measured. It is possible to calculate by *ab initio* techniques the energies of ordered compounds, which is very useful, in particular for obtaining good extrapolations of the ordered phases to higher-order systems; see section 3.3.3. It is possible using the Connolly–Williams cluster-expansion method (Connolly and Williams 1983), or by "disordered" *ab initio* calculations, to obtain estimates of the enthalpies of a disordered phase (Abrikosov *et al.* 1996). Even with very good *ab initio* data, it is still very important to have experimental data to compare them with since some effects, such as magnetism, may cause deviations between the results of *ab initio* calculations and the reality.

5.5.3 The reference state for the Gibbs energy

There is no absolute value for the Gibbs energy, so it is necessary to refer the Gibbs energy in all phases to the same reference point for each element. Thus in Eq. (5.35) and in the following the notation $°G_I^\theta$ means the Gibbs energy of formation of the constituent array I in θ from the reference states of the elements included in the constituents and is a function of temperature and pressure,

$$°G_I^\theta - \sum_k b_{ij} H_j^{\mathrm{SER}} = f(T, p) \tag{5.40}$$

where b_{ij} is the stoichiometry factor for the component j in the constituent array I and H_j^{SER} is the enthalpy of the component j in its reference state, which is normally the stable state for the component at 298.15 K and 1 bar. This reference state will normally be omitted from the equations below, but, when calculating chemical potentials, it is important to know what the reference state for each component is.

5.5.4 The ideal-substitutional-solution model

An ideal substitutional solution means a solution of non-interacting constituents mixing randomly with each other. A special case of this is when the constituents are the same as the components. The Gibbs energy is then

$$G_{\mathrm{m}} = \sum_{i=1}^{n} x_i\,°G_i + RT \sum_{i=1}^{n} x_i \ln(x_i) \tag{5.41}$$

Usually, however, the concept of an ideal model is extended also to the case when there are constituents other than the components, for example the ideal gas, which usually has many more constituents than components. The model above should then be called the *simple* ideal-substitutional-solution model. The Gibbs energy of this model is

$$G_{\mathrm{m}} = \sum_{i=1}^{n} y_i\,°G_i + RT \sum_{i=1}^{n} y_i \ln(y_i) \tag{5.42}$$

In this case the constituent fractions, y_i, defined in section 5.3.3, are used rather than the mole fractions, x_i, because there can be more constituents than components and because the mole fractions defined in section 5.3 are defined only for the components.

In the rest of the book, the term *ideal model* will mean a model in which the excess Gibbs energy, ${}^{E}G_m$, is zero. This means that its Gibbs energy depends only on the end members and that the entropy of mixing of the constituents is random. The physical contributions are included in the ideal model.

5.5.5 The gas phase

The phase most often described as an ideal substitutional model is the gas phase, which usually has many more constituents than components. For example, a model for gas with the components O and H should include at least the six constituents H, H_2, H_2O, O, O_2, and O_3:

$$G_m^{gas} = y_H\,{}^{\circ}G_H + y_{H_2}\,{}^{\circ}G_{H_2} + y_{H_2O}\,{}^{\circ}G_{H_2O} + y_O\,{}^{\circ}G_O + y_{O_2}\,{}^{\circ}G_{O_2} + y_{O_3}\,{}^{\circ}G_{O_3} + RT[y_H \ln(y_H) + y_{H_2}\ln(y_{H_2}) + y_{H_2O}\ln(y_{H_2O}) + y_O\ln(y_O) + y_{O_2}\ln(y_{O_2}) + y_{O_3}\ln(y_{O_3})] \tag{5.43}$$

The sum of the constituent fractions is unity. If there are more constituents than components, there are internal degrees of freedom in the phase and one can obtain the values of the constituent fractions only by minimizing the Gibbs energy for the current set of conditions.

Comparing Eq. (5.42) with the general form of the Gibbs energy, Eq. (5.1), gives

$${}^{srf}G_m = \sum_{i=1}^{n} y_i\,{}^{\circ}G_i \tag{5.44}$$

$${}^{cnf}S_m = -R\sum_{i=1}^{n} y_i \ln(y_i) \tag{5.45}$$

$${}^{E}G_m = 0 \tag{5.46}$$

$${}^{phys}G_m = 0 \tag{5.47}$$

The ${}^{\circ}G_i$ parameters can, at least in principle, be determined separately for each constituent i, and thus there are no adjustable parameters in the ideal-substitutional-solution model which describes the properties of the solution. For a gas phase the constituents are molecules or ions and their ${}^{\circ}G_i$ values can usually be calculated accurately from the vibrational and rotational modes of the molecule (Allendorf 2006).

It is interesting to note that it would require a very complicated excess Gibbs energy to describe the gas mentioned above with H and O using a simple ideal model based on the mixing of the atoms H and O. It is thus important to select a good "ideal" model in order to have a simple excess Gibbs energy. The shape of the Gibbs energy versus composition curve for a binary simple ideal substitutional solution is given in Fig. 5.5(a).

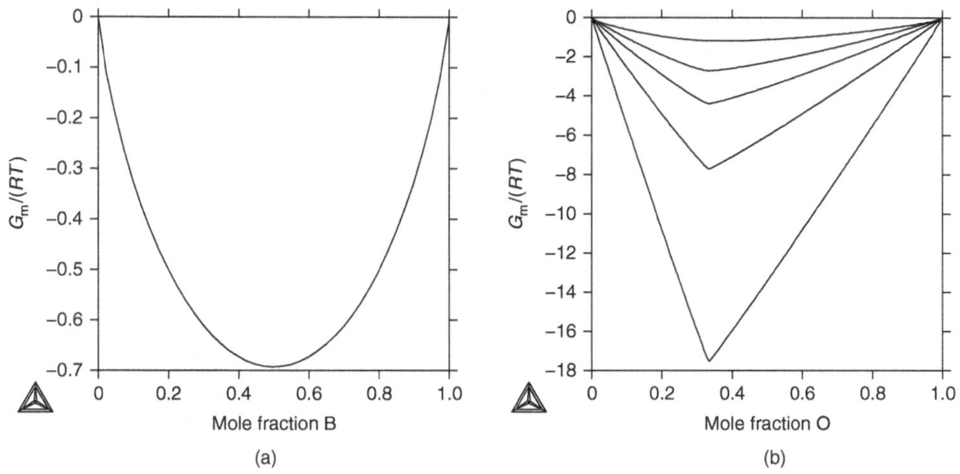

Figure 5.5 Gibbs-energy curves for ideal solutions. (a) The Gibbs energy for a simple ideal binary substitutional model divided by RT. Note that the slopes at each endpoint are infinite. (b) The Gibbs energies divided by RT for an ideal gas with the components H and O at various temperatures (500, 1000, 1500, 2000, and 3000 K) including all possible molecules (i.e. constituents) that can form. In (a) there are only A and B as constituents; in (b) there are several constituents. The strong "V" shape at low temperature is due to the formation of the molecule H_2O. At high temperature the gas becomes more like a monatomic ideal solution with H and O as major constituents.

The total Gibbs energy of one mole of "formula units" of the gas phase is

$$G_m^{gas} = \sum_i y_i \left[{}^\circ G_i^{gas} - \sum_j b_{ij} H_j^{SER} + RT \ln(y_i) \right] + RT \ln\left(\frac{p}{p_0}\right) \tag{5.48}$$

Equation (5.48) is an alternative formulation of the law of mass action. The constituent fractions y_i are the ratios of the partial pressure p_i of species i to the total pressure p, $y_i = p_i/p$.

The parameters ${}^\circ G_i^{gas}$ cannot be measured calorimetrically with good accuracy. A quantum-mechanical calculation, using spectroscopic measurements, is much more accurate. This type of calculation is outside the scope of this book. The "standard pressure" p_0 by convention is set to 10^5 Pa.

It has already been stated above that the gas phase usually has more constituents than components, i.e., there are stable molecules and thus internal degrees of freedom. It is interesting for the discussion of the model later to show how the Gibbs energy for a gas phase with the elements H and O depends on the composition for some temperatures. This is shown in Fig. 5.5(b). At low temperature the gas is either H_2 and H_2O or O_2 and H_2O because the Gibbs energy of formation of H_2O is very strong.

Non-ideal gases are often described in terms of a Helmholtz energy rather than a Gibbs energy, i.e., using T and V as variables instead of T and p. The reason for this choice is that the critical point for the gas–liquid transition is usually important for non-ideal gases, i.e., the gas and the liquid should have the same Helmholtz energy function. The transition cannot be described with a Gibbs-energy function because it represents a miscibility gap in the volume of the phase. Non-ideal gases are sometimes modeled using the fugacity; the fugacity f_i of a gas constituent i is simply related to the activity by the reference pressure p_0 as

$$f_i = p_0 a_i \tag{5.49}$$

and this can be modeled with the excess parameters introduced below.

5.5.6 *Ideal phases with special features*

Many features are associated with an ideal phase, i.e., a phase with non-interacting constituents. The definition above in Eq. (5.42) means that a phase treated with the ideal substititional model may exhibit a number of "non-ideal" features because there can be more constituents than components. For example, the activity versus composition curve for an ideal phase is often assumed to be a straight line, but that is certainly not true for a gas with H and O, as discussed above.

Even without the formation of stable species with two or more different components in the gas, one can have "non-ideal" shapes of the activity curves; see Fig. 6.7 later for an example. The curve depends on what is selected as components, but the components selected for a system should take into account all phases, not just the gas phase.

In models discussed later one will find that there are miscibility gaps even in "ideal systems," i.e., systems with non-interacting constituents.

5.5.7 *The origin of non-ideal behavior of solutions*

Systems with non-interacting constituents for which the surroundings of a constituent are irrelevant can be described by ideal models. In an ideal solution the bond between two unlike constituents, for example atoms in a lattice, is equal to that between two identical constituents. The simplest way to introduce a non-ideal behavior of solutions is to introduce the energy difference for these different bonds as

$$\epsilon_{ij} = E_{ij} - 0.5(E_{ii} + E_{jj}) \tag{5.50}$$

If ϵ_{ij} is negative it means that unlike constituents prefer to be together, i.e., long- and short-range order or clustering. If ϵ_{ij} is positive, there is a tendency toward phase separation and formation of a miscibility gap because the constituents prefer to surround themselves with constituents of the same kind.

In a theoretical approach various physical contributions to E_{ii}, E_{jj}, and E_{ij} can be identified, like electronic, vibrational, and configurational. In the present treatment, however, all contributions, except those that are modeled as separate physical contributions to the Gibbs energy (such as the magnetic contribution above), are included in the thermodynamic treatment as a value of ϵ_{ij} and are modeled together.

In crystalline solids the number of bonds is fixed, but in a liquid phase the constituents have no fixed environment and the number of nearest neighbors can vary. In many aspects the liquid is more similar to a gas than it is to a crystalline solid, but in all real liquids there is an interaction energy between the constituents that will create non-ideal behavior.

5.5.8 *The substitutional-regular-solution model*

In a crystalline phase the atoms have well-defined positions in a unit cell that can be repeated indefinitely in all directions. In a substitutional solution the constituents have the same probability of occupying any site in the unit cell and the Gibbs energy is given by

$$G_{\mathrm{m}} = \sum_{i=1}^{n} x_i \, {}^{\circ}G_i + RT \sum_{i=1}^{n} x_i \ln(x_i) + {}^{\mathrm{E}}G_{\mathrm{m}} \tag{5.51}$$

Since the probability is equal to the mole fraction, on limiting the excess Gibbs energy to binary interaction, we have

$${}^{\mathrm{E}}G_{\mathrm{m}} = \sum_{i} \sum_{j>i} x_i x_j L_{ij} \tag{5.52}$$

$$L_{ij} = \frac{z}{2} \epsilon_{ij} \tag{5.53}$$

where z is the number of bonds and ϵ is related to the bond energies by Eq. (5.50). The substitutional model is frequently used for metallic solutions, both liquids and crystalline solids, in which all the constituents mix on the same sites.

A phase may have crystallographically different sublattices and these may have different numbers of nearest neighbors and the constituents may prefer different sublattices. This is a form of long-range order (LRO) and must be included in the modeling, as will be described in section 5.8.

The models based on the CEF do not include any explicit short-range-order (SRO) contribution. The reason for this is that a proper treatment of SRO requires the introduction of clusters, which makes the numerical solution of the equilibrium too complicated for practical applications in multicomponent materials. Another reason is that, for many important materials like steels and aluminum alloys, the contribution to the Gibbs energy due to SRO is very small. Even for alloys in which SRO is significant, for example Ni-based superalloys, the SRO contribution can be approximated in the excess Gibbs energy; see section 5.8.4.5.

It is mainly for the liquid phase that models that take SRO into account are more frequently used also in thermodynamic databases. For the liquid the pair model, like the quasi-chemical model described in section 5.7.2.1, or the associated model described in section 5.7.1, is usually sufficient.

For crystalline phases the CVM developed by Kikuchi (1951) has been used with the techniques described in this book for an assessment of the Au–Cu system (Sundman *et al.* 1999). A simple case of approximating SRO in the disordered state is described in section 5.7.2.2.

5.5.9 *Dilute solutions, Henry's and Raoult's laws*

In the very dilute range all phases have the property that the activity of each of the constituents varies linearly with the fraction of the constituent. This is usually called "Henry's law" and all models described here obey this. In some cases, when the solubility range does not exceed the range within which the activity is linear, it can be convenient to model the phase with just one parameter describing this slope. Such an activity coefficient or "Henrian" parameter, γ_i°, for the solute i can be written

$$a_i = \exp\left(\frac{\mu_i}{RT}\right) = \gamma_i^\circ x_i \tag{5.54}$$

This model is useful for cases in which several dilute solutes in a single solvent phase are considered, for example aqueous solutions, because a single parameter is needed and simple software can be used. For the solvent "Raoult's law" is assumed, meaning that

$$a_1 = x_i \tag{5.55}$$

For a case with several solvents, or the same solvent in both liquid and solid states, one must keep in mind that γ_i° does not describe any property of pure i but only of i together with the solvent in a specific state. In such cases it is usually simpler to use a model that relates the properties of the pure solute i in the same state as the solvent, i.e., to use a model describing the complete composition range from solvent to solute. This requires an estimation of the Gibbs energy of the solute in a metastable state and such estimates are usually available from the SGTE (Dinsdale 1991). The assessment of an interaction parameter to describe the slope in the dilute range is then identical to assessing a value of γ_i°.

Raoult's law applies to the solvent in a dilute solution, stating that the activity of the solvent is equal to the fraction of the solvent when the solute is dilute. All models described here obey this.

Before the advent of computers, several models useful for thermodynamic calculations using just pen and paper were developed. One of them that is still popular is the dilute-solution model of Wagner (1952). In this model an additional coefficient is introduced into Eq. (5.54) for the chemical potential of the solvent i,

$$\mu_i/(RT) = \ln(\gamma_i^\circ x_i) + x_i \epsilon_{ii} \tag{5.56}$$

This equation is convenient for calculations using pen and paper but should never be used in computer calculations, since this model is inconsistent because it violates the Gibbs–Duhem relation, except in the trivial case of a binary solution.

One can compare it with the chemical potential calculated from the substitutional-regular-solution model, assuming that there are no interactions between the solutes:

$$\mu_i = {}^\circ G_i + RT\ln(x_i) + (1 - 2x_i + x_i^2)L_{1i} \tag{5.57}$$

The difference is the squared term of x_i in Eq. (5.57), without which chemical potentials described by Eq. (5.56) cannot be added to form an integral Gibbs energy

$$G_{\mathrm{m}} = \sum_i x_i \mu_i \tag{5.58}$$

except for the trivial binary case.

For very dilute solutions the fraction of solvent, x_i^2, can be set to zero in Eq. (5.57) and the sum of the terms $°G_i + L_{1j}$ is equivalent to $RT \ln(\gamma_i^{\circ})$ and L_{1i} is equal to $0.5RT\epsilon_{ii}$.

Darken improved this model by introducing his quadratic formula (Darken, 1967). Since large amounts of data have been obtained using the ϵ model, in particular for the liquid phase, it can be interesting to note that both Pelton (Pelton and Bale 1986) and Hillert (1986) have developed methods by which to convert γ_i° and ϵ_{ii} to a regular-solution model. If the solvent is denoted as "1," the partial Gibbs energies of the solutes "j" are, using Hillert's method,

$$\begin{aligned} G_j = {}& {}^{\circ}G_j + M_j + L_{1j} + RT\ln(x_j) + \sum_{k=2}^{n}(L_{kj} - L_{1j} - L_{1k})x_k \\ & + 0.5\sum_{k=2}^{n}\sum_{l=2}^{n}(L_{1k} + L_{1l} - L_{kl})x_k x_l \end{aligned} \tag{5.59}$$

where the L parameters are the interaction parameters in a substitutional regular solution, Eq. (5.52). The $°G_j$ is the reference state for pure j in this phase and M_j is a modification of this reference state when "1" is a solvent. The relations connecting M_j, the L parameters, and the ϵ parameters are

$$\begin{aligned} L_{1j}/(RT) &= -0.5\epsilon_{jj} \\ L_{kj}/(RT) &= \epsilon_{kj} - (\epsilon_{jj} + \epsilon_{kk}) \\ M_j/(RT) &= \gamma_j^{\circ} + \epsilon_{jj} \end{aligned} \tag{5.60}$$

With the last formula one may use the ϵ parameters in a regular-solution model. Note that, even converted to a regular solution, the model is limited to the dilute range of all the solutes j in Eq. (5.59) due to M_j.

5.6 Models for the excess Gibbs energy

The excess Gibbs energy is the Gibbs energy that is "in excess" of what can be described by ${}^{\mathrm{srf}}G_{\mathrm{m}}$, ${}^{\mathrm{cnf}}S_{\mathrm{m}}$ and ${}^{\mathrm{phys}}G_{\mathrm{m}}$. The binary, ternary, and higher-order interactions in the excess Gibbs energy will be described separately. All interactions will be written for a multicomponent system in order to avoid using special features that are valid in a lower-order system only:

$${}^{\mathrm{total}}G_{\mathrm{m}} = {}^{\mathrm{srf}}G_{\mathrm{m}} + {}^{\mathrm{phys}}G_{\mathrm{m}} - T\,{}^{\mathrm{cnf}}S_{\mathrm{m}} + {}^{\mathrm{E}}G_{\mathrm{m}} \tag{5.61}$$

$$\begin{aligned} {}^{\mathrm{E}}G_{\mathrm{m}} &= {}^{\mathrm{total}}G_{\mathrm{m}} - {}^{\mathrm{srf}}G_{\mathrm{m}} - {}^{\mathrm{phys}}G_{\mathrm{m}} + T\,{}^{\mathrm{cnf}}S_{\mathrm{m}} \\ &= {}^{\mathrm{bin.E}}G_{\mathrm{m}} + {}^{\mathrm{tern.E}}G_{\mathrm{m}} + {}^{\mathrm{high.E}}G_{\mathrm{m}} \end{aligned} \tag{5.62}$$

Note that the excess Gibbs energy includes all contributions not included in the ideal part, i.e., excess vibrational, excess configurational, etc.

5.6.1 The Gibbs energy of mixing

One should not confuse the excess Gibbs energy and the Gibbs energy of mixing. The Gibbs energy of mixing is defined as

$$^{\mathrm{mix}}G_{\mathrm{m}} = G_{\mathrm{m}} - \sum_{i}^{n} x_i\, {}^{\circ}G_i \tag{5.63}$$

where the summation i is for all components of the system. The total Gibbs energy of mixing, $^{\mathrm{mix}}G_{\mathrm{m}}$, is the difference between the Gibbs energy of the phase and that of a mechanical mixture of the components. The excess Gibbs energy, $^{\mathrm{E}}G_{\mathrm{m}}$ in Eq. (5.1), is a modeling quantity and equal to the difference between the Gibbs energy of the real phase and that of whatever ideal model that has been selected.

5.6.2 The binary excess contribution to multicomponent systems

The basic form of the binary excess-Gibbs-energy contributions to a multicomponent substitutional solution is

$$^{\mathrm{bin.E}}G_{\mathrm{m}} = \sum_{i=1}^{n-1} \sum_{j=i+1}^{n} x_i x_j L_{ij} \tag{5.64}$$

This equation is a sum over binary interactions in all i–j systems. Note that the mole fraction of each constituent is used, i.e., there is no assumption that $x_i + x_j = 1$, since that is true only for the binary subsystem. The same equation can be used to describe the binary excess contribution for each sublattice of a phase. The constituent fractions of the multicomponent system should then be used in Eq. (5.64).

There are several ways to justify type of binary interaction parameter from a physical point of view, for example by invoking the difference in bond energy between like and unlike atoms in Eq. (5.50). If the number of nearest neighbors is z, then $L_{ij} = (z/2)\epsilon_{ij}$. If the regular-solution parameter L_{ij} is a temperature-independent constant, the model is called *strictly regular*.

There are many extensions of this model including temperature and composition dependences and considering interactions between constituents on different sublattices.

In Fig. 5.6 the effect of the regular-solution parameter (in units of $\mathrm{J\,mol^{-1}}$) on the phase diagram for a binary system is shown. The system has only two phases, liquid and solid. In each row the liquid has the same regular-solution parameter, +20 000 in the first row, +10 000 in the second, 0 in the third, and −10 000 in the last row. The solid has a regular-solution parameter equal to +20 000, +10 000, and 0 in the first and second rows, and +10 000, 0, and −10 000 in the third and fourth rows. With a negative regular-solution parameter one can expect ordering, so a tentative second-order

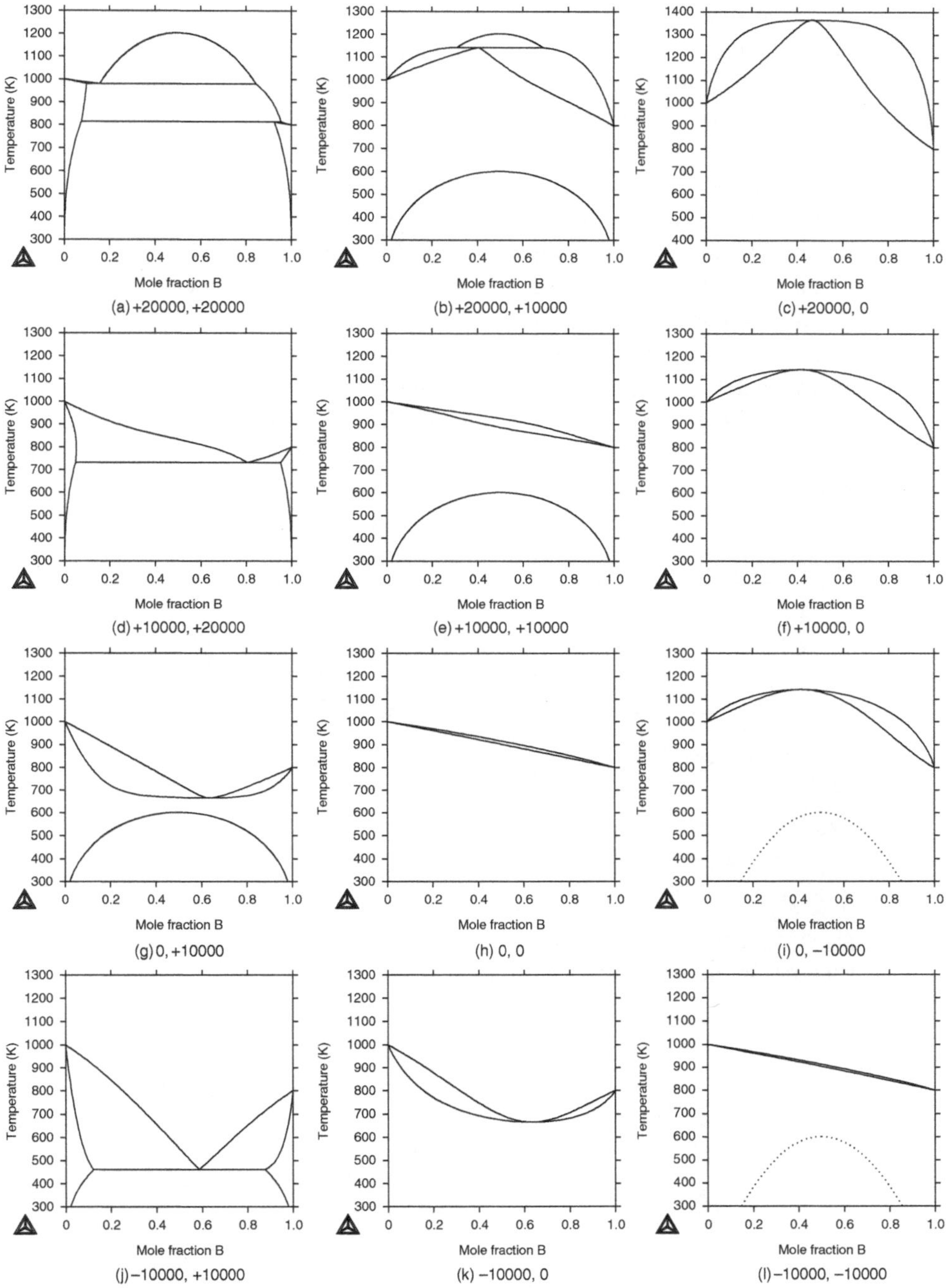

Figure 5.6 Phase diagrams for binary systems with differing values (in units of J mol^{-1}) of the regular-solution parameter. In the first row the interaction is +20 000 for the liquid and +20 000, +10 000, and 0, respectively, for the solid. In the second row the interaction is +10 000 for the liquid and +20 000, +10 000, and 0, respectively, for the solid. In the third row the interaction is 0 for the liquid and +10 000, 0, and −10 000, respectively, for the solid. In the fourth row the interaction is −10 000 for the liquid and +10 000, 0, and −10 000, respectively, for the solid. The dotted curves in diagrams (i) and (l) represent a possible second-order ordering transition.

transition line is indicated by a dotted line. The ordering has been assumed to be of A2/B2 type.

In Fig. 5.7 the Gibbs-energy curves for the liquid and solid phases are plotted for three different temperatures for the case when the interaction parameter is $-10\,000\,\text{J}\,\text{mol}^{-1}$ for the liquid and $+10\,000\,\text{J}\,\text{mol}^{-1}$ for the solid. At 800 K the liquid is stable except at the limits and the Gibbs-energy curve for the solid phase is very flat but has no inflexion points. At 500 K the Gibbs-energy curve for the solid phase has a maximum and two minima, but the miscibility gap is metastable since the liquid phase is stable in the middle of the diagram. At 400 K the miscibility gap in the solid phase dominates the phase diagram and the liquid is no longer stable.

In the assessment of real systems one should expect to obtain interaction energies approximately of the same order of magnitude in all phases modeled with a regular-solution model. The reason for this is that the regular-solution parameter depends on the difference in bond energies as given by Eq. (5.50), which remains approximately the same independently of the structure of the phase.

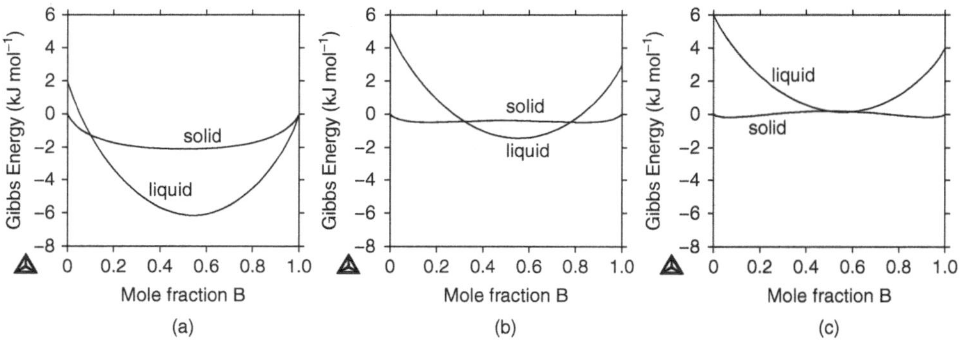

Figure 5.7 Gibbs-energy curves for the phase diagram in Fig. 5.6(j) at three different temperatures, (a) 800, (b) 500, and (c) 400 K.

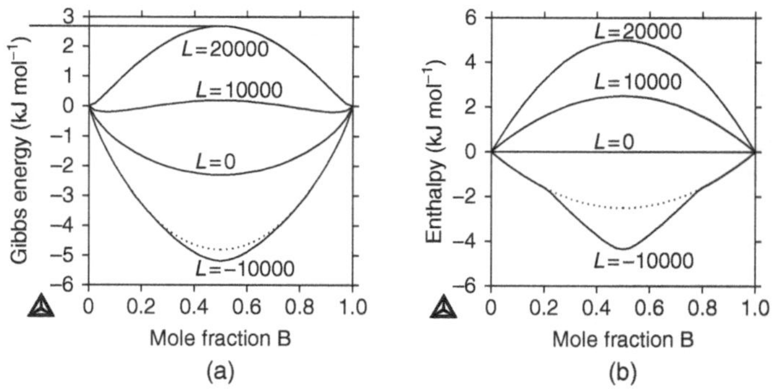

Figure 5.8 The Gibbs energy (a) and the enthalpy (b) of mixing at 400 K for the values of the regular-solution parameter used to calculate the phase diagrams in Fig. 5.6. The dotted curves represent a metastable extrapolation without ordering for the case with negative interaction.

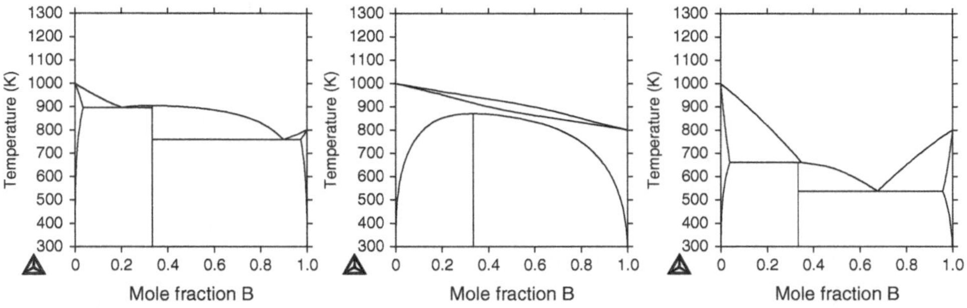

Figure 5.9 Phase diagrams with a compound A_2B for varying interaction parameters in the liquid and solid phases; the compound has the same Gibbs energy in all three diagrams. The interaction parameter in the liquid is $+10\,000\,\mathrm{J\,mol^{-1}}$ in the left and middle diagrams and $-10\,000\,\mathrm{J\,mol^{-1}}$ in the diagram on the right. The solid-phase interaction parameter is $+20\,000\,\mathrm{J\,mol^{-1}}$ in the diagram on the left and $+10\,000\,\mathrm{J\,mol^{-1}}$ in the other two. In the phase diagram the compound melts congruently in the diagram on the left, has a solid decomposition in the diagram in the middle, and undergoes peritectic melting in the diagram on the right.

In Fig. 5.8(a) the Gibbs energy of mixing and in Fig. 5.8(b) the enthalpy of mixing for the solid phase at 400 K are shown for some values of the regular-solution parameter. The dotted line indicates the metastable disordered phase for the case with negative regular-solution parameter.

In most binary systems there are also intermetallic phases, but, if these are ignored, the general shape of the phase diagram should be close to one of those plotted. In Fig. 5.9 three phase diagrams with a compound are shown. It shows that the type of transformation for the compound in the phase diagram depends on the parameters of all phases; it is not a property of the compound.

5.6.2.1 The Redlich–Kister binary excess model

Equation (5.64) gives just a single interaction parameter for the phase in the binary i–j system and that is often not enough to describe the available experimental data. One may extend the binary regular-solution term in the composition by using differences between the fractions of i and j:

$$L_{ij} = \sum_{\nu=0}^{k}(x_i - x_j)^{\nu} \cdot {}^{\nu}L_{ij} \tag{5.65}$$

This form of the composition dependence for a binary interaction was first suggested by Guggenheim (1937). It was recommended for use in modeling multicomponent systems by Redlich and Kister (1948) and is generally known as a Redlich–Kister (RK) power series. It is evident that, for a binary system, this equation can be written in many different forms using the relation $x_2 = 1 - x_1$. Other well-known expressions for the composition dependence are Legendre polynomials, and those of Margules, Borelius, etc. However, the form given above is recommended for use in multicomponent systems since it preserves the shape of the excess Gibbs energy of the binary system in the multicomponent system and only the magnitude decreases with decreasing $x_i + x_j$.

The curves in Fig. 5.10 may help the reader to understand the contributions of the various terms in an RK series to the excess Gibbs energy. Note that the regular-solution term is the only one that has a non-zero value in the middle of the system. The contributions from the odd coefficients change sign in the middle whereas the contributions from the even coefficients have the same sign throughout the whole system. All curves are calculated for the same value of the parameter, 20 000 J mol^{-1}. The magnitude of the curves decreases because the difference in fractions is raised to a higher power.

For reasons discussed later, one should avoid using many coefficients in an RK series. There should be a special reason for using more than the first three. The special name "subregular-solution model" is sometimes used when just two RK coefficients are employed, and "subsubregular" indicates using three.

The order of i and j is important in the RK series, since the sign of the coefficients depends on it. Various types of software may use different ways to denote the order.

A single positive regular-solution parameter will always give a symmetrical miscibility gap, thus at least two RK coefficients are needed in order to have an asymmetrical miscibility gap.

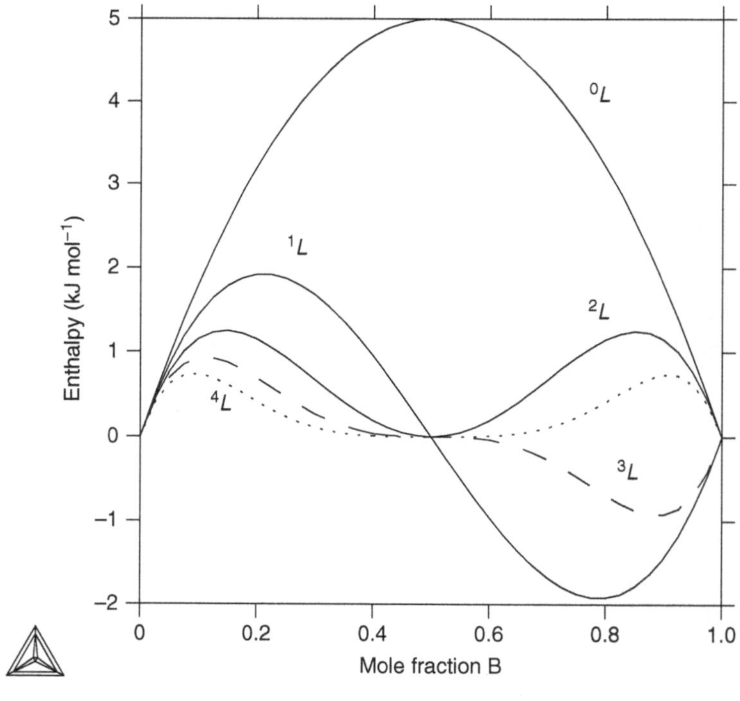

Figure 5.10 The contributions to the enthalpy of mixing for the first five terms in the RK power series. The ^3L curve is dashed and that for ^4L is dotted to make them easier to identify.

It is quite common that a phase is stable only within a narrow composition range. This is usually modeled with sublattices with various types of defects based on crystallographic and other experimental information. However, the model usually covers a much larger composition range than the stability range of the phase. It may be tempting to limit the composition range by using many RK coefficients, but one must be aware of the effect of these coefficients also outside the stability range of the phase. In particular, one may have serious problems when extrapolating a binary system with many RK coefficients to a ternary or higher-order system because the phase may then appear at quite different composition ranges, where the higher-order binary RK coefficients do not give a reasonable description. This is further described in chapter 6 on assessment methodology and in case studies in chapter 9.

The parameters ${}^{\nu}L_{ij}$ in the RK series can, of course, be temperature-dependent. Normally a linear temperature dependence is enough and only when heat-capacity data are available may one use more:

$$ {}^{\nu}L_{ij} = {}^{\nu}a_{ij} + {}^{\nu}b_{ij}T \tag{5.66} $$

The composition dependence of the excess enthalpy is described by ${}^{\nu}a_{ij}$ and the excess entropy by ${}^{\nu}b_{ij}$. If measured excess-heat-capacity data are available, then and only then may one introduce a higher-order temperature dependence into the RK coefficients. The next higher term after the linearly temperature-dependent one should be a $T\ln(T)$ term to represent a constant excess heat capacity.

The Redlich–Kister power series is useful only when the excess enthalpy is a smooth function of the constitution like for the Cu–Ni system in Fig. 5.11 from Jansson (1987). When the enthalpy varies rapidly, like in the Mg–Sn system shown in Fig. 5.16(b) later, some other modeling feature must be used.

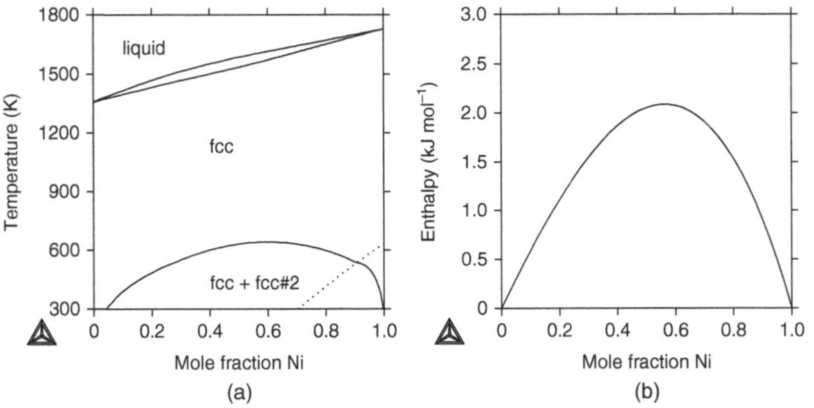

Figure 5.11 In (a) the phase diagram for Cu–Ni is shown. The temperature for the ferromagnetic transition for Ni (dashed line) makes the miscibility gap skewed. In (b) the enthalpy of mixing in the fcc phase is shown. The enthalpy curve is rather smooth and two RK coefficients were sufficient.

5.6.2.2 Other binary excess parameter models

The Redlich–Kister series is the recommended excess model because it is symmetrical and thus extrapolates well to ternary and higher-order systems. There are many other excess models for binary systems, such as the simple polynomial, Lagrange series, etc., some of which may also be modified to be symmetrical series. In a binary system one may always convert from any series to another, but the ternary extrapolations will depend on the series used, as well as on the extrapolation method; see section 5.6.6. Combining several different binary excess series and different extrapolation methods makes it very difficult to understand the behavior of a multicomponent system; therefore, it is not advisable.

5.6.3 *Two versions of a phase diagram assessed without thermochemical data*

One may find published assessments using little or no thermodynamic data. Such assessments should be considered with great skepticism, since it is quite easy to describe the same phase diagram with quite different values of the thermodynamic data. See for example the two phase diagrams in Fig. 5.12. The diagrams are very similar but the phase boundaries represent only differences between the Gibbs energies of the phases and thus the enthalpy curves from the two assessments are very different, as shown in Fig. 5.13.

The difference in the assessment is that in (a) an ordered model with sublattices has been used for the intermetallic phase, whereas in (b) the intermetallic compound has been assumed to be a substitutional model. In (a) the ordering has been assumed to be strong at the equiatomic composition, but a large deviation from stoichiometry has been allowed using anti-site atoms. In (b) the phases have been modeled with excess parameters as small as possible.

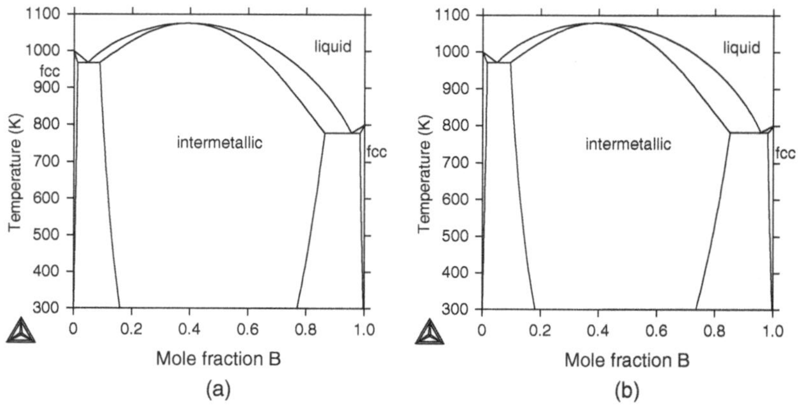

Figure 5.12 Two calculated phase diagrams fitted to the same experimental data on the phase diagram but without any thermochemical data. Different models were used for the intermediate phase. The difference between the two phase diagrams is very small.

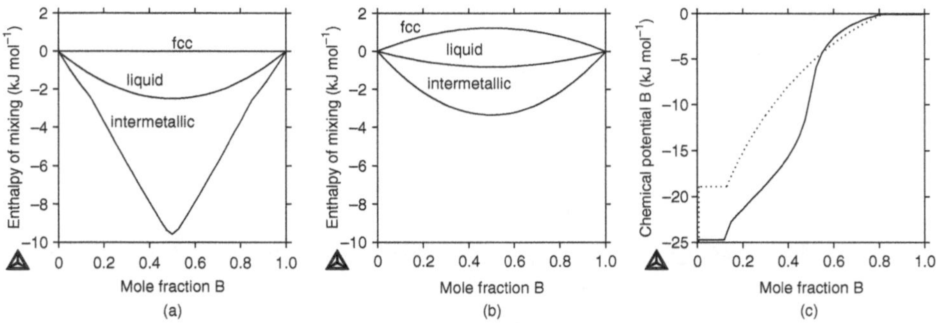

Figure 5.13 In (a) and (b) the enthalpy of mixing for all phases has been calculated for the two phase diagrams in Fig. 5.12 using ordered and disordered models, respectively. As can be seen the enthalpy curve for the intermetallic compound in (a) is much more negative and has a pronounced "V"-shape. The activity curves calculated at 600 K for the two assessments are shown in (c) (the full line is for the ordered model (a); the dotted line is for the disordered model (b)) and there are significant differences.

There are hardly any differences between the phase diagrams in Fig. 5.12 and, if there are no thermochemical measurements or crystallographic information about the type and degree of ordering, there is no way of knowing which assessment is correct.

The sharp increase in activity at $x(\mathrm{B}) = 0.5$ from the ordered model is shown in Fig. 5.13(c) (full line). It is due to the ordering and cannot be reproduced in the dashed curve from the assessment in which the intermetallic compound was treated as a substitutional solution. The diagrams show the simple relation between a sharp "V"-shaped enthalpy curve and a strong increase of the chemical potential at the same composition.

The extrapolations into a ternary system from these two different assessments will behave very differently. Thus ternary data are often a useful complement to binary assessments with no or little data.

5.6.4 The ternary excess contributions

The excess ternary interaction contribution requires one more summation:

$$^{\mathrm{tern.E}}G_{\mathrm{m}} = \sum_{i=1}^{n-2} \sum_{j=i+1}^{n-1} \sum_{k=j+1}^{n} x_i x_j x_k L_{ijk} \tag{5.67}$$

In some cases one would require that the ternary interaction parameter is also composition-dependent and, as in the binary case, several methods have been suggested as ways to handle this. However, in order to provide a symmetrical extension into higher-order systems, Hillert (1980) suggested the following type of composition dependence in the ternary excess parameter:

$$L_{ijk} = v_i \cdot {}^{i}L_{ijk} + v_j \cdot {}^{j}L_{ijk} + v_k \cdot {}^{k}L_{ijk} \tag{5.68}$$

where

$$v_i = x_i + (1 - x_i - x_j - x_k)/3$$
$$v_j = x_j + (1 - x_i - x_j - x_k)/3 \quad (5.69)$$
$$v_k = x_k + (1 - x_i - x_j - x_k)/3$$

The advantage of introducing the v_i fractions is that their sum will always be unity, even in a multicomponent system. The ternary parameter will thus behave symmetrically when extrapolated to higher-order systems. If all ${}^iL_{ijk}$ are the same then that is identical to having a composition-independent ternary parameter. A composition-independent term in Eq. (5.68) is thus redundant.

A composition-independent ternary term would have its largest contribution where the three fractions are as large as possible, i.e., in the middle of the constitutional triangle. From Eq. (5.68) it is evident that the ${}^iL_{ijk}$ will have its largest contribution toward the corner for pure i.

Before introducing ternary or higher-order interactions, one should consider modification of the binary systems if the extrapolation to a higher-order system exhibits a significant difference from the experimental information. Since the experimental information is usually scarce and uncertain, it is possible to obtain equivalent descriptions of a binary system. The binary descriptions will have different extrapolations to higher-order systems and often ternary information is crucial for obtaining a valid binary description.

5.6.5 *Higher-order excess contributions*

As for binary and ternary interactions, more summations can be added:

$$^{\text{high.E}}G_{\text{m}} = \sum_{i=1}^{n-3} \sum_{j=i+1}^{n-2} \sum_{k=j+1}^{n-1} \sum_{l=k+1}^{n} x_i x_j x_k x_l L_{ijkl} + \cdots \quad (5.70)$$

Few quaternary or higher-order parameters have been evaluated, but they would fit into Eq. (5.70). It has not been necessary to consider composition dependences of these parameters because there are not enough experimental data and, if there are discrepancies in higher-order systems, one can normally correct this by reassessment of the lower-order systems.

5.6.6 *Extrapolation methods for binary excess models*

In the derivation of the excess contributions to the Gibbs energy in Eq. (5.62) a multicomponent system considering the contribution from each of the binary, ternary, and higher-order terms was used. In the equations one should use the fraction of each constituent from the multicomponent system. One must not replace the fraction of one constituent with one minus the sum of the rest, since that would affect the contribution to higher-order systems. This is the simplest method of adding contributions from many lower-order systems together and is not related to a particular solution model. However, several methods for how to add together contributions from binary systems have been proposed on the basis of geometrical reasoning.

It may be important to point out that the choice of extrapolation model is important only for higher-order parameters in a Redlich–Kister series. The composition-independent regular parameter will extrapolate in the same way independently of the extrapolation model.

The method used in Eq. (5.62) above has been derived in a geometrical way also and is called the Muggianu method (Muggianu *et al.* 1975). Alternative methods have been suggested by Kohler (1960), Colinet (1967), and Toop (1965). The Kohler and Colinet methods are symmetrical like the Muggianu method and treat the contributions from the three binary systems in the same way, but refer the contributions from the binaries to different compositions along the binary side as shown in Fig. 5.14.

The Toop method treats one of the three elements differently, but, if the other two elements become identical, they reproduce the binary A–B = A–C systems in each vertical section through the A corner. These two features, (a) symmetrical treatment of all elements and (b) reproducing the binary systems if two elements "become" identical, cannot be combined, except in the case of composition-independent regular solutions in all three binaries. The differences among the methods are usually small. The Muggianu method has the easiest formulation, starting from the RK formalism, and thus is preferable.

The Toop method is used when one of the constituents behaves very differently from the others. For example, when mixing carbon or oxygen with two metals, it may be tempting to use the Toop method since it treats the contribution from one of the binaries differently from those from the other two. This method thus requires that constituents be classified into different groups, which can be done with small databases, but there is no obvious way to do this for a general database.

Pelton (2001) has devised an ingenious method to classify Toop elements for each ternary and to combine this with other ternary extrapolation models in a multicomponent system. This can become very complicated, so it is not recommended unless all other methods to handle the asymmetry of a ternary system have been exhausted. First one should try different sets of constituents, taking into account size effects, short-range order, electronegativity, or charge transfer. There are several other modeling tools that can handle such effects. One possibility is to add a ternary parameter that can be evaluated

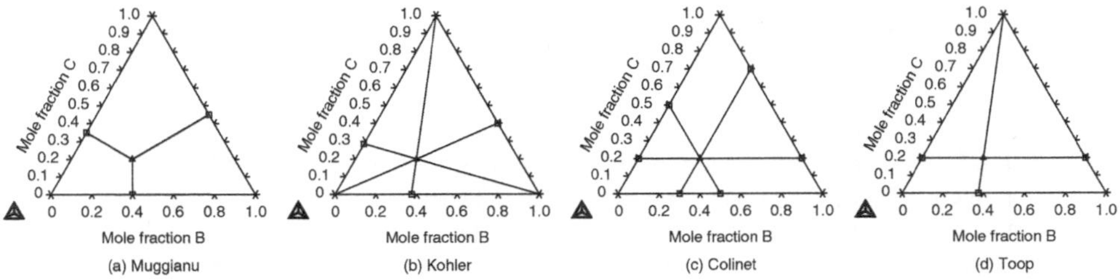

Figure 5.14 Various ternary extrapolation methods presented graphically.

from the binary excess parameters. Hillert (1980) showed that a Toop model for a ternary A–B–C system, where C is the "Toop" element and the binary A–C and B–C systems are modeled as subregular solutions, can be treated with the Muggianu model with a ternary parameter evaluated from the binary systems

$$L_{\mathrm{ABC}} = {}^{1}L_{\mathrm{AC}} + {}^{1}L_{\mathrm{BC}} \tag{5.71}$$

where ${}^{1}L_{\mathrm{AC}}$ and ${}^{1}L_{\mathrm{BC}}$ are the subregular-solution parameters for the A–C and B–C systems.

Ternary extrapolation models that can vary from "Kohler type" to "Toop type" depending on the relative values of binary excess Gibbs energies are not physical and would give strange extrapolations to higher-order systems.

5.7 Modeling using additional constituents

Any solution with a strong tendency toward ordering, i.e., those in which it is energetically favorable for unlike atoms to be close, will have a characteristic "V"-shape in its enthalpy of mixing when the composition varies. This might not be evident from the phase diagram shown in Fig. 5.15 for the Mg–Sn system from (Fries and Lukas 1993). The published assessment was later modified using a new description of the hcp phase.

The activity curves will also be affected by the ordering and the activity may change by several orders of magnitude over a small composition range. At the ordering composition the configurational entropy will have a minimum. All these effects are shown in Fig. 5.16 for the liquid phase in the Mg–Sn system. The low entropy of mixing and the sharp increase of the activity occur at the same composition. At higher temperatures the curves reveal a more regular behavior of the liquid. It should be evident that this type of behavior

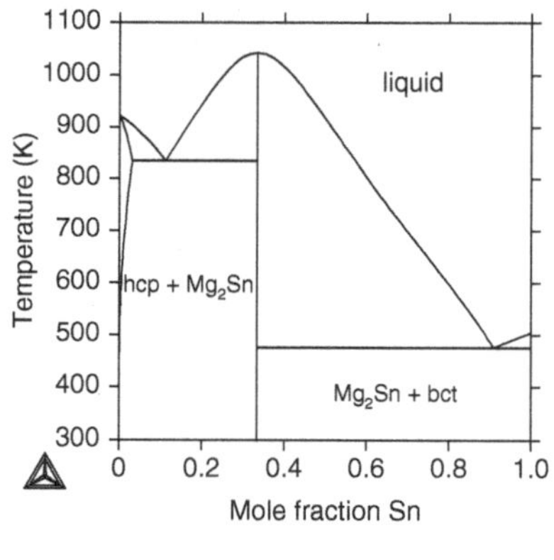

Figure 5.15 The phase diagram for Mg–Sn does not indicate that the liquid has any unusual properties. Only when one has access to thermochemical data does one find, see Fig. 5.16, that there is a strong short-range ordering in the liquid phase. The existence of the very stable compound Mg_2Sn can be taken as an indication that there may be a tendency for the liquid also to have strong ordering around the same composition, but there are cases with stable compounds but no short-range order in the liquid.

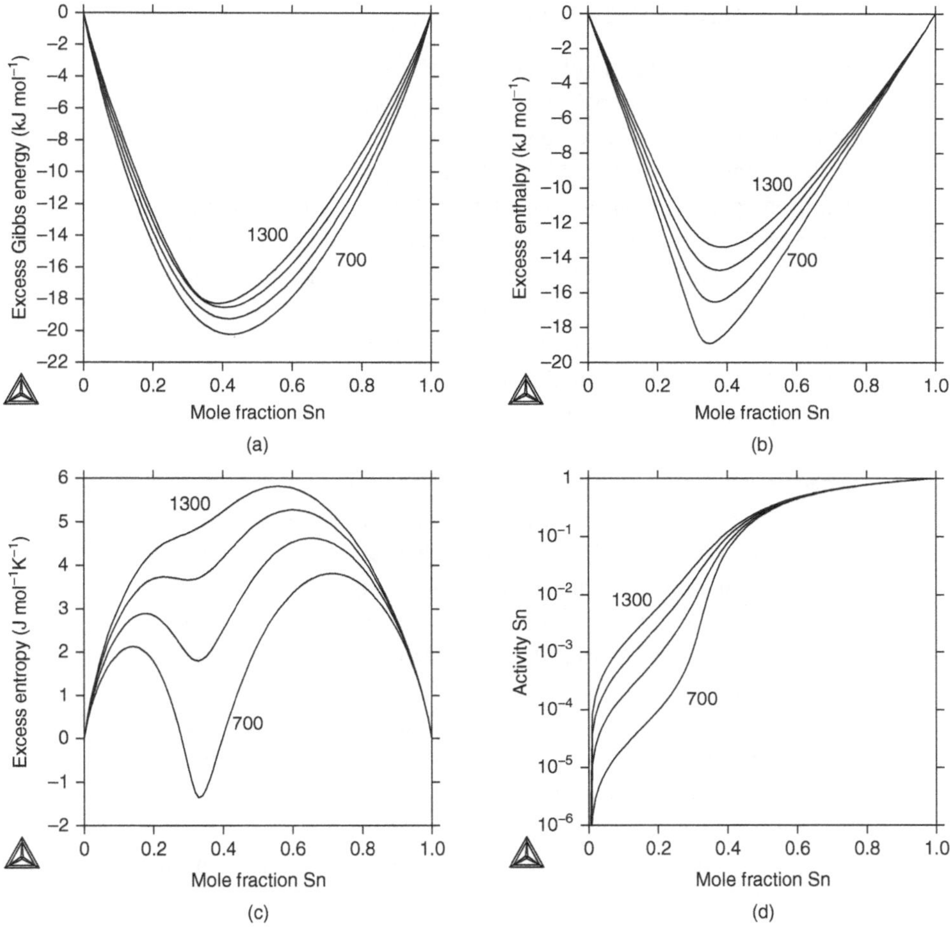

Figure 5.16 The effect of ordering on various thermodynamic properties for the Mg–Sn system: (a) the Gibbs energy, (b) the enthalpy, (c) the entropy, and (d) the activity. The curves have been calculated for the temperatures 700, 900, 1100, and 1300 K.

cannot be treated by the excess-Gibbs-energy formalism described in Eq. (5.62), which describes a smooth change in curvature over the whole composition range.

If the excess Gibbs energy for a substitutional-solution model requires too many coefficients to describe this behavior, it is preferable to modify the ${}^{\mathrm{srf}}G_{\mathrm{m}}$ and ${}^{\mathrm{cnf}}S_{\mathrm{m}}$ terms of Eq. (5.1). For solid phases, one may use crystallographic information to introduce sublattices, as described in the next section. For liquids and for cases in which sublattices are not possible, one may introduce fictitious constituents, usually called associates.

Normally, fictitious constituents are used to describe short-range order (SRO), i.e., the local arrangement of atoms, and the sublattice model to describe long-range order (LRO), i.e., the atoms form a periodic arrangement over long distances, but there are intermediate cases for which neither method fully describes the reality.

5.7.1 *The associate-solution model*

In systems with SRO, unlike atoms tend to stay together for a shorter or longer length of time. The term "associate" was introduced to denote an association between unlike atoms when the attractive forces between the atoms are not strong enough to form a stable chemical molecule. If a molecule is formed, this should be included unproblematically as a constituent in the solution. Consider, for example, the H–O system plotted in Fig. 5.5(b), which shows the variation of the Gibbs energy with composition. The pronounced minimum in the Gibbs energy is due to formation of H_2O molecules in the gas and there is no excess Gibbs energy.

For a system with a similar Gibbs energy or enthalpy versus composition curve to that of a system with formation of stable molecules, the introduction of a fictitious constituent may be reasonable. Thus the associate-solution model formally introduces an "associate" as a constituent in the solution as a modeling tool. The associate may be considered as a molecule not stable enough to allow its isolation, but having a lifetime still significantly larger than the mean time between two thermal collisions.

Fictitious constituents, sometimes called "associates" and sometimes "clusters," can be used in the same form of the equation for the surface of reference of the Gibbs energy and the configurational entropy as in Eq. (5.42). In this way one introduces a new constituent and thus creates an internal degree of freedom. The Gibbs energy of formation of the new constituent can be used to fit experimental data. An experimentally determined sharp minimum in the enthalpy curve is mainly described in terms of the enthalpy of formation of the associate. The stoichiometry of the associate must correspond to the composition of the minimum. At that composition the constituent fraction of the associate is high and thus the configurational entropy is low.

The contributions of the surface of reference and the configurational entropy to the Gibbs-energy expression, Eq. (5.1), for a substitutional associated solution are

$$ {}^{\mathrm{srf}}G_{\mathrm{m}} = \sum_{i=1}^{n} y_i\,{}^{\circ}G_i \tag{5.72} $$

$$ {}^{\mathrm{cnf}}S_{\mathrm{m}} = -R\sum_{i=1}^{n} y_i \ln(y_i) \tag{5.73} $$

where the summation over i is for all constituents. The site fraction y_i is used here to denote that the constituent fractions are not the same as the mole fractions of the components. The excess Gibbs energy can be modeled in the same way as a substitutional solution, as described in section 5.6, treating the interaction between each pair of constituents as an independent parameter. Physical properties such as magnetism can be added as usual.

5.7.2 *Non-random configurational entropy*

If the interactions between the constituents are large, the random entropy of mixing used for modeling Eq. (5.42) is not appropriate. If the interactions create a long-range ordering of the constituents, that can be modeled with sublattices as described in section 5.8. If the interactions create short-range ordering, its contribution to the Gibbs energy must be modeled differently. There is interest in predicting phase diagrams using more or less complicated, non-ideal, entropy expressions based on the "quasi-chemical" approach or on the CVM. In these models pairwise bonds between atoms or clusters with three or more atoms are introduced as fictitious constituents in calculating the entropy by use of Eq. (2.11). The models include a correction term to give the correct random configurational entropy when there is no tendency toward SRO.

5.7.2.1 The quasi-chemical model

Guggenheim proposed a model for chemical SRO called the quasi-chemical model (Guggenheim 1952). To derive this model, one may start by assuming that the bonds between A and B atoms, giving molecules AA, BB, and AB, are distributed randomly. The formation of the bonds is described by a simple chemical-reaction formula that can be written as a Gibbs energy of reaction for the molecules AA, BB, and AB:

$$\mathrm{AA} + \mathrm{BB} \rightleftharpoons \mathrm{AB} + \mathrm{BA} \tag{5.74}$$

As with the associate model, this can be treated by introducing additional "fictitious" constituents into a substitutional-solution model, in this case both AB and BA. The reason for having two constituents with the same stoichiometry is due to the fact that they do not represent free molecules as in the gas or liquid, but bonds in a crystalline solid, where the orientation of the bond is important. The exchange of an A and a B atom between two neighboring sites affects the entropy because other bonds are changed by this exchange.

For a phase with the number of bonds per atom equal to z and ignoring the fact that the bonds should agree on the atom placed at each site, the Gibbs-energy expression for the "bonds" AA, BB, AB, and BA is

$$^{\mathrm{srf}}G_{\mathrm{m}} = y_{\mathrm{AA}}{}^{\circ}G_{\mathrm{AA}} + y_{\mathrm{BB}}{}^{\circ}G_{\mathrm{BB}} + y_{\mathrm{AB}}{}^{\circ}G_{\mathrm{AB}} + y_{\mathrm{BA}}{}^{\circ}G_{\mathrm{BA}} \tag{5.75}$$

$$^{\mathrm{cnf}}S_{\mathrm{m}} = -R\frac{z}{2}[y_{\mathrm{AA}}\ln(y_{\mathrm{AA}}) + y_{\mathrm{BB}}\ln(y_{\mathrm{BB}}) + y_{\mathrm{AB}}\ln(y_{\mathrm{AB}}) + y_{\mathrm{BA}}\ln(y_{\mathrm{BA}})] \tag{5.76}$$

The mass-balance constraints give the mole fractions of A and B as

$$x_{\mathrm{A}} = y_{\mathrm{AA}} + 0.5(y_{\mathrm{AB}} + y_{\mathrm{BA}}) \tag{5.77}$$

$$x_{\mathrm{B}} = y_{\mathrm{BB}} + 0.5(y_{\mathrm{AB}} + y_{\mathrm{BA}}) \tag{5.78}$$

In the expression for the surface of reference, ${}^{\circ}G_{\mathrm{AB}} = {}^{\circ}G_{\mathrm{BA}}$ due to symmetry. If $y_{\mathrm{AB}} = y_{\mathrm{BA}}$ there is no LRO and this "degeneracy" in the disordered state, due to its distinguishing the fractions of y_{AB} and y_{BA}, gives an additional term $RTy_{\mathrm{AB}}\ln 2$ compared with the case on ignoring this degeneracy. In the expression for $^{\mathrm{cnf}}S_{\mathrm{m}}$ the configurational entropy is overestimated considerably because the number of bonds is $z/2$ times larger than the number of atoms per mole and, for the case ${}^{\circ}G_{\mathrm{AB}} = 0$, the entropy should become identical to that of an ideal solution of the components A and B.

The origin of this overestimate can be understood if $z = 2$ in Eq. (5.76) because it is then identical to the entropy of a gas phase with AA, BB, BA, and AB. In a gas phase the molecules are independent and can rotate freely, but in a crystalline phase the nearest neighbors to a site must agree what atom is placed in that site. In order to correct for the overestimation of the entropy in Eq. (5.76), the following modified entropy expression was suggested by Guggenheim (1952):

$$\begin{aligned} {}^{\mathrm{cnf}}S_{\mathrm{m}} &= -R\frac{z}{2}\left[y_{\mathrm{AA}}\ln\left(\frac{y_{\mathrm{AA}}}{x_{\mathrm{A}}^2}\right)+y_{\mathrm{BB}}\ln\left(\frac{y_{\mathrm{BB}}}{x_{\mathrm{B}}^2}\right)+y_{\mathrm{AB}}\ln\left(\frac{y_{\mathrm{AB}}}{x_{\mathrm{A}}x_{\mathrm{B}}}\right)+y_{\mathrm{BA}}\ln\left(\frac{y_{\mathrm{BA}}}{x_{\mathrm{A}}x_{\mathrm{B}}}\right)\right] \\ &\quad - R[x_{\mathrm{A}}\ln(x_{\mathrm{A}})+x_{\mathrm{B}}\ln(x_{\mathrm{B}})] \end{aligned} \tag{5.79}$$

For the case of no SRO this expression gives the same configurational entropy as a random mixing of the atoms A and B. This is the final entropy expression according to the quasi-chemical model and also to the simplest Ising model, but, even with this correction, Eq. (5.79) is valid only when the degree of SRO is small. If the ordering is strong and $z > 2$, one may even have negative values of ${}^{\mathrm{cnf}}S_{\mathrm{m}}$ from Eq. (5.79); see Fig. 5.17(b) later.

When the SRO is very strong, it can be treated as LRO, which can be described with the sublattices introduced in section 5.8. A connection between the quasi-chemical model and the sublattice model is described in section 5.8.4.3.

One can easily verify that Eq. (5.79) is identical to a random-entropy model when there is no SRO. For this case the constituent fractions can be calculated from the mole fractions:

$$\begin{aligned} y_{\mathrm{AA}} &= x_{\mathrm{A}}^2 \\ y_{\mathrm{AB}} = y_{\mathrm{BA}} &= x_{\mathrm{A}}x_{\mathrm{B}} \\ y_{\mathrm{BB}} &= x_{\mathrm{B}}^2 \end{aligned} \tag{5.80}$$

On inserting Eq. (5.80) into Eq. (5.79), this becomes identical to an ideal solution in Eq. (5.41). This is in contrast to the associate model, in which an ideal configurational entropy of all constituents is assumed and the configurational entropy will never be identical to that of an ideal solution without the associate.

5.7.2.2 The cluster-variation method

In the CVM formalism developed by Kikuchi (1951), one introduces clusters with three, four, and more atoms and, depending on the crystal structure, one can derive corrections to the entropy expression taking into account the fact that the clusters are not independent. However, the principle is the same as for the quasi-chemical formalism, namely that these clusters can be treated as independent constituents with a unique fraction, but, for the configurational entropy, one must take into account the fact that the clusters share corners, edges, surfaces, etc. The configurational entropy is different for each different crystal structure and the CVM configurational entropy for an fcc lattice in the tetrahedron approximation is given below. For the case in which there is no *LRO*, there are five constituents or "clusters," A, $A_{0.75}B_{0.25}$, $A_{0.5}B_{0.5}$, $A_{0.25}B_{0.75}$, and B. One may use the stoichiometry as fraction indices as long as there is no LRO. These five clusters all represent tetrahedra in the fcc structure and are the *end members* of the phase. The configurational enthalpy is included in their Gibbs energies of formation. Without LRO the surface of reference is

$$^{\mathrm{srf}}G_{\mathrm{m}} = y_{\mathrm{A}}\,{}^{\circ}G_{\mathrm{A}} + y_{\mathrm{A_{0.75}B_{0.25}}}\,{}^{\circ}G_{\mathrm{A_{0.75}B_{0.25}}} + y_{\mathrm{A_{0.5}B_{0.5}}}\,{}^{\circ}G_{\mathrm{A_{0.5}B_{0.5}}} + y_{\mathrm{A_{0.25}B_{0.75}}}\,{}^{\circ}G_{\mathrm{A_{0.25}B_{0.75}}} + y_{\mathrm{B}}\,{}^{\circ}G_{\mathrm{B}} \quad (5.81)$$

The ideal configurational entropy for a system with these constituents is

$$\begin{aligned} ^{\mathrm{id.cnf}}S_{\mathrm{m}} &= -R[y_{\mathrm{A}}\ln(y_{\mathrm{A}}) + y_{\mathrm{A_{0.75}B_{0.25}}}\ln(y_{\mathrm{A_{0.75}B_{0.25}}}) \\ &\quad + y_{\mathrm{A_{0.5}B_{0.5}}}\ln(y_{\mathrm{A_{0.5}B_{0.5}}}) + y_{\mathrm{A_{0.25}B_{0.75}}}\ln(y_{\mathrm{A_{0.25}B_{0.75}}}) + y_{\mathrm{B}}\ln(y_{\mathrm{B}})] \end{aligned} \quad (5.82)$$

In a crystal the clusters are not independent, but share edges and corners. They must thus agree on the type of atom placed in each site and that reduces the configurational entropy compared with that of an ideal gas. This correction can only be approximate and, with the CVM tetrahedron approximation, the following expression is used:

$$\begin{aligned} ^{\mathrm{CVM.cnf}}S_{\mathrm{m}} &= 2\,^{\mathrm{id.cnf}}S_{\mathrm{m}} + 2\,^{\mathrm{deg}}S_{\mathrm{m}} \\ &\quad + 6R[p_{\mathrm{AA}}\ln(p_{\mathrm{AA}}) + p_{\mathrm{AB}}\ln(p_{\mathrm{AB}}) + p_{\mathrm{BA}}\ln(p_{\mathrm{BA}}) + p_{\mathrm{BB}}\ln(p_{\mathrm{BB}})] \\ &\quad - 5R[x_{\mathrm{A}}\ln(x_{\mathrm{A}}) + x_{\mathrm{B}}\ln(x_{\mathrm{B}})] \end{aligned} \quad (5.83)$$

The term $^{\mathrm{deg}}S_{\mathrm{m}}$ is due to the fact that the 5 clusters above are degenerate cases of the 16 clusters needed to describe LRO (4 different $A_{0.75}B_{0.25}$, 6 different $A_{0.5}B_{0.5}$, and 4 different $A_{0.25}B_{0.75}$). This means adding

$$^{\mathrm{deg}}S_{\mathrm{m}} = -R[(y_{\mathrm{A_{0.75}B_{0.25}}} + y_{\mathrm{A_{0.25}B_{0.75}}})\ln(4) + y_{\mathrm{A_{0.5}B_{0.5}}}\ln(6)] \quad (5.84)$$

The variable p_{AA} is a pair probability that is equal to the bond fraction in the quasi-chemical-entropy expression. The values of p_{AA} etc. can be calculated from the cluster fractions as

$$\begin{aligned} p_{\mathrm{AA}} &= y_{\mathrm{A}} + 0.5y_{\mathrm{A_{0.75}B_{0.25}}} + \frac{1}{6}y_{\mathrm{A_{0.5}B_{0.5}}} \\ p_{\mathrm{AB}} = p_{\mathrm{BA}} &= 0.25y_{\mathrm{A_{0.75}B_{0.25}}} + \frac{1}{3}y_{\mathrm{A_{0.5}B_{0.5}}} + 0.25y_{\mathrm{A_{0.25}B_{0.75}}} \\ p_{\mathrm{BB}} &= y_{\mathrm{B}} + 0.5y_{\mathrm{A_{0.25}B_{0.75}}} + \frac{1}{6}y_{\mathrm{A_{0.5}B_{0.5}}} \end{aligned} \quad (5.85)$$

and the mole fractions from the pair probabilities as

$$\begin{aligned} x_{\mathrm{A}} &= p_{\mathrm{AA}} + p_{\mathrm{AB}} \\ x_{\mathrm{B}} &= p_{\mathrm{BB}} + p_{\mathrm{BA}} \end{aligned} \quad (5.86)$$

The equations for p_{AA} etc. are easily understood if one realizes that, in a tetrahedron cluster, there are six bonds and that in an $A_{0.75}B_{0.25}$ cluster half of these bonds are between A atoms, whereas in an $A_{0.5}B_{0.5}$ cluster only one is between A atoms.

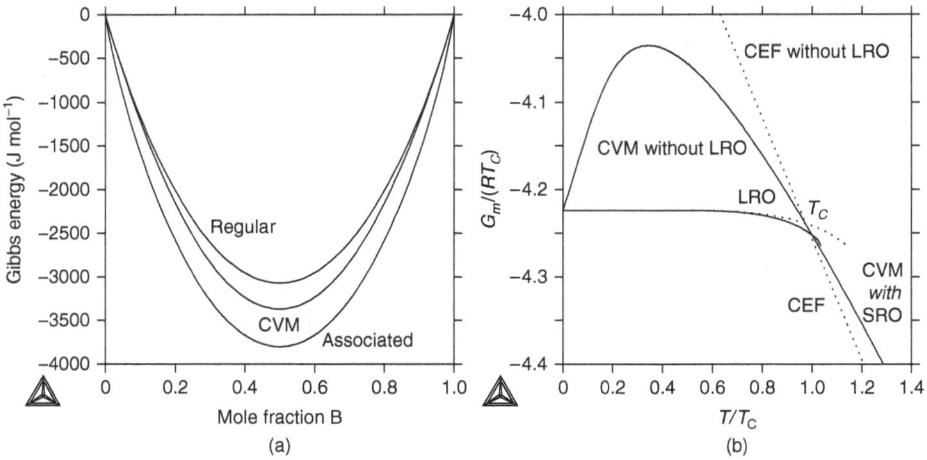

Figure 5.17 Comparisons of the Gibbs energy for tetrahedron CVM and other models using the same bond energy. T_C is the critical temperature for long-range ordering at equiatomic composition. (a) The Gibbs energy for a regular-solution model, associate model, and tetrahedron CVM model, all using the same nearest-neighbor bond energy at $T/T_C = 1.06$. (b) Gibbs energy versus T for a tetrahedron CVM model (full lines) and a four-sublattice CEF model (dashed lines) at equiatomic composition.

In Fig. 5.17(a) the Gibbs-energy curves for three related models have been calculated for an A–B system using the same nearest-neighbor bond energy just above the temperature for LRO. One curve is for a substitutional regular-solution model, Eq. (5.64), with ideal configurational entropy Eq. (5.41); one for an associated-solution model with five constituents, A, $A_{0.75}B_{0.25}$, $A_{0.5}B_{0.5}$, $A_{0.25}B_{0.75}$, and B, Eqs. (5.72) and (5.73); and one for a CVM tetrahedron model, Eqs. (5.81) and (5.83) with the same clusters as the associates. In the second and third of these models no excess Gibbs energy is used, but the same constant Gibbs energies are used for the associates or clusters. It is clear that the associate model overestimates the contribution of SRO to the Gibbs energy.

In Fig. 5.17(b) the Gibbs energies at equiatomic composition for the CVM tetrahedron model (full lines) and a four-sublattice CEF model (dashed lines) with approximate SRO contribution according to Eq. (5.151) are plotted versus temperature from 0 to 1.4 times the ordering temperature, T_C. The Gibbs energy has been normalized by dividing it by RT_C. The Gibbs energies for these two models are almost identical when there is LRO. Above T_C the LRO state is metastable within a short temperature interval both for the CVM and for the CEF. The metastable state without LRO has been extrapolated all the way down to 0 K for both models. The CVM extrapolation has an unphysical negative entropy at low temperatures. This means that one must be careful using models like the modified quasi-chemical model for liquids that cannot undergo a transition to LRO; see section 5.9.3.1.

In general the cluster energies in the CVM depend on the composition, whereas the end-member energies used in the CEF have fixed energies representing a phase with exactly the composition and structure of the end member. In the CEF the dependence on composition is modeled with the excess Gibbs energies and this method requires fewer composition variables than does the CVM. As a comparison, a CVM tetrahedron model for fcc with eight elements must have at least $8^4 = 4096$ clusters or constituents whereas a four-sublattice CEF model, which can describe the LRO just as well as CVM can and with a reasonable SRO approximation, requires only $8 \times 4 = 32$ constituents. This means that a CEF model will be several orders of magnitude faster than a CVM model to implement in three-dimensional software for simulations of phase transformations, for example.

5.7.3 The excess Gibbs energy with additional constituents

Even with fictitious constituents, Eq. (5.1) can be applied with an excess term given by Eq. (5.62) and the summations are made including the fictitious constituents. Introducing a fictitious constituent into a binary substitutional solution allows one to treat this formally as a ternary system and evaluate binary interactions of all three constituents. From the mathematical point of view, this is not a problem, but of course the software for calculation of the equilibrium must handle the mass-balance conditions properly.

A simple case of introducing associates is the model for the Mg–Sn system. The liquid exhibits a high degree of SRO at the Mg_2Sn composition and, in order to describe this,

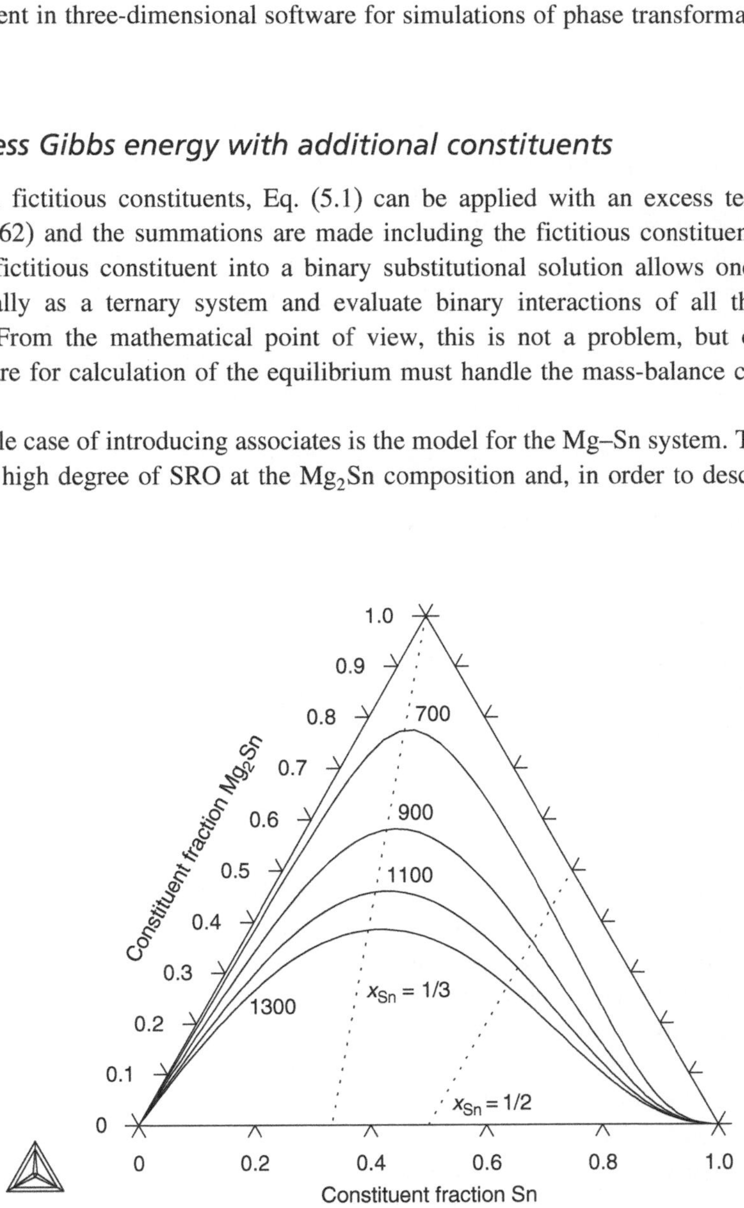

Figure 5.18 The variation of the constituent fractions in liquid Mg_2Sn with composition at temperatures of 700, 900, 1100, and 1300 K. The dotted lines represent two constant amounts of Sn as indicated.

one may introduce Mg_2Sn as a constituent of the liquid. The terms in the Gibbs-energy expression will be

$$ {}^{\mathrm{srf}}G_{\mathrm{m}}^{\mathrm{liq}} = y_{\mathrm{Mg}}\,{}^{\circ}G_{\mathrm{Mg}} + y_{\mathrm{Sn}}\,{}^{\circ}G_{\mathrm{Sn}} + y_{\mathrm{Mg_2Sn}}\,{}^{\circ}G_{\mathrm{Mg_2Sn}} \quad (5.87) $$

$$ {}^{\mathrm{cnf}}S_{\mathrm{m}}^{\mathrm{liq}} = -R[y_{\mathrm{Mg}}\ln(y_{\mathrm{Mg}}) + y_{\mathrm{Sn}}\ln(y_{\mathrm{Sn}}) + y_{\mathrm{Mg_2Sn}}\ln(y_{\mathrm{Mg_2Sn}})] \quad (5.88) $$

$$ {}^{\mathrm{E}}G_{\mathrm{m}} = y_{\mathrm{Mg}}y_{\mathrm{Sn}}L_{\mathrm{Mg,Sn}} + y_{\mathrm{Mg_2Sn}}y_{\mathrm{Sn}}L_{\mathrm{Mg_2Sn,Sn}} + y_{\mathrm{Mg}}y_{\mathrm{Mg_2Sn}}L_{\mathrm{Mg,Mg_2Sn}} \quad (5.89) $$

The parameter ${}^{\circ}G_{\mathrm{Mg_2Sn}}$ determines the fraction of Mg_2Sn in the liquid. There are three excess interaction parameters that describe each "binary" side of the constitutional triangle shown in Fig. 5.18. In this diagram, which looks like a ternary system, for the Mg–Sn system, taken from an assessment by Fries and Lukas (1993), the full lines show how the constitution of the liquid phase varies with composition at several temperatures. In these cases the parameter $L_{\mathrm{Mg,Sn}}$ is usually zero because the fractions of Mg and Sn are never high simultaneously. One may include a "ternary" interaction $L_{\mathrm{Mg,Sn,Mg_2Sn}}$.

Note that the selection of the associate is very important, the model will behave differently if one selects as a model (Mg, Sn, $Mg_{2/3}Sn_{1/3}$).

5.8 Modeling using sublattices

In crystalline solids it is important to take into account that the atoms may occupy different types of sublattices with different coordination numbers, bond lengths, etc. Sublattices represent LRO, which will modify both the entropy expression and the excess Gibbs energy. The simplest case of a sublattice model would be two sublattices with two constituents in each. A shorthand notation of this would be

$$ (\mathrm{A,B})_m(\mathrm{C,D})_n \quad (5.90) $$

where the subscripts m and n give the ratio of sites on the two sublattices. In a crystalline solid m and n are fixed numbers, but one may chose to extract a common factor determining the size of the formula unit of the phase. It is advisable to use the smallest possible integer numbers for the site ratios. The constituents A, B, C, and D can represent atoms, ions, anti-site atoms, vacancies, etc. The most general formalism for describing thermodynamic properties of phases with two or more sublattices is the CEF, which is given by Eqs. (5.35)–(5.37).

The very simplest case of using two or more sublattices is when there is a single component on either sublattice. That represents a stoichiometric phase and is not different from having a single species as constituent. The case of mixing on one sublattice and a single constituent on the other is identical to mixing species in the substitutional model, although some care has to be taken about the stoichiometry of the species in order to obtain a reasonable configurational entropy.

5.8.1 Reciprocal solutions

A simple case, which cannot be modeled as a substitutional model, is that of a phase with two sublattices and two constituents on each sublattice. This is known as a reciprocal system because of the reaction

$$A_mC_n + B_mD_n = A_mD_n + B_mC_n \tag{5.91}$$

The reciprocal energy represented by this reaction is

$$\Delta G = {}^{\circ}G_{A:C} + {}^{\circ}G_{B:D} - {}^{\circ}G_{A:D} - {}^{\circ}G_{B:C} \tag{5.92}$$

An example of this type of reaction occurs on mixing the salts NaCl and KBr:

$$\mathrm{NaCl} + \mathrm{KBr} = \mathrm{NaBr} + \mathrm{KCl} \tag{5.93}$$

In the reciprocal model, just like in the CEF, one assumes ideal configurational entropy on each sublattice separately and the total entropy is weighted with respect to the number of sites on each sublattice. The sublattice model thus gives a different behavior of the configurational entropy from that given by a substitutional model. The first two terms of Eq. (5.1) will be

$$^{\mathrm{srf}}G_{\mathrm{m}} = y'_{\mathrm{A}}y''_{\mathrm{C}}{}^{\circ}G_{\mathrm{A:C}} + y'_{\mathrm{A}}y''_{\mathrm{D}}{}^{\circ}G_{\mathrm{A:D}} + y'_{\mathrm{B}}y''_{\mathrm{C}}{}^{\circ}G_{\mathrm{B:C}} + y'_{\mathrm{B}}y''_{\mathrm{D}}{}^{\circ}G_{\mathrm{B:D}} \tag{5.94}$$

$$^{\mathrm{cnf}}S_{\mathrm{m}} = -R\{m[y'_{\mathrm{A}}\ln(y'_{\mathrm{A}}) + y'_{\mathrm{B}}\ln(y'_{\mathrm{B}})] + n[y''_{\mathrm{C}}\ln(y''_{\mathrm{C}}) + y''_{\mathrm{D}}\ln(y''_{\mathrm{D}})]\} \tag{5.95}$$

where y'_i and y''_j are the constituent fractions on sublattices 1 and 2, respectively. The reference energy expression has four terms representing the possible four compounds obtainable by combining the constituents on the two sublattices. The Gibbs energy of formation of these compounds ${}^{\circ}G_{i:j}$ is multiplied by the fractions of one constituent from each sublattice. Note that a colon is used to separate constituents on different sublattices in the parameter expression. For clarity, two constituents on the same sublattice will sometimes be separated by a comma.

The configurational entropy is calculated from Eq. (2.11) assuming random mixing on each sublattice and multiplying this by the site ratios. The excess term, ${}^{E}G_{\mathrm{m}}$, will be discussed in the next section. Already without any excess terms, i.e., an ideal reciprocal model, there are some specific features.

In Fig. 5.19(a) the constitutional square is shown and the four corner compounds are indicated. If the reciprocal model is used for a binary system, the constituents on the two sublattices can be the components, species or ions formed by the components or vacancies. The Wagner–Schottky model described in section 5.8.2.3 is one of the most important examples of the use of this model. Many examples of the application of the two-sublattice model will be described in other sections.

For the reciprocal model one may be able to calculate partial Gibbs energies not for the components but only for end members as defined in section 5.5.1. For a model with

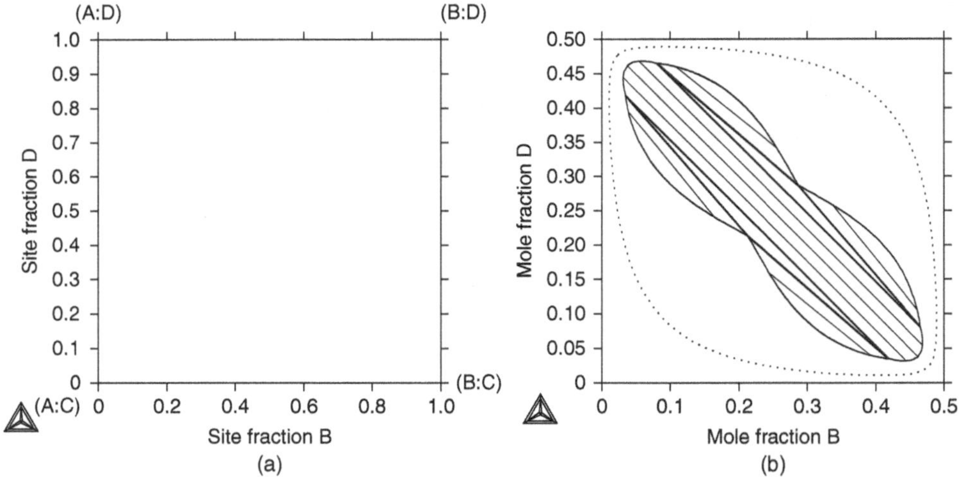

Figure 5.19 Diagrams for reciprocal systems. (a) The constitutional square for a reciprocal system (A,B)(C,D). The lower left corner represents the compound AC, the upper right the compound BD. (b) The reciprocal miscibility gap in a system (A,B)(C,D). The dotted curve represents the miscibility gap without any excess parameters; the full line, which actually has five miscibility gaps, represents the stable phase diagram with the reciprocal parameter according to Eq. (5.102) added.

the Gibbs energy given by Eqs. (5.94) and (5.95), the partial Gibbs energy for the end member AD can be calculated using Eq. (5.38),

$$G_{\mathrm{A:D}} = {}^{\circ}G_{\mathrm{A:D}} + (1 - y'_{\mathrm{A}})(1 - y''_{\mathrm{D}})\Delta G + mRT\ \ln(y'_{\mathrm{A}}) + nRT\ \ln(y''_{\mathrm{D}}) \qquad (5.96)$$

where ΔG is the reciprocal energy according to Eq. (5.92).

5.8.1.1 Excess Gibbs energy for the reciprocal solution

In the reference energy term there are already two fractions, one from each sublattice, multiplied by the parameter for formation of the compound. In the excess parameter one will thus start with three fractions multiplied by each other, two from one sublattice and one from the other. Additionally there is a reciprocal excess parameter that is multiplied by all four fractions:

$${}^{\mathrm{E}}G_{\mathrm{m}} = y'_{\mathrm{A}}y'_{\mathrm{B}}y''_{\mathrm{C}}L_{\mathrm{AB:C}} + y'_{\mathrm{A}}y'_{\mathrm{B}}y''_{\mathrm{D}}L_{\mathrm{AB:D}} + y'_{\mathrm{A}}y''_{\mathrm{C}}y''_{\mathrm{D}}L_{\mathrm{A:CD}} + y'_{\mathrm{B}}y''_{\mathrm{C}}y''_{\mathrm{D}}L_{\mathrm{B:CD}} + y'_{\mathrm{A}}y'_{\mathrm{B}}y''_{\mathrm{C}}y''_{\mathrm{D}}L_{\mathrm{AB:CD}} \qquad (5.97)$$

There are two "binary" interaction parameters for each sublattice depending on the constituent on the other sublattice and these are related to the four sides of the constitutional square in Fig. 5.19(a). The reciprocal parameter will have its largest influence in the center of the square. In Fig. 5.26 in the section 5.9.4 the surface of reference for a reciprocal system is shown.

The binary interaction parameters can be expanded in an RK formula, Eq. (5.65), in terms of the two fractions from the same sublattice, for example

$$L_{\mathrm{AB:C}} = \sum_{\nu=0}^{n} (y'_{\mathrm{A}} - y'_{\mathrm{B}})^{\nu} \cdot {}^{\nu}L_{\mathrm{AB:C}} \tag{5.98}$$

It may also be necessary in some cases to use a composition dependence for the reciprocal parameter. The following expression has been adopted:

$$L_{\mathrm{AB:CD}} = {}^{0}L_{\mathrm{AB:CD}} + (y'_{\mathrm{A}} - y'_{\mathrm{B}}) \cdot {}^{1}L_{\mathrm{AB:CD}} + (y''_{\mathrm{C}} - y''_{\mathrm{D}}) \cdot {}^{2}L_{\mathrm{AB:CD}} \tag{5.99}$$

So far no more than two composition-dependent terms have been used and it is recommended that one try to do without any composition dependence.

For the models with more sublattices, the excess Gibbs energy is expanded in the same way as for the reciprocal system. With three or more sublattices, however, the excess Gibbs energy is less important since the major part of the Gibbs energy is described by the parameters in the surface of reference, ${}^{\circ}G_I$.

5.8.1.2 The reciprocal miscibility gap

The reciprocal sublattice model has an inherent tendency to form a miscibility gap in the middle of the system. When the reciprocal Gibbs energy, ΔG in Eq. (5.92), is sufficiently large this will create a phase separation with the tie-lines parallel to the diagonal and with the largest energy difference between the corner compounds. This miscibility gap can appear without any excess parameters and it cannot be suppressed completely even by adding excess parameters. Even the reciprocal parameter might not suppress the miscibility gap, but may just separate it into two smaller ones.

When there are experimental data on the Gibbs energies of the corner compounds, the appearance of the miscibility gap is usually predicted reasonably well by the reciprocal model with no or small excess energy. However, if one or two of the corners represent metastable or purely fictitious compounds that must be estimated, it is often difficult to control the appearance of this miscibility gap.

When there are no data for an end member of a reciprocal solution, it is often useful to estimate using the assumption that the ΔG in Eq. (5.92) is zero, i.e., the end member (A:C) representing a fictitious compound can be estimated by

$${}^{\circ}G_{\mathrm{A:C}} = {}^{\circ}G_{\mathrm{A:D}} + {}^{\circ}G_{\mathrm{B:C}} - {}^{\circ}G_{\mathrm{B:D}} \tag{5.100}$$

In this way one can avoid problems with reciprocal miscibility gaps. If several end members are fictitious and cannot be estimated in any other way, one should start by estimating the one which is most significant. One should also take care about compound energies that might not be specific to a particular system. In particular, the parameters for compounds with some sublattices filled with vacancies may be part of other systems and it is then important to estimate the values of such parameters by incorporating all available information, just as when determining the lattice stabilities for the elements.

To avoid the presence of a reciprocal miscibility gap, Hillert suggested the use of a special reciprocal parameter,

$$^{\mathrm{E}}G_{\mathrm{m}} = \sqrt{y_{\mathrm{A}} y_{\mathrm{B}} y_{\mathrm{C}} y_{\mathrm{D}}} L_{\mathrm{A,B:C,D}} \tag{5.101}$$

This can completely suppress the miscibility gap and can be added to the normal excess Gibbs energy, for the expression Eq. (5.97). Since there is no physical reason for this composition dependence, it must be considered as a curve-fitting parameter.

The reciprocal miscibility gap is often decreased by short-range ordering in the system. It is possible to derive an approximation to this contribution, as was first done by Pelton and Blander (1986) for molten salts. It was later extended to solid phases by Sundman *et al.* (1998). The reciprocal parameter is expressed as a function of the reciprocal energy, ΔG in Eq. (5.92), as

$$L_{\mathrm{A,B:C,D}} = -\frac{\Delta G^2}{zRT} \tag{5.102}$$

The factor z is the number of nearest neighbors. This parameter has also been used successfully by Frisk *et al.* (2001) for modeling carbides and nitrides, for which many end members are not stable and their Gibbs energies thus must be estimated.

In Fig. 5.19(b) the reciprocal miscibility gap in a phase modeled as (A, B)(C, D) is shown as a dotted line. This miscibility gap is calculated with a ΔG parameter according to Eq. (5.92) only and no excess parameters. Adding a reciprocal parameter according to Eq. (5.102) changes the miscibility gap into five different ones, which are shown by the full lines.

5.8.2 *Models using two sublattices*

In this section we consider a number of models that can be described with the two-sublattice CEF. When there are several constituents in each sublattice, one can write the model

$$(A, B, \ldots)_m(U, V, \ldots)_n \tag{5.103}$$

In some cases the same constituent can be on both sublattices; in others the constituents are different. The Gibbs energy for this model is

$$G_{\mathrm{m}} = \sum_i \sum_j y'_i y''_j {}^{\circ}G_{i:j} + {}^{\mathrm{phys}}G_{\mathrm{m}} + RT\left(m \sum_i y'_i \ln(y'_i) + n \sum_j y''_j \ln(y''_j)\right) + {}^{\mathrm{E}}G_{\mathrm{m}} \tag{5.104}$$

5.8.2.1 Interstitial solutions

A common application of the two-sublattice model is to interstitial solutions of carbon or nitrogen in metals. Owing to the difference in size, these elements prefer to occupy the interstitial sites in the metallic sublattices. In the austenite phase with fcc lattice the interstitial sites form an interwoven fcc lattice and the structure type for Fe with C is thus B1.

The main characteristic of an interstitial solution is that one of the constituents in Eq. (5.103) is the vacancy. This was already introduced in Eq. (5.28) and the vacancy can be treated just like any other "real" constitent if its chemical potential is always equal to zero, and it is excluded from the summation to calculate the mole fractions, Eq. (5.29).

Some carbides, nitrides, and borides have the B1 structure (TiC, VN, etc.) and they can be treated with the same model as austenite, i.e., the fcc phase. A model for B1 as both metal and carbon–nitride–boride phases in a multicomponent steel is

$$(\mathrm{Fe, Cr, Ni, Ti, \ldots})_1(\mathrm{Va, C, N, B})_1 \tag{5.105}$$

In Thermo-Calc two "composition sets" of the fcc phase can be used if both the metallic phase (with mainly vacancies in the interstitial sublattice) and the carbide are stable in the same system. Usually the carbide is the second composition set, but, after a calculation, one may have to list the compositions of the phases in order to know which composition set is metallic or carbide.

5.8.2.2 Models for phases involving metals and non-metals

Many phases involving a metallic element and a non-metal like carbon, nitrogen, or oxygen are modeled with one sublattice for the metallic elements and another for the non-metal. Some examples are the cubic carbo-nitride mentioned above, and the M_7C_3 carbide. In some cases there is more than one crystallographically different sublattice for the metallic element, but there are normally not enough experimental data to describe this. One case for which two metallic sublattices have been used is for the $M_{23}C_6$ carbide,

$$(\mathrm{Cr, Fe, \ldots})_{20}(\mathrm{Cr, Fe, Mo, W, \ldots})_3(\mathrm{C})_6 \tag{5.106}$$

In the model above one has used the experimental information that some of the elements do not enter all sublattices, in this example Mo and W. Even if there is a third sublattice with just carbon, this is in principle a two-sublattice model because there is mixing on only two sublattices.

An interesting case is the spinel phase. In this case the constituents must be treated as ions:

$$(\mathrm{Fe^{2+}, Fe^{3+}})_1(\mathrm{Fe^{2+}, Fe^{3+}, Va})_2(\mathrm{O^{2-}})_4 \tag{5.107}$$

There are additional constraints when a phase has charged constituents. This model will be discussed further in section 6.2.5.8.

5.8.2.3 The Wagner–Schottky defect model

The Wagner–Schottky model (Wagner and Schottky 1930) describes the variation of the Gibbs energy of formation of a compound within a small composition range as a function of various types of defects. The model contains only three parameters, the Gibbs energy of formation of the *ideal compound* and the Gibbs energies of formation of two *defects*, allowing a deviation from stoichiometry on both sides of the *ideal composition*. This model can be described in the CEF as a reciprocal model with four end members. If the ideal constituents are called A and B and the defects X and Y, the model will be

$$(\mathrm{A}, \mathrm{X})_a(\mathrm{B}, \mathrm{Y})_b \tag{5.108}$$

where a and b are the stoichiometry ratios. The defects can be

(1) anti-site atoms, i.e., B atoms on the sublattice for A and A atoms on the sublattice for B;
(2) vacancies;
(3) interstitials; or
(4) a mixture of the above defects.

If interstitial defects are important, one must in CEF add a sublattice for these and specify how many sites there are in this interstitial sublattice and that the normal constituent on the interstitial sublattice is the vacancy. For example, if the main type of defect is that both A and B prefer to appear as interstitials on the same interstitial sublattice, one has the model

$$(\mathrm{A})_a(\mathrm{B})_b(\mathrm{Va}, \mathrm{A}, \mathrm{B})_c \tag{5.109}$$

The crystal structure for the phase provides additional information. For example, in some compounds with the B2 structure type, i.e., ordered bcc with two identical sublattices, one often has anti-site atoms on one side of the ideal composition and vacancies on the other. This means that it is tempting to use a model of the form

$$(\mathrm{A}, \mathrm{B})_1(\mathrm{B}, \mathrm{Va})_1 \tag{5.110}$$

but the two sublattices are crystallographically identical, which means that, if the atoms in the two sublattices are interchanged, the crystal is still the same and the ideal compound can equally well be written BA as AB. For this reason one must include all defects on both sublattices; allowing both anti-site atoms and vacancies the model should be

$$(\mathrm{A}, \mathrm{B}, \mathrm{Va})_1(\mathrm{B}, \mathrm{A}, \mathrm{Va})_1 \tag{5.111}$$

One should take into account that B2 is an ordered form of the bcc lattice. The recommended model for B2 is discussed further in section 5.8.2.4.

The Wagner–Schottky defect model considers only one defect for each sublattice and three parameters are needed, the energy of formation of the compound and the energy needed to create a defect on either sublattice. On comparing this with a reciprocal CEF model like Eq. (5.104), the energy of formation is equal to the $^{\circ}G_{\mathrm{A:B}}$ parameter, the energy needed to create defects on the first sublattice is related to the $°G_{X:B}$ parameter, and the energy needed to create defects on the second sublattice is related to $°G_{A:Y}$.

The mathematical identification of the Wagner–Schottky model with the CEF leads to

$$°G_{A:B} = °G_{A_aB_b} \tag{5.112}$$

$$°G_{A:Y} = °G_{A_aB_b} + \frac{b}{a+b}\,\Delta°G_{A:Y} \tag{5.113}$$

$$°G_{X:B} = °G_{A_aB_b} + \frac{a}{a+b}\,\Delta°G_{X:B} \tag{5.114}$$

$$\begin{aligned} °G_{X:Y} &= °G_{A_aB_b} + \frac{b}{a+b}\,\Delta°G_{A:Y} + \frac{b}{a+b}\,\Delta°G_{X:B} \\ &= °G_{A:Y} + °G_{X:B} - °G_{A:B} \end{aligned} \tag{5.115}$$

where $°G_{A_aB_b}$, $\Delta°G_{A:Y}$, and $\Delta°G_{X:B}$ are the parameters of the WS model

$$G_{ms} = °G_{A_aB_b} + \frac{n_X}{n}\,\Delta°G_{X:B} + \frac{n_Y}{n}\,\Delta°G_{A:Y} \tag{5.116}$$

where n_X is the number of X on A sites, n_Y is the number of Y on B sites, and n is the total number of sites considered.

The parameter $°G_{X:Y}$ from the CEF represents a phase consisting only of defects, which should be very unstable. Equation (5.115) relating this parameter and the other end members is equivalent to setting the reciprocal energy, Eq. (5.92), to zero.

On combining assessments of the same phase from different systems in a database, one may find that the parameter $°G_{X:B}$ occurs in several systems, not just in the A–B system. It is thus necessary to agree on a recommendation for the value of $°G_{X:B}$, similarly to what was done for the metastable states of the pure elements; see section 5.2.3.

When the values of $°G_{X:B}$ are fixed, one has to use the excess parameters to describe the properties of the individual binaries. It is illustrative to show how to determine which parameters in the CEF are most important. This is done below for the case in which the defects are anti-site atoms on both sublattices, i.e.,

$$(\mathrm{A},\mathrm{B})_a(\mathrm{B},\mathrm{A})_b \tag{5.117}$$

The differential of the Gibbs energy with respect to any variation of the site fractions can be calculated from the partial Gibbs energies, Eq. (5.38):

$$\mathrm{d}G = \sum_i \frac{\partial G_m}{\partial y_i}\,\mathrm{d}y_i \tag{5.118}$$

This differential should be zero at equilibrium and, for a fixed composition, the relations between the $\mathrm{d}y_i$ are

$$\begin{aligned} \mathrm{d}y'_A &= -\mathrm{d}y'_B \\ \mathrm{d}y''_A &= -\mathrm{d}y''_B \\ a\,\mathrm{d}y'_A &= -b\,\mathrm{d}y''_A \end{aligned} \tag{5.119}$$

This gives the following equation for the equilibrium fraction of defects in the single-phase region:

$$\frac{\partial G_{\rm m}}{\partial y'_{\rm A}} - \frac{\partial G_{\rm m}}{\partial y'_{\rm B}} - \frac{a}{b}\frac{\partial G_{\rm m}}{\partial y''_{\rm A}} + \frac{a}{b}\frac{\partial G_{\rm m}}{\partial y''_{\rm B}} = 0 \tag{5.120}$$

The Gibbs-energy expression from Eq. (5.104) with five parameters in the excess Gibbs energy is

$$^{\rm srf}G_{\rm m} = y'_{\rm A}y''_{\rm A}{}^\circ G_{\rm A:A} + y'_{\rm A}y''_{\rm B}{}^\circ G_{\rm A:B} + y'_{\rm B}y''_{\rm A}{}^\circ G_{\rm B:A} + y'_{\rm B}y''_{\rm B}{}^\circ G_{\rm B:B} \tag{5.121}$$

$$^{\rm cnf}S_{\rm m} = RT\{a[y'_{\rm A}\ln(y'_{\rm A}) + y'_{\rm B}\ln(y'_{\rm B})] + b[y''_{\rm A}\ln(y''_{\rm A}) + y''_{\rm B}\ln(y''_{\rm B})]\} \tag{5.122}$$

$$^{\rm E}G_{\rm m} = y'_{\rm A}y'_{\rm B}(y''_{\rm A}L_{\rm A,B:A} + y''_{\rm B}L_{\rm A,B:B}) + y''_{\rm A}y''_{\rm B}(y'_{\rm A}L_{\rm A:A,B} + y'_{\rm B}L_{\rm B:A,B}) + y'_{\rm A}y'_{\rm B}y''_{\rm A}y''_{\rm B}L_{\rm A,B:A,B} \tag{5.123}$$

As usual a colon ":" is used to separate constituents in different sublattices and for clarity a comma "," has been used between interacting constituents in the same sublattices, but that will normally be suppressed.

The parameter ${}^\circ G_{\rm A:B}$ is the Gibbs energy of formation of the ideal compound A_aB_b; ${}^\circ G_{\rm A:A}$ is the Gibbs energy of formation of pure A with the same structure as A_aB_b, i.e., all B sites filled with anti-site atoms. The parameter ${}^\circ G_{\rm B:B}$ is the equivalent for pure B and the parameter ${}^\circ G_{\rm B:A}$ is the Gibbs energy of formation of the compound consisting entirely of defects; its value will be discussed further below. As mentioned above, ${}^\circ G_{\rm A:A}$ and ${}^\circ G_{\rm B:B}$ should not be used to fit this phase in the binary A–B system because they may occur for the same phase in other binary or ternary systems.

Inserting the Gibbs-energy expressions into Eq. (5.120) gives, after simplification,

$$\begin{aligned}
RT\ln\left(\frac{y''_{\rm A}y'_{\rm B}}{y'_{\rm A}y''_{\rm B}}\right) &= \left(\frac{y''_{\rm A}}{a} - \frac{y'_{\rm A}}{b}\right){}^\circ G_{\rm A:A} - \left(\frac{y''_{\rm B}}{a} - \frac{y'_{\rm B}}{b}\right){}^\circ G_{\rm B:B} + \left(\frac{y''_{\rm B}}{a} + \frac{y'_{\rm A}}{b}\right){}^\circ G_{\rm A:B} \\
&\quad - \left(\frac{y''_{\rm A}}{a} + \frac{y'_{\rm B}}{b}\right){}^\circ G_{\rm B:A} - \left(\frac{y'_{\rm A}y'_{\rm B}}{b^2} - \frac{y'_{\rm B}y''_{\rm A}}{ab} + \frac{y'_{\rm A}y''_{\rm A}}{ab}\right)L_{\rm A,B:A} \\
&\quad + \left(\frac{y'_{\rm A}y'_{\rm B}}{b^2} + \frac{y'_{\rm B}y''_{\rm B}}{ab} - \frac{y'_{\rm A}y''_{\rm B}}{ab}\right)L_{\rm A,B:B} \\
&\quad + \left(\frac{y''_{\rm A}y''_{\rm B}}{a^2} - \frac{y'_{\rm A}y''_{\rm B}}{ab} + \frac{y'_{\rm A}y''_{\rm A}}{ab}\right)L_{\rm A:A,B} - \left(\frac{y''_{\rm A}y''_{\rm B}}{a^2} - \frac{y'_{\rm B}y''_{\rm B}}{ab} + \frac{y'_{\rm B}y''_{\rm A}}{ab}\right)L_{\rm B:A,B} \\
&\quad + y'_{\rm A}y'_{\rm B}y''_{\rm A}y''_{\rm B}\left(\frac{1}{ay'_{\rm A}} - \frac{1}{ay'_{\rm B}} - \frac{1}{by''_{\rm A}} + \frac{1}{by''_{\rm B}}\right)L_{\rm A,B:A,B}
\end{aligned} \tag{5.124}$$

The Wagner–Schottky model is applicable when the fraction of defects is very small and the defects are non-interacting. At the ideal composition with no defects, one has

$$\begin{aligned}
y'_{\rm A} = y''_{\rm B} &\Rightarrow 1 \\
y'_{\rm B} = y''_{\rm A} &\Rightarrow 0
\end{aligned} \tag{5.125}$$

and inserting this into the right-hand side of Eq. (5.124) (and the first equation also in the logarithm) gives

$$RT\ \ln(y''_{\rm A}y'_{\rm B}) = \left(\frac{1}{a}+\frac{1}{b}\right){}^{\circ}G_{\rm A:B} - \frac{1}{b}{}^{\circ}G_{\rm A:A} - \frac{1}{a}{}^{\circ}G_{\rm B:B} - \frac{1}{ab}L_{\rm A,B:B} - \frac{1}{ab}L_{\rm A:A,B} \tag{5.126}$$

For very small defect fractions, their product can thus be expressed by the following equation:

$$y''_{\rm A}y'_{\rm B} = \exp\left(\frac{(a+b){}^{\circ}G_{\rm A:B} - a{}^{\circ}G_{\rm A:A} - b{}^{\circ}G_{\rm B:B} - L_{\rm A,B:B} - L_{\rm A:A,B}}{abRT}\right) \tag{5.127}$$

In the case in which ${}^{\circ}G_{\rm A:A}$ and ${}^{\circ}G_{\rm B:B}$ formally are unary parameters, which are already defined in the database, the parameters $L_{\rm A,B:B}$ and $L_{\rm A:A,B}$ can be used for adjustment to the defect fractions. These two parameters are equally important at the ideal composition, but $L_{\rm A,B:B}$ will be most important for deviations toward B and $L_{\rm A:A,B}$ will be most important for deviations toward A. Note that the parameter ${}^{\circ}G_{\rm B:A}$ does not appear in Eq. (5.127) because the fractions by which it is multiplied were set to zero. Even when interaction parameters are used, it is commonly restricted by Eq. (5.115).

The other two interaction parameters, $L_{\rm A,B:A}$ and $L_{\rm B:B,A}$, are for one sublattice filled with defects and are not significant close to the ideal stoichiometry. It is recommended that they are set equal to the former two:

$$\begin{aligned} L_{\rm A,B:A} &= L_{\rm A,B:B} = L_{\rm A,B:*} \\ L_{\rm B:B,A} &= L_{\rm A:A,B} = L_{\rm *:B,A} \end{aligned} \tag{5.128}$$

which is equivalent to assuming that the interaction on each sublattice is independent of the occupation of the other sublattice. Equation (5.128) is just an estimate to prevent selection of arbitrary numbers and to keep insignificant parameters within suitable ranges.

If the defect fractions are larger, one may use other parameters in the Gibbs-energy expression, as a first step by expanding the interaction parameters using RK coefficients, Eq. (5.98). In fact, the CEF is not restricted to the dilute range; as has already been described, it can be extended to a sublattice full of defects.

5.8.2.4 A model for B2 ordering of bcc

The B2 ordering of the A2 structure type means that the atom in the center of the unit cell is different from those at the corners, see Fig. 5.20. A typical phase diagram with B2 ordering is that for Fe–Si or Cu–Zn from Seiersten and Tibballs (1993) and Kowalski and Spencer (1993), respectively. In these systems there is a *second-order* transition between B2 and A2 and this enforces the treatment of the ordered and disordered phases by use of the same Gibbs-energy expression.

In many other systems a phase with B2 structure appears as an intermetallic phase with more or less restricted solubility, for example Fe–Ti. In such cases it is frequently treated as a separate phase from the disordered A2 phase, using, for example, a Wagner–Schottky model; see section 5.8.2.3. However, this is not recommended because, on

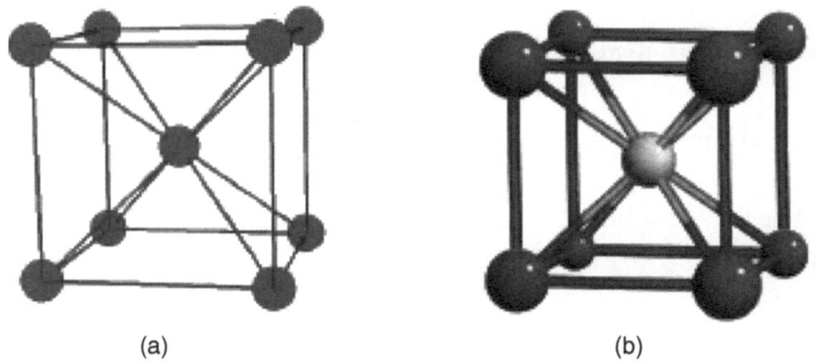

Figure 5.20 The disordered A2 (a) and ordered B2 (b) variants of the bcc lattice. From the Crystal Lattice Structures web page (http://cst-www.nrl.navy.mil/lattice).

extending into a ternary, the B2 phase may form a continuous solution to another binary system for which there is a second-order A2/B2 transition, for example in Al–Fe–Ni, see section 3.3.3.

The B2 ordering requires a two-sublattice model in which the sublattices have equal numbers of sites and must be treated as identical for crystallographic reasons. It is thus irrelevant for the Gibbs energies of formation whether element A is in the first sublattice and element B in the second or vice versa; one has

$$ {}^{\circ}G_{i:j} = {}^{\circ}G_{j:i} \tag{5.129} $$

All interaction parameters on the two sublattices must also be symmetrical, thus

$$ L_{i,j:k} = L_{k:i,j} \tag{5.130} $$

$$ L_{i,j,k:l} = L_{l:i,j,k} \tag{5.131} $$

Note that these parameters will give a contribution also in the disordered state, i.e., when $y_i' = y_i'' = x_i$. In section 5.8.4.1 this feature is discussed in detail.

As discussed in section 5.8.2.3, there are both anti-site atoms and vacancy defects in B2. Thus, since the two sublattices are crystallographically equivalent, a binary model would be

$$ (\mathrm{A,B,Va})_1(\mathrm{A,B,Va})_1 \tag{5.132} $$

This means that one has nine end members for a B2 phase with two different defects on each sublattice. However, many of these can be eliminated using the requirement that the model must be symmetrical with respect to exchanging the constituents on the sublattices, thus

$$ {}^{\circ}G_{\mathrm{A:B}} = {}^{\circ}G_{\mathrm{B:A}} \tag{5.133} $$

$$ {}^{\circ}G_{\mathrm{A:Va}} = {}^{\circ}G_{\mathrm{Va:A}} \tag{5.134} $$

$$ {}^{\circ}G_{\mathrm{B:Va}} = {}^{\circ}G_{\mathrm{Va:B}} \tag{5.135} $$

Using the fact that B2 is an ordered bcc phase, one also has that

$$°G_{\mathrm{A:A}} = °G_{\mathrm{A}}^{\mathrm{bcc}} \tag{5.136}$$

$$°G_{\mathrm{B:B}} = °G_{\mathrm{B}}^{\mathrm{bcc}} \tag{5.137}$$

Finally there is a parameter $°G_{\mathrm{Va:Va}}$ and there is no well-defined value to assign for this. The parameter $°G_{\mathrm{Va}}^{\mathrm{bcc}}$ is related to the energy needed to create thermal vacancies, but, since this is dependent on the element, it is better to use the parameter $L_{\mathrm{A,Va}}^{\mathrm{bcc}}$ to determine this. There has been some argument that this parameter represents empty space and its value should thus be zero, but this reasoning is doubtful and it creates numerical problems since the parameter is for zero amount of real atoms. When the number of real atoms approaches zero the Gibbs energy per mole of atoms becomes zero divided by zero, which can give very large negative values and thus make a bcc phase consisting mainly of vacancies very stable. Other arguments are that one should consider it as a phase full of defects and use a very large positive number to suppress the thermal vacancies. The present recommendation is to use a reasonably large positive value like $30T$ in order to avoid numerical problems and fit an interaction parameter to describe the real vacancy fraction.

It is important to model the B2 phase in a compatible way in each binary since one often has solubilities of ternary elements across the whole composition range, for example in Al–Fe–Ni; see section 3.3.3. In this ternary system the model for the B2 phase must include the disordered bcc phase as a special case. More about modeling ordered phases can be found in section 5.8.4.

5.8.2.5 A model for $\mathrm{L1_2}$/A1 ordering

The A1 structure type is shown in Fig. 5.21(a) together with some ordered structures. The $\mathrm{L1_2}$ ordering means that the eight corner atoms are of a different kind from the atoms on the sides. This gives an ideal stoichiometry $\mathrm{A_3B}$ and initially this was modeled with two sublattices, $(\mathrm{A,B})_3(\mathrm{A,B})_1$. In this model the constituents in the sublattice with three sites have eight nearest neighbors in the same sublattice. This means that there is a relation between the Gibbs energy of formation $°G_{\mathrm{A:B}}$ and the interaction parameters $L_{\mathrm{AB:*}}$ since both are related to the bonding energy of the nearest neighbors. The recommended way to model ordering on the fcc lattice is to use four sublattices, as described in section 5.8.4.4. A problem with using the two-sublattice model in an assessment of an $\mathrm{L1_2}$ ordering is that it may predict an ordered $\mathrm{L1_0}$ ordering if the parameters are converted to a four-sublattice model.

The $\mathrm{L1_2}$ phase is the most important ordered fcc phase and it may be practical to use the two-sublattice model for multicomponent systems. There are rather complicated restrictions on the parameters that are needed in the two-sublattice model sublattice. The method employed to convert from a four-sublattice model is described in section 5.8.4.6.

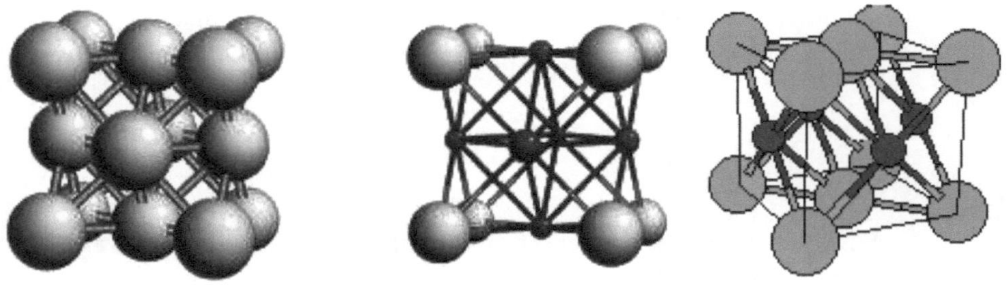

(a) (b) (c)

Figure 5.21 The disordered A1 (a) and ordered $L1_2$ (b) and $L1_0$ (c) variants of the fcc lattice. From the Crystal Lattice Structures web page (http://cst-www.nrl.navy.mil/lattice).

5.8.3 *Models with three or more sublattices*

Many crystalline phases have more than two sublattices, but it may be necessary to simplify the models to fewer sublattices since there might not be enough experimental data. Nonetheless, it is fairly common to use three or four sublattices to describe real phases. The model for a phase with three sublattices is

$$(\mathrm{A}, \mathrm{B}, \ldots)_{a_1}(\mathrm{K}, \mathrm{L}, \ldots)_{a_2}(\mathrm{U}, \mathrm{V}, \ldots)_{a_3} \tag{5.138}$$

When the same constituent enters more than one sublattice, it may be useful to remind the reader about Eq. (5.29) concerning how to obtain the mole fractions from the constituent fractions.

The Gibbs-energy expression for a phase with three sublattices is

$$\begin{aligned} G_{\mathrm{m}} &= \sum_i \sum_j \sum_k y'_i y''_j y'''_k \, {}^{\circ}G_{i:j:k} + {}^{phys}G_{\mathrm{m}} \\ &\quad + RT\left(a_1 \sum_i y'_i \ln(y'_i) + a_2 \sum_j y''_j \ln(y''_j) + a_3 \sum_j y'''_k \ln(y'''_k)\right) + {}^{\mathrm{E}}G_{\mathrm{m}} \end{aligned} \tag{5.139}$$

where

$$\begin{aligned} {}^{\mathrm{E}}G_{\mathrm{m}} &= \sum_i \sum_j \sum_k \sum_{l>i} y'_i y''_j y'''_k y'_l L_{i,l:j:k} \\ &\quad + \sum_i \sum_j \sum_k \sum_{l>j} y'_i y''_j y'''_k y''_l L_{i:j,l:k} \\ &\quad + \sum_i \sum_j \sum_k \sum_{l>k} y'_i y''_j y'''_k y'''_l L_{i:j:k,l} \\ &\quad + \sum_i \sum_j \sum_k \sum_{l>k} \sum_{m>i} y'_i y''_j y'''_k y'_l y'_m L_{i,l,m:j:k} + \cdots \\ &\quad + \sum_i \sum_j \sum_k \sum_{l>k} \sum_{m>j} y'_i y''_j y'''_k y'_l y''_m L_{i,l:j,m:k} + \cdots \end{aligned} \tag{5.140}$$

The number of possible excess parameters is very large, but normally very few are used. The most important parameters to optimize are those in the surface of reference, ${}^{\circ}G_{i:j:k}$.

Phases are not very commonly modeled with four or more sublattices and several constituents in each because it is difficult to acquire enough experimental data to evaluate the necessary number of parameters. An exception is when there are crystallographic relations between the parameters, as in the case of chemical ordering. In the next section a four-sublattice model for ordering on the fcc lattice will be described.

5.8.3.1 Models for intermetallic phases

Some intermetallic phases have been modeled with three sublattices, the most well known perhaps being the σ and the μ phases (Fig. 5.22). In both cases the crystallographic information has been simplified in the thermodynamic modeling. The σ phase has five different crystallographic sublattices, but only three are used in the modeling. The main reason for this is that there is not enough experimental information to evaluate all necessary parameters for a five-sublattice model. Such information would be the distributions of the various components on the sublattices, which are rarely known. In the σ phase the first sublattice is mainly occupied by fcc-type elements, the second by mainly bcc-type elements, and the third is a mixture of all.

The composition range of the intermetallic phase is important to consider when deciding on the sublattices. If the intermetallic phase appears in several systems, all of these should be considered before selecting a model. The constituents of a sublattice would normally be those which can be the only constituent on that sublattice, but, in order to describe the composition range, it may be necessary to include some components as "defects" in sublattices that they normally would not enter. However, this may increase the number of parameters significantly in multicomponent systems and should be done with care. Sometimes one may accept an error of a few percent in the composition range in a lower-order system in order to simplify the multicomponent description.

It is important for the development of a general thermodynamic database that all agree to use the same set of sublattices when modeling a phase. In the second Ringberg report Ansara *et al.* (1997a) presented a survey and recommendations regarding the modeling of many intermetallic compounds. For example, they recommended that the σ phase be modeled with three sublattices and the site ratios 10 : 4 : 16, not 8 : 4 : 18 as had been

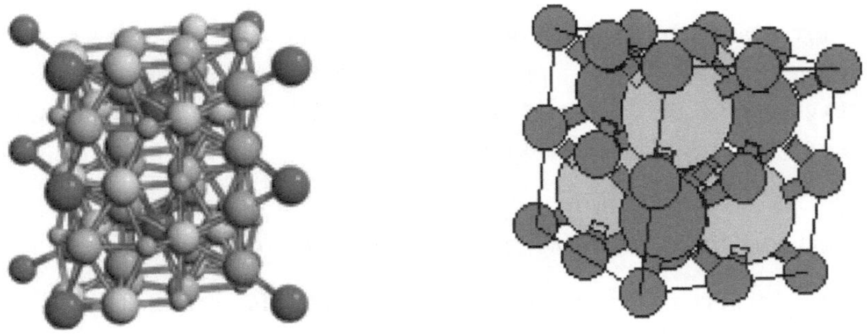

Figure 5.22 The crystal structures of σ (left) and $L2_1$ (right) phases. From the Crystal Lattice Structures web page (http://cst-www.nrl.navy.mil/lattice).

used previously. The reason for the earlier model was to simplify the description of the composition range of the σ phase in the Ni–V system and to describe the composition range of the σ phase in the Cr–Fe–Mo system.

The possibility of using *ab initio* calculations to obtain values for metastable end members in intermetallic phases is rapidly becoming an important source of information, for example on the Re–W (Fries and Sundman 2002) and Cr–Fe (Korzhavyi *et al.* 2005) system. Since the CEF can use the energies calculated for the ordered end members directly without any cluster-expansion approximation and because SRO is insignificant in these phases, the combination of the CEF and *ab initio* methods amounts to an important step forward for the development of new materials.

The availability of improved experimental techniques such as Rietveld refinement (Joubert 2002) also means that additional information on site occupancies and defect fractions has become available to use in the modeling of phases. The experimental determination of the site occupancy of the μ phase in the Nb–Ni system showed that none of the previous models, which had been selected arbitrarily or on the basis of coordination-number arguments, coincides with the one revealed by the experimental results (Joubert and Feutelais 2002).

5.8.3.2 Models for metal–non-metal phases

The $M_{23}C_6$ phase is an example of a phase that has been modeled with two sublattices for the metallic elements and one for the carbon, in order to treat the limited solubility of the heavy elements W and Mo in a proper way. The crystallographic information would give $(\mathrm{Cr, Fe, \ldots})_{21}(\mathrm{Cr, Fe, W, Mo, \ldots})_2\mathrm{C}_6$, but it was not followed entirely since that would not have described the full composition range. Instead a model of the form $(\mathrm{Cr, Fe, \ldots})_{20}(\mathrm{Cr, Fe, W, Mo, \ldots})_3\mathrm{C}_6$ was adopted.

Many oxide phases have different types of metallic sublattices, for example the spinel phase. In this phase the oxygen ions form an fcc lattice and the metallic ions enter some of the tetrahedral and octahedral interstitial sites. A four-sublattice model has been adopted for this phase, which can also describe the wide stability range of the spinel phase in the $MgO–Al_2O_3$ system.

An interesting application of the three-sublattice model is to the Fe–Al–C system. In the ferrite phase, carbon dissolves interstitially and Fe–Al can form a B2 ordered structure. This means a combination of two sublattices for ordering of the metallic elements with one extra sublattice for interstitials. Some care must be taken with the model parameters to ensure that the disordered phase remains stable, but, using the partitioning as described in section 5.8.4.1, that is no problem.

5.8.4 *Models for phases with order–disorder transitions*

The sublattice model can describe phases that exhibit ordering of the metallic constituents on different sublattices that may disappear above a certain temperature or beyond a particular composition, for example the B2/A2 transition already mentioned in section 5.8.2.4, the $L1_2/A1$ transition in section 5.8.2.5, and the four-sublattice models in sections 5.8.4.4

and 5.8.4.7. If one restricts the model parameters to just the ideal model without excess parameters, the sublattice model is identical to the Bragg–Williams–Gorsky treatment of ordering, but with the excess parameters in the CEF one may reproduce the real properties of the ordering transformation better.

In the disordered state the constituents are distributed equally but randomly on the sublattices. In the ordered state the constituents have different fractions. This degeneration of the two or more ordered sublattices into a single one enforces several restrictions on the possible parameters in the CEF, otherwise the disordered state would never be stable. These restrictions are based on the requirement that the Gibbs energy of the phase must have an extremum against variations in the ordering variables at the disordered composition. In the disordered state such a requirement is self-evident and the extremum must be a minimum, otherwise the phase would undergo ordering. Such requirements were first derived by Ansara *et al.* (1988).

In some cases the ordered phase is described with a different Gibbs-energy expression from that used for the disordered phase. The ordered phase will then normally never be completely disordered. Such a treatment of ordered phases is not recommended in general, although some ordered structure types, such as $D0_{22}$ and $L2_1$, see Fig. 5.22 and section 5.8.4.7, may require too many sublattices for it to be practically reasonable to treat them with a single mathematical expression.

The configurational entropy for a phase with an order–disorder transition may deviate considerably from the ideal entropy used in the sublattice model. However, from a practical point of view these differences can be modeled in terms of the excess Gibbs energy without losing any agreement with experimental data, although one may suspect that the extrapolations based on these simplifications may be less accurate than a proper treatment of the configurational entropy. For future work it is recommended that models making use of the CVM, see section 5.7.2.2, or similar techniques should be considered for phases with order–disorder transitions.

5.8.4.1 The disordered state of an ordered phase

For phases modeled with an order–disorder transition like the B2/A2 case in section 5.8.2.4 both the ordered and the disordered state can be described with a single Gibbs-energy function.

Applying $y_i' = y_i'' = x_i$ to the G_m expression (5.104) for an ordered binary A–B phase and a regular-solution interaction parameter gives

$$\begin{aligned} G_m &= y_A' y_A'' \,{}^\circ G_{A:A} + y_A' y_B'' \,{}^\circ G_{A:B} + y_B' y_A'' \,{}^\circ G_{B:A} + y_B' y_B'' \,{}^\circ G_{B:B} \\ &\quad + 0.5RT\,[y_A' \ln(y_A') + y_B' \ln(y_B') + y_A'' \ln(y_A'') + y_B'' \ln(y_B'')] + y_A' y_B' L_{AB:*} + y_A'' y_B'' L_{*:AB} \\ &= x_A^2 \,{}^\circ G_{A:A} + 2x_A x_B \,{}^\circ G_{A:B} + x_B^2 \,{}^\circ G_{B:B} + RT\,[x_A \ln(x_A) + x_B \ln(x_B)] + 2x_A x_B L_{AB:*} \\ &= x_A^2 \,{}^\circ G_{A:A} + x_A x_B \,{}^\circ G_{A:A} + x_B^2 \,{}^\circ G_{B:B} + x_A x_B \,{}^\circ G_{B:B} \\ &\quad + x_A x_B (2\,{}^\circ G_{A:B} - {}^\circ G_{A:A} - {}^\circ G_{B:B}) + RT\,[x_A \ln(x_A) + x_B \ln(x_B)] + 2x_A x_B L_{AB:*} \\ &= x_A \,{}^\circ G_A + x_B \,{}^\circ G_B + RT\,[x_A \ln(x_A) + x_B \ln(x_B)] + x_A x_B L_{AB} \end{aligned} \tag{5.141}$$

The star, *, used instead of a constituent index in the parameters above indicates that the parameter is independent of the constitution of that sublattice. In the derivations of the relations above, ${}^{\circ}G_{\mathrm{A:B}} = {}^{\circ}G_{\mathrm{B:A}}$ and $L_{\mathrm{AB:*}} = L_{\mathrm{*:AB}}$ have been used and in the last line also the substitutions

$$\begin{aligned} {}^{\circ}G_{\mathrm{A}} &= {}^{\circ}G_{\mathrm{A:A}} \\ {}^{\circ}G_{\mathrm{B}} &= {}^{\circ}G_{\mathrm{B:B}} \\ L_{\mathrm{AB}} &= 2{}^{\circ}G_{\mathrm{A:B}} - {}^{\circ}G_{\mathrm{A:A}} - {}^{\circ}G_{\mathrm{B:B}} + 2L_{\mathrm{AB:*}} \end{aligned} \tag{5.142}$$

The fact that the parameters used to describe the ordered state also influence the disordered state is a disadvantage when the ordering is added to an already-assessed disordered phase. Also when extrapolating binary systems with and without ordering to a ternary system this model requires that the parameters derived for the substitutional model for the disordered phase be converted to the two-sublattice model. This can always be done by replacing the mole fractions by the appropriate site fractions,

$$\begin{aligned} x_{\mathrm{A}} &= 0.5(y'_{\mathrm{A}} + y''_{\mathrm{A}}) \\ x_{\mathrm{B}} &= 0.5(y'_{\mathrm{B}} + y''_{\mathrm{B}}) \end{aligned} \tag{5.143}$$

but it is rather cumbersome for multicomponent systems.

These were the reasons for introducing a "partitioning" of the Gibbs energy for an ordered phase into one part that is independent of the ordering and one that describes the contribution to the Gibbs energy due to ordering.

5.8.4.2 Partitioning of the Gibbs energy for ordered phases

For phases with order–disorder transitions the partitioning of the Gibbs-energy expression is into two parts, a general part, which depends only on the composition of the phase, expressed as mole fractions x, and another part giving the contribution due to long-range ordering, depending on the site fractions y:

$$G_{\mathrm{m}} = G_{\mathrm{m}}^{\mathrm{dis}}(x) + \Delta G_{\mathrm{m}}^{\mathrm{ord}}(y) \tag{5.144}$$

The first term, $G_{\mathrm{m}}^{\mathrm{dis}}(x)$, is independent of the ordering state of the phase. The second term, $\Delta G_{\mathrm{m}}^{\mathrm{ord}}$, gives the contribution due to long-range ordering and must be zero when the phase is disordered. The simplest way to ensure that $\Delta G_{\mathrm{m}}^{\mathrm{ord}}$ is zero in the disordered state is to calculate this contribution as

$$\Delta G_{\mathrm{m}}^{\mathrm{ord}}(y) = G_{\mathrm{m}}^{\mathrm{ord}}(y) - G_{\mathrm{m}}^{\mathrm{ord}}(y \text{ replaced by } x) \tag{5.145}$$

In Eq. (5.145) the parameter describing the ordering, $G_{\mathrm{m}}^{\mathrm{ord}}$, is first calculated with the original site fractions, y, which describe the ordering. The site fractions are then set equal to the mole fractions, x, which means that each constituent has the same site fraction in all sublattices, and the value of the expression is calculated again. The difference is the contribution to the Gibbs energy due to ordering. If the phase was originally disordered, the two terms will be equal and the difference will be zero.

Ordered phases become less ordered with increasing temperature. Above a certain temperature LRO may disappear completely, but some SRO will remain even in the disordered state. If the ordered phase has no order–disorder transition within the experimental range, for example the ordered phase may melt before it disorders, there is a certain arbitrariness in the distribution of the Gibbs energy between $G_{\rm m}^{\rm dis}(x)$ and $G_{\rm m}^{\rm ord}(x)$. It may be possible to resolve this using *ab initio* calculations.

An important advantage of this partitioning is that it is much easier to combine assessments, including assessments of both ordered and disordered descriptions of a phase. If a system has a disordered contribution only, that is added to the $G_{\rm m}^{\rm dis}$ part and no parameters are needed for the $G_{\rm m}^{\rm ord}$ part.

It is important to understand that the partitioning of the Gibbs energy into an ordered and a disordered part is not a new model; the Gibbs-energy expression can always be written using the site fractions only. It is a simplification that is useful for calculations of ordered phases in multicomponent systems.

5.8.4.3 The quasi-chemical model and long-range order

An interesting connection with the quasi-chemical model for SRO is that one can calculate site fractions for a two-sublattice model from the "bond fractions" $y_{\rm AA}, y_{\rm AB}, y_{\rm BA}$, and $y_{\rm BB}$ as

$$
\begin{aligned}
y'_{\rm A} &= y_{\rm AA} + y_{\rm AB} \\
y'_{\rm B} &= y_{\rm BA} + y_{\rm BB} \\
y''_{\rm A} &= y_{\rm AA} + y_{\rm BA} \\
y''_{\rm B} &= y_{\rm AB} + y_{\rm BB}
\end{aligned}
\tag{5.146}
$$

Note that it is essential to treat $y_{\rm AB}$ and $y_{\rm BA}$ as different, otherwise one cannot describe the long-range ordering. This means that a quasi-chemical model can be formally treated as a two-sublattice model, but in addition it includes a contribution from SRO. It is not possible to calculate the bond fractions from the site fractions without introducing an additional variable ϵ:

$$
\begin{aligned}
y_{\rm AA} &= y'_{\rm A} y''_{\rm A} + \epsilon \\
y_{\rm AB} &= y'_{\rm A} y''_{\rm B} - \epsilon \\
y_{\rm BA} &= y'_{\rm B} y''_{\rm A} - \epsilon \\
y_{\rm BB} &= y'_{\rm B} y''_{\rm B} + \epsilon
\end{aligned}
\tag{5.147}
$$

It is this ϵ variable that describes the SRO. The quasi-chemical model is not used for assessment since it is limited to crystalline structures that can be described with a two-sublattice model with all nearest neighbors on the opposite sublattice, but it has been applied to liquids, see section 5.9.3. Note the problem with negative entropies for large SRO shown in Fig. 5.17(b) if one uses a quasi-chemical model without allowing for LRO. Models for crystalline phases with explicit SRO are based on the CVM, see section 5.7.2.2; models with implicit SRO are described in section 5.8.4.5.

5.8.4.4 Simultaneous $L1_2$ and $L1_0$ ordering

With four sublattices one may model both $L1_2$ and $L1_0$ ordering on the fcc lattice. Actually, there are many more ordered structures based on the fcc lattice, but most of them can be treated as separate phases with different Gibbs-energy expressions. The restriction that these phases will never undergo disordering is usually a minor problem.

In the four-sublattice model the restrictions on the parameters can be derived fairly easily from the symmetry of the lattice. For example,

$$\begin{aligned}
{}^{\circ}G_{A:A:A:B} &= {}^{\circ}G_{A:A:B:A} = {}^{\circ}G_{A:B:A:A} = {}^{\circ}G_{B:A:A:A} \\
{}^{\circ}G_{A:A:B:B} &= {}^{\circ}G_{A:B:A:B} = {}^{\circ}G_{B:A:A:B} = {}^{\circ}G_{A:B:B:A} = {}^{\circ}G_{B:A:B:A} = {}^{\circ}G_{B:B:A:A} \\
{}^{\circ}G_{B:B:B:A} &= {}^{\circ}G_{B:B:A:B} = {}^{\circ}G_{B:A:B:B} = {}^{\circ}G_{A:B:B:B} \\
{}^{\circ}L_{A,B:A:A:A} &= {}^{\circ}L_{A:A,B:A:A} = {}^{\circ}L_{A:A:A,B:A} = {}^{\circ}L_{A:A:A:A,B} \\
&\ldots
\end{aligned} \tag{5.148}$$

The $L1_2$ and $L1_0$ ordering and also the disordered A1 phase can be described with the same model. Normally the Gibbs energy is partitioned into a disordered and an ordered part as described in section 5.8.4.2. The L parameters above are then normally not used since they can be described in the disordered part. If one assumes that the bond energy between AB pairs, u_{AB}, depends only slightly on the composition, one can write the G parameters as follows:

$$\begin{aligned}
{}^{\circ}G_{A:A:A:B} = G_{A_3B} &= 3u_{AB} + \Delta u_1 \\
{}^{\circ}G_{A:A:B:B} = G_{A_2B_2} &= 4u_{AB} \\
{}^{\circ}G_{A:B:B:B} = G_{AB_3} &= 3u_{AB} + \Delta u_2
\end{aligned} \tag{5.149}$$

The factors of 3 and 4 above come from the number of AB bonds in each end member. The bond energy may be different when the overall composition is different and thus the terms Δu_1 and Δu_2 can be used as corrections to fit the experimental data.

It is a critical test that one has used a correct set of parameters that the disordered phase can really be disordered, i.e., that all site fractions on all sublattices are equal at some high temperature. If there is a deviation, even a small one, then probably one or more parameters are missing or have wrong values.

5.8.4.5 Approximation of the short-range-order contribution to the Gibbs energy

The SRO contribution to the Gibbs energy of the fcc phase can be approximated with a reciprocal parameter as shown in Eq. (5.102). For a four-sublattice phase one can have three different such parameters:

$$\begin{aligned} L_{\mathrm{A,B:A,B:A:A}} &= L_{\mathrm{A.B:A:A,B:A}} = \cdots = L_{\mathrm{AA}} \\ L_{\mathrm{A,B:A,B:A:B}} &= L_{\mathrm{A.B:A:A,B:B}} = \cdots = L_{\mathrm{AB}} \\ L_{\mathrm{A,B:A,B:B:B}} &= L_{\mathrm{A.B:B:A,B:B}} = \cdots = L_{\mathrm{BB}} \end{aligned} \tag{5.150}$$

Unless there is a lot of experimental information on the system, one may set all of these equal and just use

$$L_{\mathrm{A,B:A,B:*:*}} = L_{\mathrm{A,B:*:A,B:*}} = \cdots = L_{**} = u_{\mathrm{AB}} + \Delta u_3 \tag{5.151}$$

A reasonable initial value is obtained, according to Abe and Sundman (2003), by setting this parameter equal to the bond energy, u_{AB}, and, if necessary, using a small correction term, Δu_3. The SRO should vanish at high temperatures, which can be achieved by multiplying u_{AB} by a factor like $\exp\left[(T_{\mathrm{C}} - T)/(2T_{\mathrm{C}})\right]$, where T_{C} is the ordering temperature for the equiatomic composition.

The "prototype" fcc phase diagram shown in Fig. 5.23(a) was calculated with a single constant bond energy $u_{\mathrm{AB}} = -10\,000\,\mathrm{J\,mol^{-1}}$ used in Eqs. (5.149) and (5.151) with all $\Delta u_i = 0$. This diagram is almost identical to the prototype phase diagram for fcc ordering calculated using a CVM-based tetrahedron model, as shown by Sundman and Mohri (1990). The thermodynamic functions at 700 K for the system are plotted in Fig. 5.23(b). The entropy curve is like teeth on a saw and, for ordered compositions, the entropy is almost zero. The heat-capacity curve is even more irregular, as discussed by Kusoffsky and Sundman (1998).

To describe a real binary phase diagram with fcc ordering, one has to investigate the parameters of the Gibbs-energy function that can be adjusted. The Gibbs energy has been partitioned between the disordered fcc and the ordered phases according to Eq. (5.145):

$$G_{\mathrm{m}}^{\mathrm{tot}} = G_{\mathrm{m}}^{\mathrm{A1}}(x_i) + \Delta G_{\mathrm{m}}^{\mathrm{ord}}(y_i) \tag{5.152}$$

$$G_{\mathrm{m}}^{\mathrm{A1}} = \sum_{i=\mathrm{A,B}} x_i\,{}^{\circ}G_i + RT \sum_{i=\mathrm{A,B}} x_i \ln(x_i) + x_{\mathrm{A}} x_{\mathrm{B}} \sum_{\nu=0}^{4} (x_{\mathrm{A}} - x_{\mathrm{B}})^{\nu} \cdot {}^{\nu}L_{\mathrm{A,B}}^{\mathrm{A1}} \tag{5.153}$$

$$\Delta G_{\mathrm{m}}^{\mathrm{ord}} = G_{\mathrm{m}}^{\mathrm{ord}}(y_i) - G_{\mathrm{m}}^{\mathrm{ord}}(x_i) \tag{5.154}$$

$$G_{\mathrm{m}}^{\mathrm{ord}} = \sum_{i=\mathrm{A,B}} \sum_{j=\mathrm{A,B}} \sum_{k=\mathrm{A,B}} \sum_{l=\mathrm{A,B}} y_i^{(1)} y_j^{(2)} y_k^{(3)} y_l^{(4)}\,{}^{\circ}G_{ijkl} + RT \sum_{s=1}^{4} \sum_{i=\mathrm{A,B}} y_i^{(s)} \ln(y_i^{(s)}) + {}^{\mathrm{E}}G_{\mathrm{m}}^{\mathrm{ord}} \tag{5.155}$$

$${}^{\mathrm{E}}G_{\mathrm{m}}^{\mathrm{ord}} = \sum_{s=1}^{3} \sum_{t=s+1}^{4} y_{\mathrm{A}}^{(s)} y_{\mathrm{B}}^{(s)} y_{\mathrm{A}}^{(t)} y_{\mathrm{B}}^{(t)} L_{**} \tag{5.156}$$

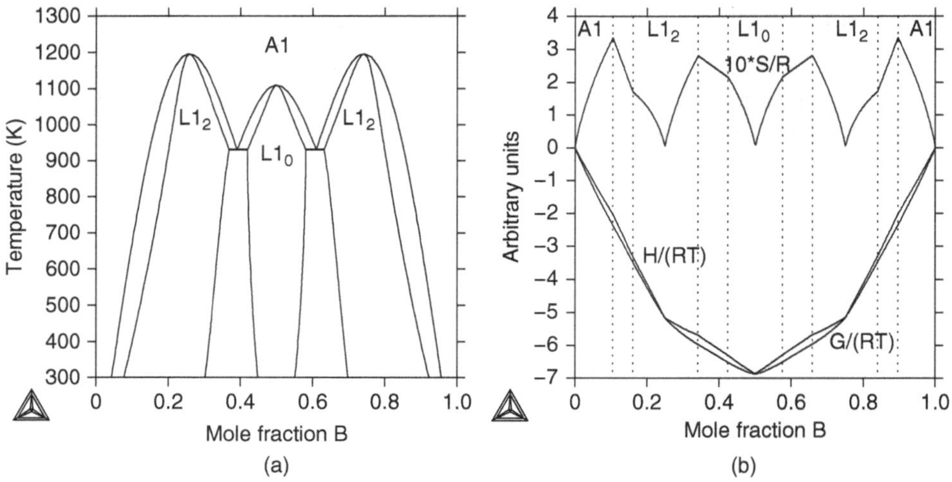

Figure 5.23 A prototype phase diagram for fcc ordering with two $L1_2$ ordered phases and one $L1_0$ ordered phase is shown in (a). The diagram is calculated using a single Gibbs-energy function, Eq. (5.152), with a single bond energy value $u_{AB} = -10\,000\,J\,mol^{-1}$. In (b) the dimensionless thermodynamic functions at 700 K are plotted (the entropy is scaled by a factor of 10, to make it more easily visible). The vertical dotted lines indicate the two-phase regions.

In the disordered part, Eq. (5.153), the contribution due to SRO must be included and the following parameters must be set using the notation from Eqs. (5.149) and (5.150):

$$\begin{aligned}
{}^{0}L^{A1}_{A,B} &= G_{A_3B} + 1.5G_{A_2B_2} + G_{AB_3} + 0.75L_{AA} + 0.75L_{BB} + l_0 \\
{}^{1}L^{A1}_{A,B} &= 2G_{A_3B} - 2G_{AB_3} + 0.75L_{AA} - 0.75L_{BB} + l_1 \\
{}^{2}L^{A1}_{A,B} &= G_{A_3B} - 1.5G_{A_2B_2} + G_{AB_3} - 1.5L_{AB} + l_2 \\
{}^{3}L^{A1}_{A,B} &= -0.75L_{AA} + 0.75L_{BB} \\
{}^{4}L^{A1}_{A,B} &= -0.75L_{AA} + 1.5L_{AB} - 0.75L_{BB}
\end{aligned} \tag{5.157}$$

These parameters can be derived from the four-sublattice model by setting all site fractions equal to the mole fractions and identifying the interaction parameters in the substitutional model. The coefficients l_0 to l_2 are zero in Fig. 5.23 but can be optimized to fit experimental data. It is necessary to include the ordered parameters in the disordered part because otherwise one would not have a separation of the three maxima for the order–disorder transitions.

One may think that including the SRO contribution in the disordered part, G^{dis}_m, would give twice the SRO contribution when the phase is ordered. However, according to Eq. (5.154), the SRO contribution from the ordered part will be subtracted using mole fractions and thus cancel out the contribution from the disordered part when the phase is ordered.

The phase diagrams in Fig. 5.24 show the influence of varying the Δu and l_i parameters. In both diagrams $u_{AB} = -10\,000\,J\,mol^{-1}$, $\Delta u_1 = -1000\,J\,mol^{-1}$, and

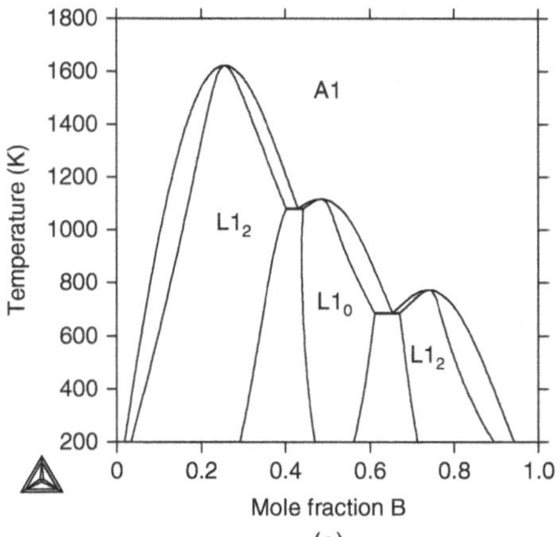

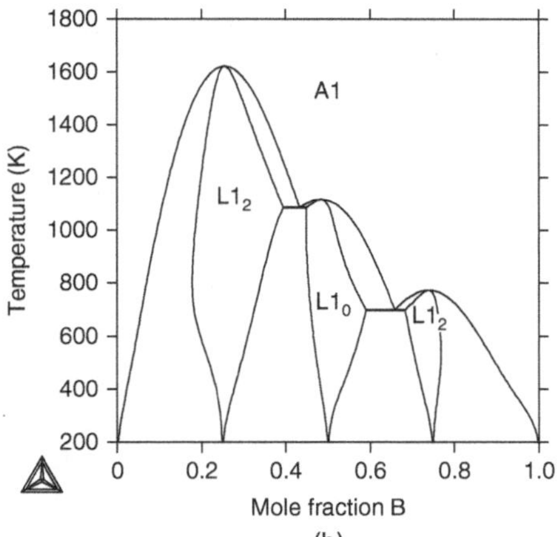

Figure 5.24 Ordered phase diagrams, with (a) $l_0 = 0$ and (b) $l_0 = 40\,000\,\mathrm{J\,mol^{-1}}$, for fcc with two $\mathrm{L1_2}$ ordered phases and one $\mathrm{L1_0}$ ordered phase calculated using a single Gibbs-energy function, Eq. (5.152), with different parameter values, according to Eqs. (5.149) and (5.153); the values are given in the text.

$\Delta u_2 = +1000\,\mathrm{J\,mol^{-1}}$, and in (a) the disordered l_0 parameter is zero whereas in (b) it is $40\,000\,\mathrm{J\,mol^{-1}}$.

The same four-sublattice model can also be applied to ordering in hcp phases. It is also possible to add a fifth interstitial sublattice to the ordered model, but all parameters for the interstitial component should be in the disordered part.

It is of historical interest to mention that the first order–disorder phase diagram for fcc was calculated by Shockley (1938) using a Bragg–Williams–Gorsky model without the reciprocal parameters, Eq. (5.151), and he could then not separate the three ordering maxima. It was a success of the CVM when Sanchez *et al.* (1982) showed that a CVM tetrahedron model could describe the topology of the ordered fcc phase diagram with three separate maxima.

5.8.4.6 Transforming a four-sublattice ordered fcc model to the two-sublattice model

The relation between the parameters for the two-sublattice $\mathrm{L1_2}$ model can be derived from a four-sublattice model in which the site fractions on three sublattices are set equal and related to the normal parameters in the two-sublattice model, or by using the criterion that $\partial G/\partial y_i = 0$ for the disordered state for all independent y_i. The reason for using a two-sublattice model is that calculations with it are significantly faster, but it means that one cannot calculate a possible $\mathrm{L1_0}$ transformation.

Higher-order systems with an $\mathrm{L1_2}$ phase modeled with two sublattices require a lot of ternary and some quaternary interaction parameters dependent on the binary ones in order to make the disordered state stable. For a phase $(\mathrm{A,B,C,D})_3(\mathrm{A,B,C,D})$ the list below shows some of these parameters as derived by Dupin (1995), which can also be found in Kusoffsky *et al.* (2001):

$$\begin{aligned}
{}^0L^{\mathrm{ord}}_{\mathrm{A,B:A}} &= -1.5G_{\mathrm{AB_3}} + 1.5G_{\mathrm{A_2B_2}} + 1.5G_{\mathrm{A_3B}} \\
{}^1L^{\mathrm{ord}}_{\mathrm{A,B:A}} &= 0.5G_{\mathrm{AB_3}} - 1.5G_{\mathrm{A_2B_2}} + 1.5G_{\mathrm{A_3B}} \\
L^{\mathrm{ord}}_{\mathrm{A,B,C:A}} &= 6G_{\mathrm{A_2BC}} - 1.5G_{\mathrm{AB_2C}} - 1.5G_{\mathrm{ABC_2}} - 1.5G_{\mathrm{A_3B}} - 1.5G_{\mathrm{A_2B_2}} \\
&\quad + G_{\mathrm{AB_3}} - 1.5G_{\mathrm{A_3C}} - 1.5G_{\mathrm{A_2C_2}} + G_{\mathrm{AC_3}} \\
L^{\mathrm{ord}}_{\mathrm{A,B,D:A}} &= 6G_{\mathrm{A_2BD}} - 1.5G_{\mathrm{AB_2D}} - 1.5G_{\mathrm{ABD_2}} - 1.5G_{\mathrm{A_3B}} - 1.5G_{\mathrm{A_2B_2}} \\
&\quad + G_{\mathrm{AB_3}} - 1.5G_{\mathrm{A_3D}} - 1.5G_{\mathrm{A_2D_2}} + G_{\mathrm{AD_3}} \qquad (5.158) \\
{}^0L^{\mathrm{ord}}_{\mathrm{A,C:A}} &= -1.5G_{\mathrm{AC_3}} + 1.5G_{\mathrm{A_2C_2}} + 1.5G_{\mathrm{A_3C}} \\
{}^1L^{\mathrm{ord}}_{\mathrm{A,C:A}} &= 0.5G_{\mathrm{AC_3}} - 1.5G_{\mathrm{A_2C_2}} + 1.5G_{\mathrm{A_3C}} \\
L^{\mathrm{ord}}_{\mathrm{A,C,D:A}} &= 6G_{\mathrm{A_2CD}} - 1.5G_{\mathrm{AC_2D}} - 1.5G_{\mathrm{ACD_2}} - 1.5G_{\mathrm{A_3C}} - 1.5G_{\mathrm{A_2C_2}} \\
&\quad + G_{\mathrm{AC_3}} - 1.5G_{\mathrm{A_3D}} - 1.5G_{\mathrm{A_2D_2}} + G_{\mathrm{AD_3}} \\
{}^0L^{\mathrm{ord}}_{\mathrm{A,D:A}} &= -1.5G_{\mathrm{AD_3}} + 1.5G_{\mathrm{A_2D_2}} + 1.5G_{\mathrm{A_3D}} \\
{}^1L^{\mathrm{ord}}_{\mathrm{A,D:A}} &= 0.5G_{\mathrm{AD_3}} - 1.5G_{\mathrm{A_2D_2}} + 1.5G_{\mathrm{A_3D}} \\
{}^\circ G^{\mathrm{ord}}_{\mathrm{B:A}} &= G_{\mathrm{AB_3}} \\
&\ldots
\end{aligned}$$

It is not necessary to give a complete list since software can generate all these parameters. The parameters from the four-sublattice model are denoted as in Eqs. (5.149) and (5.159). The symbols $G_{\mathrm{AB_3}}$, etc. have the same meanings as for the respective binary systems according to Eq. (5.149). The new symbols introduced are defined as

$$\begin{aligned}
G_{\mathrm{ABC_2}} &= u_{\mathrm{AB}} + 2u_{\mathrm{AC}} + 2u_{\mathrm{BC}} + \Delta u_4 \\
G_{\mathrm{AB_2C}} &= 2u_{\mathrm{AB}} + u_{\mathrm{AC}} + 2u_{\mathrm{BC}} + \Delta u_5 \qquad (5.159) \\
G_{\mathrm{A_2BC}} &= 2u_{\mathrm{AB}} + 2u_{\mathrm{AC}} + u_{\mathrm{BC}} + \Delta u_6 \\
G_{\mathrm{ABCD}} &= u_{\mathrm{AB}} + u_{\mathrm{AC}} + u_{\mathrm{AD}} + u_{\mathrm{BC}} + u_{\mathrm{BD}} + u_{\mathrm{CD}} + \Delta u_7
\end{aligned}$$

where u_{AB} etc. are from the binary systems but the terms Δu_4 to Δu_7 can be optimized to fit data in the ternary system.

5.8.4.7 B32, $D0_3$, and $L2_1$ ordering

The B32, $D0_3$, and $L2_1$ phases are ordered forms of the A2 structure type and require four sublattices for their modeling. The ideal composition of a B32 ordered phase is AB, as for B2, but not all nearest neighbors are different. The ideal composition of a $D0_3$ ordered phase is A_3B, as for $L1_2$, but, in contrast to $L1_2$ ordering, the $D0_3$ ordering does not have identical surroundings in the three sublattices with the majority constituent.

One difference between the four-sublattice models for ordering in bcc and fcc phases is that the bcc ordering requires two bond energies, both nearest and next-nearest neighbors are important. In B2 ordering it is sufficient to have two sublattices, one with the central atom, the other with the eight corner atoms. In $D0_3$ ordering one must consider eight bcc unit cells and each of the previous sublattices is split into two new ones in such a way that the atoms on the same sublattice are arranged diagonally. The four sublattices must be grouped two and two since all nearest neighbors are on two of the sublattices and all next-nearest neighbors on the third. If sublattices 1 and 2 have no nearest neighbors and neither do sublattices 3 and 4, the compound energies are

$$\begin{aligned} {}^{\circ}G_{\mathrm{A:A:A:B}} &= {}^{\circ}G_{\mathrm{A:A:B:A}} = {}^{\circ}G_{\mathrm{A:B:A:A}} = {}^{\circ}G_{\mathrm{B:A:A:A}}\, 2\epsilon_{\mathrm{AB}} + \eta_{\mathrm{AB}} \\ {}^{\circ}G_{\mathrm{A:A:B:B}} &= {}^{\circ}G_{\mathrm{B:B:A:A}}\, 4\epsilon_{\mathrm{AB}} \quad (2) \\ {}^{\circ}G_{\mathrm{A:B:A:B}} &= {}^{\circ}G_{\mathrm{A:B:B:A}} = {}^{\circ}G_{\mathrm{B:A:A:B}} = {}^{\circ}G_{\mathrm{B:A:B:A}}\, 2\epsilon_{\mathrm{AB}} + 2\eta_{\mathrm{AB}} \end{aligned} \tag{5.160}$$

where η_{AB} is the next-nearest-neighbor bond energy.

The $L2_1$ phase is also known as the Heusler phase with the ideal composition A_2BC and can appear only in ternary systems; the lattice is shown in Fig. 5.22. It has the same arrangement of sites as $D0_3$, but two sublattices have the same element (A) while the other two have different elements (B and C).

5.8.5 *Partitioning of the Gibbs energy for phases that never undergo disordering*

Phases with many sublattices may become very complicated in multicomponent systems since the number of "end members" is the product of the number of constituents in each sublattice. Many of these "end members" may represent compositions inside the single-phase region or far outside the phase's stability region, thus it can be impossible to assess or evaluate any parameter value for these. The technique of setting the parameters equal to the mean value of the Gibbs energies of formation of each constituent, like in Eq. (5.39), is one possible way to handle this.

Another recent model involves partitioning the Gibbs energy into a substitutional and a sublattice description also for phases with sublattices that never disorder. For such a case there is no need to make the ordered contribution equal to zero and for those phases one simply has

$$G_{\mathrm{m}} = G_{\mathrm{m}}^{\mathrm{dis}}(x) - T^{\mathrm{cnf}}S_{\mathrm{m}}^{\mathrm{dis}} + \Delta G_{\mathrm{m}}^{\mathrm{ord}} \tag{5.161}$$

$$\Delta G_{\mathrm{m}}^{\mathrm{ord}}(y) = G_{\mathrm{m}}^{\mathrm{ord}}(y) \tag{5.162}$$

$G_{\mathrm{m}}^{\mathrm{dis}}$ is described with a substitutional model and $G_{\mathrm{m}}^{\mathrm{ord}}$ includes the sublattices according to the crystalline structure. In Eq. (5.161) the ideal configurational entropy ${}^{\mathrm{cnf}}S_{\mathrm{m}}^{\mathrm{dis}}$ is subtracted from the disordered part and the configurational entropy is calculated for the ordered part only.

The physical reason for adding a disordered contribution to a phase like σ is that the interaction energies need not depend strongly on the sublattices where the constituents are located. In particular, for a phase with a narrow composition range the disordered contribution makes it possible to have a smoother enthalpy curve far away from the composition range of stability and avoid unphysical curvatures. It also seems possible to model phases with small composition ranges with fewer parameters, using disordered interaction parameters.

This partitioning for phases that are always ordered is a very new model and little practical experience has been obtained. For the Laves phase the usual model has all constituents on all sublattices, but the σ phase is normally modeled with a restricted set of constituents on the sublattices. With a disordered part it may be advantageous if all constituents were allowed to enter all sublattices. In the first case one would use for Fe–Cr

$$(\mathrm{Fe})_{10}(\mathrm{Cr})_4(\mathrm{Cr},\mathrm{Fe})_{16} \tag{5.163}$$

and in the second case

$$(\mathrm{Cr},\mathrm{Fe})_{10}(\mathrm{Cr},\mathrm{Fe})_4(\mathrm{Cr},\mathrm{Fe})_{16} \tag{5.164}$$

An advantage of the second method would be that one could assign a value just to the ${}^{\circ}G^{\sigma}_{\mathrm{Fe}}$ representing pure σ Fe and would not have to bother about multicomponent parameters like ${}^{\circ}G^{\sigma}_{\mathrm{Fe:Mo:Ni}}$ that cannot be determined from experimental data. The value of ${}^{\circ}G^{\sigma}_{\mathrm{Fe}}$ have to be estimated or determined from *ab initio* calculations. This partitioning is used in the case study of the Re–W system in section 9.3.

5.8.6 *Partitioning of parameters in physical models*

Physical properties that are modeled separately, such as magnetism, may also have their parameters partitioned between the ordered and the disordered part. The individual parameters such as the Curie temperature from both parts will be added together before they are used to calculate contributions to the Gibbs energy of the phase. In Fig. 5.4 the assessed critical temperatures for magnetic ordering in the bcc phases in the Fe–Cr system and for fcc and bcc phases in the Fe–Ni system are plotted.

5.9 Models for liquids

The substitutional-solution model is the most commonly used model for liquids. The associate model, described earlier, is often applied to liquids that exhibit a tendency toward SRO. The description of these models will not be repeated here.

5.9.1 *Metallic liquids*

Most metallic liquids can be well described in terms of a substitutional solution of the elements with a Redlich–Kister excess Gibbs energy, Eq. (5.51), which is repeated here:

$$G_{\mathrm{m}} = \sum_{i=1}^{n} x_i {}^{\circ}G_i + RT\sum_{i=1}^{n} x_i \ln(x_i) + {}^{\mathrm{E}}G_{\mathrm{m}} \tag{5.165}$$

where x_i is equal to the mole fraction. The binary excess Gibbs energy, using the Redlich–Kister series explained in section 5.6.2.1, is

$$^{\mathrm{E}}G_{\mathrm{m}} = \sum_i \sum_{j>i} x_i x_j L_{ij} \tag{5.166}$$

$$L_{ij} = \sum_{\nu=0}^{k} (x_i - x_j)^{\nu} \cdot {}^{\nu}L_{ij} \tag{5.167}$$

Ternary and higher-order terms can be added as for solid solutions, as described in section 5.6.4. For the liquid phase it is also sometimes useful to introduce other ternary extrapolation models, as described in section 5.6.6.

5.9.2 Liquids with strong short-range order

Some liquids have a tendency to exhibit SRO around certain compositions. This is usually apparent also in the solid state with a stable compound of that composition. The heat of mixing of the liquid should have a pronounced "V"-shape and it can be modeled with associates, see section 5.7.1, or, for cases in which the SRO is due to electronic charge transfer, the particular models for liquids described below.

5.9.3 Quasi-chemical entropy for liquids

Short-range order is particularly important in liquids and several models have been proposed for various types of liquids. The already-mentioned associated model is particularly useful for the liquid phase when different atoms like to be close to each other but still do not form stable molecules. Since the associated solution still uses the ideal configurational entropy, several attempts have been made to improve the agreement with experimental data by taking into account a quasi-chemical entropy expression. Two such models will be described here because they can be used in the PARROT program.

5.9.3.1 The modified quasi-chemical model

Various modifications of the quasi-chemical model have been used with considerable success by the FACT (Bale *et al.* 2002) group in Montreal to describe oxide and sulfide liquids. In their assessments they have not used the same model for the metallic liquid, which makes it impossible to describe some systems like Fe–S and Zr–O for which there is no miscibility gap between the metallic and ionic liquids.

This model has been modified several times and it has not been used for any assessments with the software described in this book, so it will not be discussed further. The interested reader is referred to the publications by Pelton *et al.* (Pelton *et al.* 2000, 2001, Pelton and Chartrand 2001, Pelton 2001).

Just to investigate the importance of the quasi-chemical configurational entropy, a quasi-chemical modification of the liquid two-sublattice model, described by Hillert *et al.* (2001), has been tested. The liquid two-sublattice model is described below in section 5.9.4. For a simple system with ions all with unit charge, all cations mix on one sublattice and all anions on the other, and the numbers of sites on the two sublattices are the same. For this simple case this model is identical to the modified quasi-chemical model of Pelton *et al.*,

$$(A^{1+}, B^{1+})_1(C^{1-}, D^{1-})_1 \tag{5.168}$$

The configurational entropy in the ionic-liquid model is given by Eq. (5.183) and, for the simple system above,

$$\begin{aligned} {}^{\text{q-chem}}G_m = {} & y_{AC}{}^{\circ}G_{AC} + y_{AD}{}^{\circ}G_{AD} + y_{BC}{}^{\circ}G_{BC} + y_{BD}{}^{\circ}G_{BD} \\ & + \frac{RTz}{2}\left[y_{AC}\ln\left(\frac{y_{AC}}{y_{A^{1+}}y_{C^{1-}}}\right) + y_{AD}\ln\left(\frac{y_{AD}}{y_{A^{1+}}y_{D^{1-1}}}\right) + y_{BC}\ln\left(\frac{y_{BC}}{y_{B^{1+}}y_{C^{1-}}}\right) + y_{BD}\ln\left(\frac{y_{BD}}{y_{B^{1+}}y_{D^{1-}}}\right)\right] \\ & + RT[y_{A^{1+}}\ln(y_{A^{1+}}) + y_{B^{1+}}\ln(y_{B^{1+}}) + y_{C^{1-}}\ln(y_{C^{1-}}) + y_{D^{1-}}\ln(y_{D^{1-}})] \end{aligned} \tag{5.169}$$

The equation above is simplified by the fact that all charges are the same. This means that it is similar to a crystalline quasi-chemical model, Eq. (5.79).

The reciprocal miscibility gap for this system has been calculated using the same values for the end members in both models and is shown in Fig. 5.25 for the normal ionic-liquid model (dashed lines) and for the ionic quasi-chemical model, Eq. (5.169) (full lines). For the quasi-chemical model the number of bonds, z, was assumed to be 12. The only difference between the models is in the configurational entropy, the effect of which is quite small.

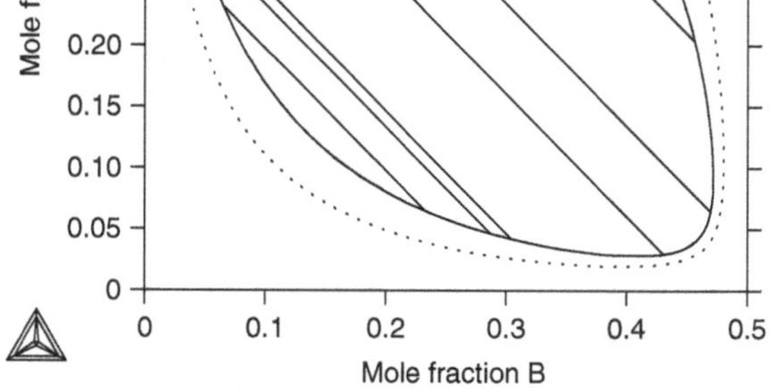

Figure 5.25 The reciprocal miscibility gap for an ionic liquid (dashed lines) and that for the ionic quasi-chemical model (full lines) calculated using the same Gibbs energies for the end members.

5.9.3.2 The cell model

A binary cell model was proposed by Kapoor *et al.* (1974) and later extended to multi-component systems by Guye and Welfringer (1984). It was first used for liquid oxide slags, but has recently been extended to sulfides and fluorides also.

The cell model has a special form of the quasi-chemical entropy. One considers cells with one anion and two cations. Originally only oxygen was considered, so the presentation here will be limited to oxides. The cells can have two identical or two different cations. The cells can be treated like constituents, but one has to be careful about the numbers of atoms in the constituents. The general rule is that the same number of anions must be provided from both cations in the constituent with two different cations. The mole fractions of the "component oxides," i.e., the cells with only one type of cation, are, for n cations,

$$x_i = \frac{y_i + \sum_{j \neq i}^{n} v_j y_{ij}}{Q} \tag{5.170}$$

$$Q = \sum_{i=1}^{n} \left(y_i + \sum_{j \neq i}^{n} v_j y_{ij} \right) \tag{5.171}$$

where v_j is the oxygen stoichiometry in the component oxide of the cation M, $M_{u_i}O_{v_i}$. As an example one may use the CaO–SiO$_2$ system, for which the constituents would be (CaO, SiO$_2$, (CaO)$_2$SiO$_2$) and the quantities above would be

$$x_{CaO} = \frac{y_{CaO} + 2y_{Ca_2SiO_4}}{Q}$$

$$x_{SiO_2} = \frac{y_{SiO_2} + y_{Ca_2SiO_4}}{Q}$$

$$Q = y_{CaO} + y_{SiO_2} + 3y_{Ca_2SiO_4} \tag{5.172}$$

The Gibbs energies of formation of the "component oxides" as well as the cell with two different cations are simply introduced into the ${}^{srf}G_m$ part of Eq. (5.1) multiplied by their constituent fractions as in the ideal-solution model, Eq. (5.42). One has to be careful that the parameters are given the right values considering the number of atoms in the constituent; the parameters for formation of a cell are for one oxygen atom whereas the constituents defined here will be for $v_i + v_j$ oxygen atoms.

The model then introduces a sum over the component oxides, D_i defined as

$$D_i = \sum_{j=i}^{n} v_j x_j \tag{5.173}$$

The important property of this sum is that the "component oxides" must be ordered in decreasing valency according to the cations. By introducing D_i the model attempts to account for the charged behavior of the cells. The expression for the configurational entropy below, Eq. (5.174), is in principle derived by first distributing the highest-charged constituents on all possible sites, then the second-highest-charged ones on the remaining sites and so on.

If two cations have the same valency, they are arranged in decreasing order of "acidity." The order is slightly arbitrary, but the proposed arrangement of cations is (P^{5+}, Si^{4+}, $(PO)^{3+}$, Cr^{3+}, Al^{3+}, Fe^{3+}, Cr^{2+}, Fe^{2+}, Mn^{2+}, Mg^{2+}, Ca^{2+}, Na^{1+}). Note that several cations have two valencies and P also appears as a complex $(PO)^{3+}$, in order to obtain a better description of the experimental information. The entropy expression of the cell model is then

$$\begin{aligned} {}^{\text{cnf}}S_{\text{m}} &= R\frac{u_i}{v_i}\sum_{i=1}^{n-1}\left[D_i \ln\left(\frac{D_i}{v_i x_i}\right) - D_{i+1}\ln\left(\frac{D_{i+1}}{v_i x_i}\right)\right] \\ &\quad - 2R\sum_{i=1}^{n} v_i x_i \ln\left(\frac{v_i x_i}{D_1}\right) - R\sum_{j=1}^{m} y_j \ln\left(\frac{y_j}{D_1}\right) \end{aligned} \tag{5.174}$$

The first two sums are over all component oxides, but the last sum over j is for all m constituents.

The excess parameters of the cell model are slightly special and will not be discussed here.

5.9.4 *The partially ionic-liquid two-sublattice model*

This model has a very long name and it is usually abbreviated as the liquid two-sublattice model, although that can be ambiguous. It is interesting to note that the sublattice model developed by Hillert and Staffanson (1970) for solid phases was based on a model suggested by Temkin (1945) to take into account the configurational entropy in molten salts. In a liquid there is no LRO and one cannot distinguish sites for anions or cations, but the mathematical formalism using mixing on two different sites gives good agreement with experimental information. This means that exchanges of a cation with an anion must not be counted in calculating the configurational entropy by use of Eq. (2.11). In particular, on mixing four salts, A_cC_a, A_dD_a, B_cC_b, and B_dD_b, one obtains a reciprocal system as shown in Fig. 5.26, with A and B in the first sublattice and C and D in the second:

$$(A^{a+}, B^{b+})_P(C^{c-}, D^{d-})_Q \tag{5.175}$$

If all ions have the same valence then $P = Q = a$ and the Gibbs-energy expression for this liquid is identical to that for a crystalline two-sublattice model. However, if the valencies of the cations or anions are not equal, one must find some method by which to maintain electro-neutrality in this liquid. One method is to use equivalent fractions defined by

$$z_A = \frac{\dfrac{N_A}{a}}{\dfrac{N_A}{a} + \dfrac{N_B}{b}} \tag{5.176}$$

$$z_C = \frac{\dfrac{N_C}{c}}{\dfrac{N_C}{c} + \dfrac{N_D}{d}} \tag{5.177}$$

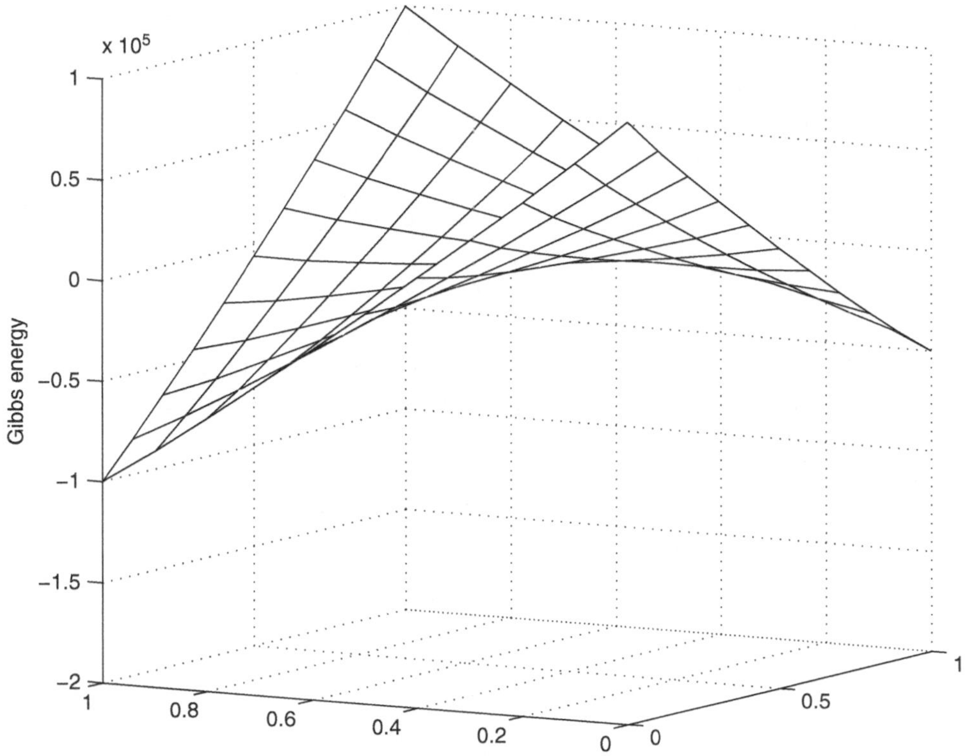

Figure 5.26 The surface of reference for a reciprocal system. Note that the surface is curved even without any configurational entropy or excess parameters. Courtesy of Qing Chen.

where $a+$, $b+$, $c-$, and $d-$ are the valences of A, B, C, and D, respectively, and $P = Q = 1$. However, the use of equivalent fractions has the drawback that it is impossible to extend the model to systems with neutral constituents.

5.9.4.1 Hypothetical vacancies and neutral species

Therefore another model has been developed, which can be extended both to systems with only cations (i.e., metallic systems) and also to non-metallic liquids, for example liquid sulfur. This model is called the "partially ionic two-sublattice liquid model" (Hillert *et al.* 1985) and it uses constituent fractions as composition variables. In order to handle a liquid with only cations, i.e., a metallic liquid, hypothetical vacancies are introduced on the anion sublattice, whereas in order to extend the model to non-metallic systems one introduces neutral species on the anion sublattice. The model can be written as

$$(\mathrm{C}_i^{\nu_i+})_P(\mathrm{A}_j^{\nu_j-}, \mathrm{Va}, \mathrm{B}_k^0)_Q \qquad (5.178)$$

where each pair of parentheses surrounds a sublattice, while C represents cations, A anions, Va hypothetical vacancies, and B neutral species. The charge of an ion is denoted

ν_i and the indices i, j, and k are used to denote a specific constituent. The superscripts ν_i on cations and anions as well as 0 for the neutral species will be omitted unless needed for clarity. The numbers of sites on the sublattices, P and Q, vary with the composition in order to maintain electro-neutrality. The values of P and Q are calculated from the following equations:

$$P = \sum_j \nu_j y_{A_j} + Q y_{Va} \tag{5.179}$$

$$Q = \sum_i \nu_i y_{C_i}$$

where y_i denotes the constituent fraction of constituent i. Equation (5.179) simply means that P and Q are equal to the average charge on the opposite sublattice. The hypothetical vacancies have an induced charge equal to Q.

The ordinary mole fractions can be calculated from the constituent fractions in the following way for the components which behave like cations:

$$x_{C_i} = \frac{P y_{C_i}}{P + Q(1 - y_{Va})} \tag{5.180}$$

For the components which behave like anions or neutral species:

$$x_{D_i} = \frac{Q y_{D_i}}{P + Q(1 - y_{Va})} \tag{5.181}$$

where D is used to denote any constituent on the anion sublattice. Equation (5.181) cannot be applied to the hypothetical vacancies, because the mole fraction of vacancies is zero, of course.

The integral Gibbs-energy expression for this model is

$$^{srf}G_m = \sum_i \sum_j y_{C_i} y_{A_j} {^\circ}G_{C_i:A_j} + Q y_{Va} \sum_i y_{C_i} {^\circ}G_{C_i} + Q \sum_k y_{B_k} {^\circ}G_{B_k} \tag{5.182}$$

$$^{cnf}S_m = -R\left[P \sum_i y_{C_i} \ln(y_{C_i}) + Q\left(\sum_j y_{A_j} \ln(y_{A_j}) + y_{Va} \ln(y_{Va}) + \sum_k y_{B_k} \ln(y_{B_k})\right)\right] \tag{5.183}$$

$$\begin{aligned} ^{E}G_m = & \sum_{i_1} \sum_{i_2} \sum_j y_{i_1} y_{i_2} y_j L_{i_1 i_2:j} + \sum_{i_1} \sum_{i_2} y_{i_1} y_{i_2} y_{Va}^2 L_{i_1 i_2:Va} \\ & + \sum_i \sum_{j_1} \sum_{j_2} y_i y_{j_1} y_{j_2} L_{i:j_1 j_2} + \sum_i \sum_j y_i y_j y_{Va} L_{i:j Va} \\ & + \sum_i \sum_j \sum_k y_i y_j y_k L_{i:jk} + \sum_i \sum_k y_i y_k y_{Va} L_{i:Va k} + \sum_{k_1} \sum_{k_2} y_{k_1} y_{k_2} L_{k_1,k_2} \end{aligned} \tag{5.184}$$

where ${^\circ}G_{C_i:A_j}$ is the Gibbs energy of formation for $\nu_i + \nu_j$ moles of atoms of liquid C_iA_j. ${^\circ}G_{C_i}$ and ${^\circ}G_{B_i}$ are the Gibbs energies of formation per mole of atoms of liquids C_i and B_i, respectively. The factor Q in front of the second and third sums comes from the variation of the number of sites with the composition. Note that G_m in Eqs. (5.182)–(5.184) is defined for a formula unit that contains $P + Q(1 - y_{Va})$ moles of atoms. $^{cnf}S_m$ is a random configurational entropy on each sublattice and $^{E}G_m$ is the excess Gibbs energy.

5.9.4.2 The excess parameters

There are many possible excess parameters in this model and in the expression above only the binary ones are included. The following list may be useful in order to relate them to real systems.

$L_{i_1 i_2:j}$ represents interaction between two cations with a common anion, for example, in the CaO–MgO system, $L_{\mathrm{Ca^{2+},Mg^{2+}:O^{2-}}}$.

$L_{i_1 i_2:\mathrm{Va}}$ represents interaction between two metallic elements, for example, in the Ca–Mg system, $L_{\mathrm{Ca,Mg}}$. The charge is irrelevant in this case. Note that this parameter is equivalent to that in a substitutional solution of i and j if the parameter is multiplied by the fraction of vacancies squared. The corresponding ternary parameter with three cations must be multiplied by the third power of the fraction of vacancies in order for the parameter to be compatible with the ternary interaction parameter in a substitutional solution with three metallic atoms. This is discussed in more detail by Sundman (1991b).

$L_{i_1 i_2:k}$ is forbidden because the number of cation sites is zero.

$L_{i:j_1 j_2}$ represents interaction between two anions in systems with a common cation, for example, in the $\mathrm{Ca(OH)_2}$–$\mathrm{CaCO_3}$ system, $L_{\mathrm{Ca^{2+}:OH^{1-},CO_3^{2-}}}$.

$L_{i:j\mathrm{Va}}$ represents interaction between a metallic atom and an anion, for example describing the Fe-rich side of the Fe–S system, $L_{\mathrm{Fe^{2+}:S^{2-},Va}}$.

$L_{i:jk}$ represents interaction between an anion and a neutral atom, for example describing the sulfur-rich side of the Fe–S system, $L_{\mathrm{Fe^{2+}:S^{2-},S}}$.

$L_{i:\mathrm{Va}k}$ represents interaction between a metal and a neutral species, for example, in the Fe–C system, $L_{\mathrm{Fe^{2+}:Va,C}}$.

L_{k_1,k_2} represents interaction between two neutral species, for example, in the $\mathrm{Si_3N_4}$–$\mathrm{SiO_2}$ system, $L_{\mathrm{Si_3N_4,SiO_2}}$. Note that the cation is irrelevant in this case because the number of cation sites is zero.

It is straightforward to include ternary interaction parameters in the model, but this will not be described here.

Equation (5.182) may look formidable in its complexity and one may wonder whether simpler models might not be equally useful. This criticism misses the point, because Eq. (5.182) is the general multicomponent expression and this model is indeed identical to simpler models in many special cases. The great advantage with Eq. (5.182) is that it allows a continuous description of a liquid that changes in character with varying composition. Equation (5.182) has successfully been used to describe oxide liquids, silicates, and sulfides as well as SRO in liquids and also molten salts and ordinary metallic liquids.

5.9.4.3 Compatibility between different liquid models

It is instructive to show how to convert from some different models to the liquid two-sublattice model. A liquid with two metallic elements like Fe and Cr would normally be treated as a substitutional-solution model, (Fe, Cr). In the liquid two-sublattice model the description would be $(Fe^{2+}, Cr^{3+})_Q(Va)_Q$ (taking the dominant valency only). For solutions with only vacancies in the anion sublattice, P and Q are always equal. All parameters for interactions between cations evaluated for a substitutional model can thus be used directly in the liquid two-sublattice model.

A liquid solution of C in Fe may also be treated as a substitutional solution, at least at high concentrations of carbon. However, it would not be realistic to assume that carbon has a positive valency and in the liquid two-sublattice formalism one would like to model it like $(Fe^{2+})_P(Va, C)_Q$. It is quite remarkable that this is also mathematically identical to a substitutional solution (Fe, C).

Another remarkable feature of Eq. (5.182) is that it becomes identical to the associate model for some simple binary systems, for example the Cu–S system modeled with an associate Cu_2S. The associate model uses a substitutional solution of (Cu, S, Cu_2S) whereas the liquid two-sublattice model uses $(Cu^{1+})_P(S^{2-}, Va, S)_Q$, but it is possible to identify the parameters in these two models with each other and they give identical results in both models. The assumptions behind the associate model and the liquid two-sublattice model are very different and it is interesting that these different assumptions can lead to the same mathematical expression. This shows clearly that one cannot make any statement about the true nature of a system just because a mathematical formalism based on some physical model of the system gives a good result. It may be possible that another physical model of the system will yield exactly the same mathematical formalism.

5.9.5 *The aqueous solution*

Model parameters for aqueous solutions are usually not stored in general databases but are often re-evaluated from various models for each particular application. However, parameters for a simplified Pitzer model and some other models used for aqueous solutions can be evaluated using the techniques described in this book. Interested users are referred to the particular software for instructions on how to handle the parameters of the models implemented.

5.9.6 *A model for polymers – the Flory–Huggins model*

This model has been proposed for polymer systems in which the constituents can be very different in size or volume. For each constituent a "size" parameter must be given. It is used in the following binary expression for the Gibbs energy of mixing:

$$ {}^{\mathrm{M}}G_{\mathrm{m}} = RT\left[x_1 \ln\left(\frac{\phi_1}{x_1}\right) + x_2 \ln\left(\frac{\phi_2}{x_2}\right)\right] + \frac{\phi_1\phi_2(\nu_1 x_1 + \nu_2 x_2)\chi_{12}}{\nu_1} \qquad (5.185) $$

where x_1 and x_2 are mole fractions, ν_1 and ν_2 are measures of the sizes of the constituents, χ_{12} is an interaction parameter, and

$$ \begin{aligned} \phi_1 &= \frac{\nu_1 x_1}{\nu_1 x_1 + \nu_2 x_2} \\ \phi_2 &= \frac{\nu_2 x_2}{\nu_1 x_1 + \nu_2 x_2} \end{aligned} \qquad (5.186) $$

It is not clear how to introduce composition-dependent binary excess parameters or how to extend this model to ternary or higher-order systems. In order to conform with the current models, the following general expression has been implemented in Thermo-Calc:

$$G_m = x_1 {}^\circ G_1 + x_2 {}^\circ G_2 + RT[x_1 \ln(\phi_1) + x_2 \ln(\phi_2)] + \frac{x_1 x_2}{\nu_1 x_1 + \nu_2 x_2} L_{12} \quad (5.187)$$

where the parameter $L_{12} = \nu_2 \chi_{12}$.

5.10 Chemical reactions and thermodynamic models

In chemistry it is common to use reactions to describe equilibria between different phases or constituents in a phase. This is very simple and useful when there is only one solution phase involved and this can be modeled as a substitutional phase. When there are many constituents or the phases involved have complex structures, there are more problems than simplifications in using reaction formulae. It is thus recommended that one use chemical reactions only as an introduction to equilibrium calculations and then switch to using a general minimization of the total Gibbs energy for equilibrium calculations as described in section 2.3.

5.10.1 The solubility product

An example of a homogeneous chemical reaction was given in section 5.3.1. Here an example of a heterogeneous chemical reaction will be given in order to show the relation between chemical reactions and the thermodynamic modeling. The chemical reaction for formation of the AlN phase from Al and N dissolved in liquid steel can be written

$$(\mathrm{Al}) + (\mathrm{N}) = \mathrm{AlN} \quad (5.188)$$

The parentheses around Al and N are used to denote that they are dissolved in an unspecified solvent and the reaction formula does not specify that AlN is a different phase, but all of this is usually evident from the context. If the liquid steel is an ideal solution, the reaction can be written

$${}^\circ G_{\mathrm{Al}}^{\mathrm{L}} + RT \ln(x_{\mathrm{Al}}^{\mathrm{L}}) + {}^\circ G_{\mathrm{N}}^{\mathrm{L}} + RT \ln(x_{\mathrm{N}}^{\mathrm{L}}) = {}^\circ G_{\mathrm{AlN}}^{\mathrm{AlN}} \quad (5.189)$$

where the superscripts now denote the phases liquid (L) and solid AlN. Rearranging this gives the solubility product

$$x_{\mathrm{Al}}^{\mathrm{L}} x_{\mathrm{N}}^{\mathrm{L}} = \exp\left(\frac{{}^\circ G_{\mathrm{AlN}}^{\mathrm{AlN}} - {}^\circ G_{\mathrm{Al}}^{\mathrm{L}} - {}^\circ G_{\mathrm{N}}^{\mathrm{L}}}{RT}\right) \quad (5.190)$$

The right-hand side is the equilibrium constant giving the solubility product. There is thus a simple relation between the thermodynamic technique described in this book and chemical-reaction formulae. The chemical reactions should never be used when two or more solution phases are involved or the model for any phase is more complicated than the substitutional-solution model. Note also that Eq. (5.190) is formally very similar to Eq. (5.127), but, for the latter equation, the defect fractions and the energy parameters belong to the same phase, a so-called homogeneous equilibrium, whereas in Eq. (5.190) the reaction concerns two different phases, a so-called heterogeneous reaction. This difference is obvious when using thermodynamic models but rather obscured by the reaction formalism.

5.10.2 *Comparison with the Kröger and Vink notation*

A method by which to describe equilibria with defects in ionic systems is to use the Kröger and Vink notation, or something similar, in chemical reactions. This technique has the same limitation as that of using chemical reactions for other equilibria as mentioned above.

For example, the reaction of a gas containing oxygen with bunsenite (NiO) to form a vacant site for Ni and two holes (denoted h•) in the conduction band to compensate for the electrons taken up forming the oxygen ion can be written as a chemical reaction using the Kröger and Vink notation:

$$0.5\mathrm{O}_2^{\mathrm{gas}} = \mathrm{V}_{\mathrm{Ni}}'' + 2\mathrm{h}^\bullet + \mathrm{O}_{\mathrm{O}}^x \tag{5.191}$$

In standard textbooks such as Birks *et al.* (2006) it is shown that the fraction of vacancies in the Ni sublattice formed in this reaction depends on the partial pressure of oxygen as $p_{\mathrm{O}_2}^{1/6}$. This relation is obtained by using the "law of mass action" on the reaction above and replacing the fraction of holes by the fraction of vacancies on the Ni^{2+} sublattice from the electro-neutrality condition.

In Chen *et al.* (1998) the modeling of semiconductors is discussed and, to formulate a physically realistic model for the reaction above, it is reasonable to assume, as shown by Grundy (2006), that each hole is associated with an Ni^{2+} ion to form an Ni^{3+} ion and the model should be

$$(\mathrm{Ni}^{2+}, \mathrm{Ni}^{3+}, \mathrm{Va})_1(\mathrm{O}^{2-})_1 \tag{5.192}$$

This model is identical to that used for wustite described in section 6.2.5.8 later. The Gibbs energy for this model would be

$$\begin{aligned} G_m = {} & y_{\mathrm{Ni}^{2+}}{}^\circ G_{\mathrm{Ni}^{2+}:\mathrm{O}^{2-}} + y_{\mathrm{Ni}^{3+}}{}^\circ G_{\mathrm{Ni}^{3+}:\mathrm{O}^{2-}} + y_{\mathrm{Va}}{}^\circ G_{\mathrm{Va}:\mathrm{O}^{2-}} \\ & + RT(y_{\mathrm{Ni}^{2+}} \ln(y_{\mathrm{Ni}^{2+}}) + y_{\mathrm{Ni}^{3+}} \ln(y_{\mathrm{Ni}^{3+}}) + y_{\mathrm{Va}} \ln(y_{\mathrm{Va}})) \end{aligned} \tag{5.193}$$

The partial Gibbs energy for oxygen in the model can be calculated as the difference between the partial Gibbs energies for the end members using Eq. (5.38),

$$
\begin{aligned}
G_{\mathrm{O}} &= 2G_{\mathrm{Ni^{3+}:O^{2-}}} + G_{\mathrm{Va:O^{2-}}} - 2G_{\mathrm{Ni^{2+}:O^{2-}}} \\
&= 2({}^{\circ}G_{\mathrm{Ni^{3+}:O^{2-}}} + RT\ln(y_{\mathrm{Ni^{3+}}})) + {}^{\circ}G_{\mathrm{Va:O^{2-}}} + RT\ln(y_{\mathrm{Va}}) \\
&\quad - 2({}^{\circ}G_{\mathrm{Ni^{2+}:O^{2-}}} + RT\ln(y_{\mathrm{Ni^{2+}}})) \\
&= \Delta G_{\mathrm{Va}} + RT(2\ln(y_{\mathrm{Ni^{3+}}}) + \ln(y_{\mathrm{Va}}) - 2\ln(y_{\mathrm{Ni^{2+}}}))
\end{aligned} \tag{5.194}
$$

In the last line $\Delta G_{\mathrm{Va}} = 2{}^{\circ}G_{\mathrm{Ni^{3+}:O^{2-}}} + {}^{\circ}G_{\mathrm{Va:O^{2-}}} - 2{}^{\circ}G_{\mathrm{Ni^{2+}:O^{2-}}}$ has been introduced for the energy needed to create an Ni^{3+} ion plus a vacancy. Electro-neutrality requires $y_{\mathrm{Ni^{3+}}} = 2y_{\mathrm{Va}}$ and, because the sum of site fractions is unity, $y_{\mathrm{Ni^{2+}}} = 1 - 3y_{\mathrm{Va}}$, which gives

$$
G_{\mathrm{O}} = \Delta G_{\mathrm{Va}} + RT(2\ln(2y_{\mathrm{Va}}) + \ln(y_{\mathrm{Va}}) - 2\ln(1 - 3y_{\mathrm{Va}})) \approx \Delta G_{\mathrm{Va}} + 2RT\ln(2) + 3RT\ln(y_{\mathrm{Va}}) \tag{5.195}
$$

where one has used $\ln(1 - 3y_{\mathrm{Va}}) \approx 0$ since the fraction of vacancies is small. Combining this with the expression for the chemical potential for oxygen in the gas at the standard pressure p_0 gives

$$
0.5\left[{}^{\circ}G^{\mathrm{gas}}_{\mathrm{O_2}} + RT\ln\left(\frac{p_{\mathrm{O_2}}}{p_0}\right)\right] = \Delta G_{\mathrm{Va}} + 2RT\ln(2) + 3RT\ln(y_{\mathrm{Va}}) \tag{5.196}
$$

Rearranging the terms gives

$$
3RT\ln(y_{\mathrm{Va}}) - 0.5RT\ln\left(\frac{p_{\mathrm{O_2}}}{p_0}\right) = 0.5\,{}^{\circ}G^{\mathrm{gas}}_{\mathrm{O_2}} - \Delta G_{\mathrm{Va}} - 2RT\ln(2) \tag{5.197}
$$

$$
\frac{(y_{\mathrm{Va}})^3}{\sqrt{p_{\mathrm{O_2}}/p_0}} = \exp\left(-\frac{\Delta G_{\mathrm{Va}} + 2RT\ln(2) - 0.5\,{}^{\circ}G^{\mathrm{gas}}_{\mathrm{O_2}}}{RT}\right) \tag{5.198}
$$

$$
y_{\mathrm{Va}} = Ap_{\mathrm{O_2}}^{1/6} \tag{5.199}
$$

The Kröger and Vink notation does not specify the model for the phase and it is quite arbitrary how one includes electrons and holes in a model based on the CEF. It can nonetheless be illuminating to transfer a reaction notation to a CEF-type model in order to understand how realistic it is.

5.11 Final remarks

In most models the excess Gibbs energy includes contributions from all kinds of physical phenomena, for example vibrational, electronic, and configurational, which are not described separately in a magnetic model, for example. It is not practical to use very detailed models for a particular contribution, for example configurational entropy, if other phenomena with equal or larger contributions to the Gibbs energy are described with a simpler excess model.

A note about model names may also be appropriate. For example, the special case of the CEF applied to chemical ordering is called the Bragg–Williams–Gorsky (BWG) model above when no excess terms are included, but some authors may call all models

Table 5.1 *A tentative classification of the different models. Note that the Gibbs energies of "clusters" and "pairs" can be composition-dependent whereas the Gibbs energies of "end members" in the CEF are not. All models with the point approximation are special cases of the CEF.*

Configurational entropy	Sublattices used	Constituents	Excess	Model
Point	No	Elements	No	Ideal substitutional
			Yes	Regular substitutional
		Species	No	Ideal (gas)
			Yes	Real gas or associate
	Yes	Elements	No	BWG
			Yes	Sublattice model
		Species	No	Ideal CEF
			Yes	CEF
Pair	No	Pairs	No	Quasi-chemical
			Yes	Modified quasi-chemical
Complex	Yes	Clusters	No	CVM
			Yes	Calphad CVM

limited to the point approximation of the configurational entropy BWG models. A possible classification of the different models, following Sundman (1990), is shown in Table 5.1. Of course, all models are special cases of the Ising model (Ising 1925).

5.11.1 *Adjustable parameters in the models*

The models and formalisms described in this chapter usually contain many more adjustable parameters than can be fitted with the available experimental values. This problem will be discussed several times in chapter 6. It is important to understand that the most critical parameters to be optimized are not those in the excess model, but those in the surface of reference part, ${}^{\mathrm{srf}}G_{\mathrm{m}}$. When a phase is modeled with associates or with sublattices there are parameters, such as ${}^{\circ}G^{\mathrm{L}}_{\mathrm{Mg_2Sn}}$ or ${}^{\circ}G^{\mathrm{B2}}_{\mathrm{Al:Fe}}$, in the surface of reference energy that must be fitted to experimental information. Since these parameters are multiplied by the lowest power of the fractions, they will have a larger influence on the behavior of the Gibbs energy than will any excess parameter. All parameters in ${}^{\mathrm{srf}}G_{\mathrm{m}}$ must be referred to the reference state of the elements, but, if they describe compounds inside the composition range of the phase, the Kopp–Neumann rule, see section 5.2.3, is often used in addition to an adjustable enthalpy and entropy term. One must never set an end-member parameter equal to zero.

In fitting a thermodynamic model to experimental values, the final value of each adjustable coefficient depends on many of the different measurements and each measured value contributes to many of the coefficients. The advantage of the least-squares method is that these influences do not need to be known quantitatively, since the strategy of the method itself is to select the best possible agreement among all the coefficients and all the experimental values. Many of the parameters and coefficients of the models, however, are not able to improve the fit between descriptions and measurements significantly. Using them in an excessive manner may lead the calculation to follow just the scatter of the experimental values, creating maxima and minima where a smooth line is physically more probable.

To judge whether a certain coefficient is well defined by the available set of measured values, the effect of each coefficient on the shape of calculated curves should be known at least qualitatively. It must be determined for each coefficient whether its influence on the shape of calculated functions really is necessary to improve the fit between calculation results and the experimental dataset. In the case studies in chapter 9 this check is described. If it is difficult to find enough arguments in a given case, the least-squares method should be applied twice, with and without the coefficient. Comparing the two results makes a decision possible in most cases. *It is better to start a calculation with fewer coefficients than with some unnecessary ones.* A systematic misfit between some series of experimental points and the corresponding calculated curve usually gives some hints regarding what parameter or coefficient should be added.

5.11.2 Models, formalisms, and curve-fitting formulae

In real phases a lot of features of the models described in this chapter may be more or less redundant. Then the "general model" becomes a "formalism," which is used to describe the Gibbs energy of the phase by application of a "particular model."

Using a formalism like the CEF, one may have many more parameters than necessary to fit the available data. One may then be forced to use some parameters purely for "curve fitting" of the experimental data, since it is not possible to relate their values to measured quantities.

Curve fitting normally means that a set of data is fitted to a mathematical function, ignoring any underlying physical or chemical relations. This can be the case also for thermodynamic modeling when just a single property is fitted. For example, if a phase diagram is fitted without using any thermochemical information, it is most unlikely that the enthalpies and entropies obtained from such an assessment will be realistic. It is mandatory to use all possible thermochemical information, even estimated values, to obtain realistic values also of the separate thermodynamic properties of a system, i.e., for the enthalpy, entropy, and heat capacity. In addition, one must take care that the model parameters for a phase do not predict unrealistic values outside the range of stability of the phase. However, the question of what is unrealistic is sometimes debatable.

The lack of data is usually the main reason for using "curve fitting," since it might not be possible to relate the model parameters to more than a single measured quantity.

5.11.3 Limitations in the models

Almost all models in commercial thermodynamic databases available today use the ideal (point) configurational entropy. This is sometimes a severe limitation since the SRO must then be modeled as an excess contribution. This excess contribution may give a bad extrapolation to higher-order systems, but the effort needed to change to the use of quasi-chemical or CVM-based models is considerable because the number of clusters increases exponentially with the number of components. Another problem with these models is that one cannot just put a number of assessed systems together and extrapolate to a higher-order system. One must add the necessary pairs or clusters for all combinations of elements and calculate or estimate energies for these. Methods intended to simplify the CVM by using fewer clusters, like the CSA (Oates and Wenzl 1996), may resolve some of these problems, but any general change of model requires that all systems in the databases today be reassessed. Thus one should be careful in selecting any new model to be of general use.

Many simplifications in the modeling of the crystal structure may be necessary in order to limit the model parameters, in particular when there is little or no experimental information about the occupancies of the constituents in the sublattices. Even with very good information about this, one must have a lot of thermodynamic information, either experimental or from *ab initio* calculations.

6 Assessment methodology

6.1 Starting the assessment

In chapter 5, various models were described in order to understand how they can be fitted to the experimental features that were described in chapter 4. In the present chapter, we start from experimental evidence and search for the model best able to describe it. Therefore many topics of chapter 5 will be revisited here.

An assessor working on a system will experience almost all the steps described and discussed here. Since a system is often reassessed many times, by the same researcher or by another, it is very important to keep records about the decisions made in order to make it easy to restart the work, for example, when new experimental evidence requires a new optimization. The process of assessing a system is made easier if an **assessment logbook** is kept. An important function of this logbook is that one should record all mistakes and failures so that one does not repeat them later. In the final paper only the sucessful modeling will be reported and there is no information about the difficulties encountered in obtaining it.

The assessment methodology described here includes a critical assessment of the available literature in the way in which it is normally done, for example, in the *Journal of Phase Equilibria and Diffusion*. By combining this with thermodynamic models, an analytical description is created and the determination of adjustable model parameters is often done using the least-squares method to obtain a description that represents best the complete set of available consistent experimental data. From this point of view that technique can be called an optimization. However, the least-squares method can work well only if the scatter of experimental data is completely random. Non-randomly distributed deviations of some data may completely destroy the utility of the least-squares method. They must be classified as systematic errors and excluded from the optimization. Therefore subjective judgments are required and decisions have to be taken on the selection of data during the optimization. From that point of view, the technique can also be called an assessment. A schematic picture of an assessment procedure is shown in Fig. 6.1.

6.1.1 Literature searching

The first step of the optimization process is the literature search. It is very important to consider all the data that are able to contribute to the optimization of the description one wants to carry out. A useful way to organize the literature is to classify it by the types of the measured quantities.

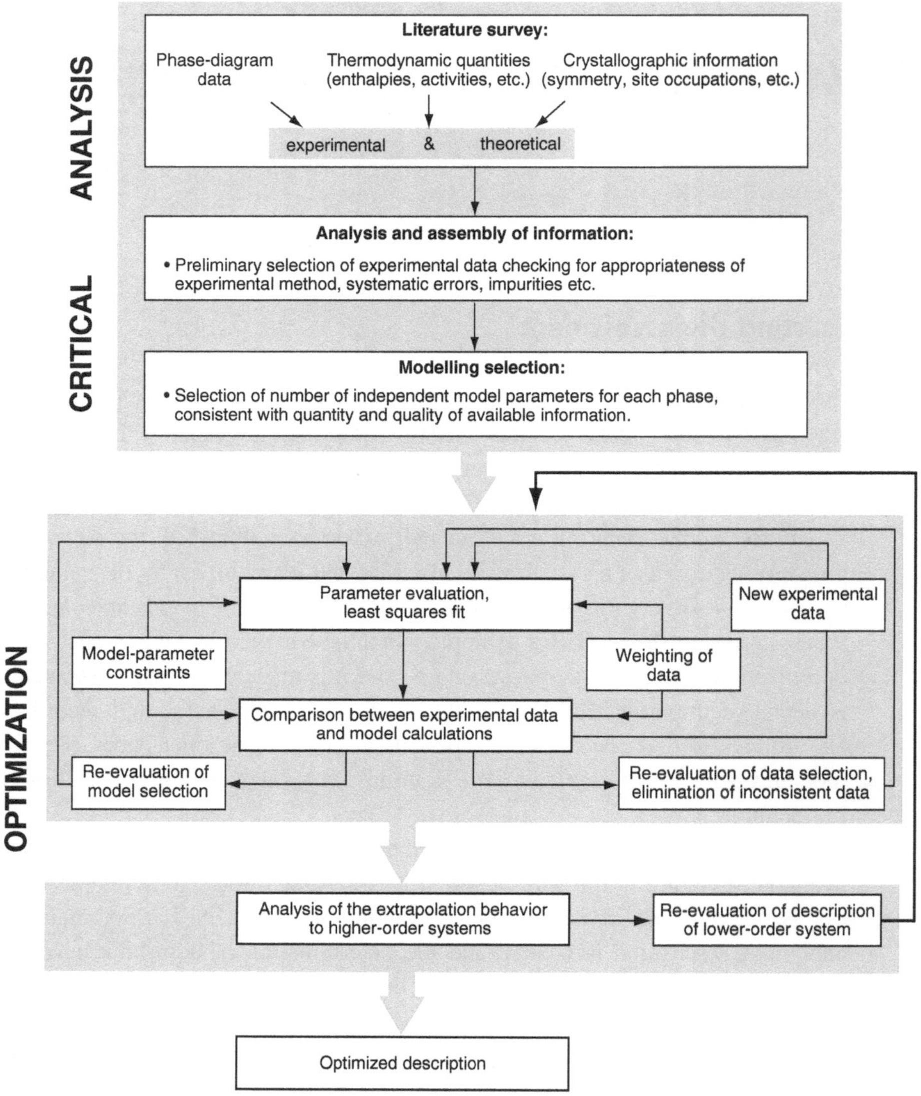

Figure 6.1 A schematic diagram of the Calphad assessment method.

A good start for the literature search is the *Journal of Phase Equilibria and Diffusion*. The last issue of each volume contains a cumulative index of alloy systems. It covers all the evaluations published under the aegis of the Alloy Phase Diagram International Commission (APDIC). The articles quoted there are a critical analysis of the literature: they give an overview of the literature and, if there are discrepancies between published measurements, the authors of the articles amass arguments to decide which values are the most reliable ones. The *Journal of Phase Equilibria and Diffusion* deals specifically with phase diagrams, so most of the literature cited there is relevant for the optimization and it usually covers much of the literature of interest. A scanning of the literature quoted in the JANAF tables (Chase 1998) may be complementarily useful for thermodynamic data such as H-, S-, and C_p-values. Auxiliary information helpful for the modeling of Gibbs energies, for example types of point defects, carrier mobilities, variation of lattice parameters with pressure, temperature, and composition, information from spectroscopic methods such as Raman and Mössbauer spectroscopies, bulk-modulus data, thermal expansion, vibrational (phonon) spectra, total-energy calculations (from first principles), and resistivity measurements to determine phase boundaries, may not be mentioned in the above-mentioned journal.

Calphad, the *Zeitschrift für Metallkunde* (now called the *International Journal of Materials Research*), the *Journal of Alloys and Compounds*, *Intermetallics*, and *Thermochimica Acta* are some of the other journals that often present assessments or data useful for assessments. The information useful for thermodynamic assessments is, however, widespread in many other journals, including *Physical Review* B, the *Philosophical Magazine*, *Acta Materialia*, *Metallurgical and Materials Transactions*, *Nature Materials*, *Acta Crystallographica*, and *Scripta Materialia*. These can provide literature more recent than the critical evaluation of the International Programme for Alloy Phase Diagram Data. With electronic literature databases used on line, searching for keywords can provide large lists of references. However, many of them may be irrelevant for the optimization.

As a rule one should obtain all the original papers. It is very important to check all the data as they are reported by the original author. Do not use "secondhand" information from the author of an assessment, unless it is impossible to order the original literature.

It may be enlightening to know a story about how important it is to have original data. In the accepted Fe–Mo phase diagram before 1980 there was a three-phase temperature for liquid $+\sigma+$ bcc at 1813 K. This was in several references reported as determined by Sykes (1926), but the thermodynamic assessment showed that this temperature was impossible, since the parameters required to get this low three-phase temperature were absolutely impossible. However, Sykes being a respected experimentalist, it was not possible just to discard his data and it had been quoted in all references since 1926. But when the original reference was finally retrieved it was found that he had not really measured this temperature. He had measured some points on the $\sigma+$ liquid solubility curve and then extrapolated this to meet the solubility line of σ in equilibrium with bcc Mo and put the three-phase temperature where these lines met. The solubility of Fe in σ he had estimated as 50% Mo, but he did not have any experimental data on this solubility; he had misinterpreted another paper that stated that the solubility limit of the μ phase was 50% Fe, mistaking μ for the σ phase.

Later the solubility limit of the σ phase was measured and a much higher solubility of Mo, almost 60%, was found. However, in the phase diagram drawn the three-phase line was not changed; instead a very strange curvature of the solubility curve of σ was imposed in order to fit the accepted three-phase temperature of 1813 K. If those who constructed this diagram had read the Sykes paper and carried out the same construction as he did, they would have put the three-phase temperature at a much higher value. In the thermodynamic assessment this was now done. At the same time some new experimental data above

1813 K were obtained, showing that σ is indeed stable at much higher temperatures, and the new assessment was confirmed.

Information can be lost. Information can be wrong. Information can have been misinterpreted by the assessor at the time of his assessments. New experimental results reported in the literature may invalidate the previous arguments and decisions.

In many cases the originally measured quantity is reported after having been transformed. For example, enthalpies of mixing are often obtained by dropping a cold sample into a calorimetric bath (section 4.1.1.1). The reported value is the measured one minus the enthalpy difference between the cold sample and the sample heated up to the calorimeter temperature. For this enthalpy difference the numerical value which is actually used is often not given, but only a reference is cited and the reference may be ambiguous in evaluating this value. In the dataset to be optimized, however, this enthalpy difference is already defined and may deviate to some extent from that used by the author. Then the value used by the author should be replaced by the enthalpy difference defined by the dataset.

Sometimes the original measurements may contain more information than is reported in tables or curves. The additional information can, however, often still be extracted from the reported data, if the experimental method is well described in the publication.

A useful way to organize the literature is to classify it by the types of the measured quantities (see chapter 4). A first classification may distinguish among the following.

- Experimental thermodynamic data.
  Here all kinds of enthalpies appear, plus their variations with temperature and composition (calorimetric data) together with all the experimental data related to chemical-potential measurements, obtained by emf or vapor-pressure methods.
- Experimental phase-diagram data.
  Phase-diagram data can be obtained by various methods and require careful research. Thermal analysis, metallography, X-ray diffraction, solidification-path experiments, microprobe measurements, and use of diffusion couples are the most common methods.
- Other experimental data, which have a quantitative relation to thermodynamic functions: bulk moduli, thermal expansions, elastic constants, etc. All these quantities are related to derivatives of the Gibbs energy.
- Crystal-structure data, point defects, densities (vacancies), ordering, resistivity, vibrations, etc. These data are very important in selecting details of the models to be used for each phase.
- Theoretical papers for calculations of total energies, at zero and/or at finite temperature, estimates, trends, thermodynamic properties, and phase diagrams of similar systems. See section 6.2.1.3.
- Review papers, critical assessments, and previous optimizations.
  These papers may contain an optimization that already satisfies all the recommendations for a good optimization. This means that the work has already been done for the system in question. If an optimization is not satisfactory, the arguments given in

that paper may nevertheless be helpful; and they can be a very useful starting point for an improved description.

- Miscellaneous
  Here one can include information that is indirectly connected to Gibbs energies, such as information on kinetics, microstructure, and solidification. It can be very useful for checking the use of the assessment in some application.

6.1.2 Analysis of the experimental data

After reading all the literature, one should have a good overview of the system. Already at this stage some contradictions between different sets of measurements may be detected. The most obvious contradictions are recognized when different values are reported for the same quantity, for example contradicting points of a solvus line of a binary system. In this case at least one of the datasets contains systematic errors and this set must be excluded from a least-squares fitting unless one can evaluate the error. It is impossible to give a general rule on how to select the best one of several contradictory sets of measurements. The descriptions of the experimental procedures must be read carefully to ascertain where it is more likely that errors could have occurred, like reaction with the crucible material or with the gas phase, evaporation of a volatile component, or not reaching equilibrium. Anyway, the arguments deployed to categorize sets of data as more or less reliable must be clearly specified in the publication of the optimization. It is possible that later measurements will throw new light on this problem and suggest a new decision regarding which set of conflicting data should be preferred.

Contradictions between quantities obtained from different experimental methods might be undetectable before running an optimization, but found after the first trials. Thus the evaluation of the reliability of the experimental data must be repeated.

For values that are unique to a system, such as temperatures of invariant equilibria and standard enthalpies of formation of stoichiometric phases, it is recommended that one select a "best" value for each of these measured or estimated quantities before the optimization. If several values of the same quantity are used, the least-squares method can do nothing else but find a mean of the reported values. The selection of the "best" value can, however, be changed if more information is obtained during the optimization procedure.

Values depending continuously on a state variable, such as temperature or composition, should be used for the optimization without prior smoothing or replacement by mean values. Every smoothing procedure has associated with it the danger of introducing a subjective preference, which might not agree well with the "reality."

Measurements excluded from the optimization due to contradictions should not be totally erased from the computer, but should be plotted together with the accepted measurements and compared with the result of the optimization.

Ternary phase-diagram data, especially indications of the temperature at which the second solid phase appears on cooling, must be interpreted with great care since the composition of the liquid phase will change while it is precipitating the first solid phase. One may use a Scheil simulation, see Fig. 8.7(b), to check such experimental data.

6.1.3 *Selection of the set of phases to be considered*

An important part of the critical assessment is the identification of the phases present in a system. The **liquid phase** should always be included, since it is usually stable across the whole system. The **terminal phases** are in general not a problem because they are usually the pure elements themselves, with homogeneity ranges varying from so small that they can be ignored to extending across the whole system. The identification of the **intermediate phases**, however, is not always straightforward. By intermediate phase one should understand any phase that does not appear as stable for any of the pure elements, for example intermetallics, oxides, carbides, etc. The early investigations might not have found all phases or they may have attributed a different composition to a phase. A phase may be metastable, despite its having been found by all investigators. A well-known example of a metastable phase is cementite, Fe_3C, in the Fe–C system. A metastable phase such as cementite must be included in the set of phases considered in the optimization of the Fe–C system.

The existence of a phase is most convincingly confirmed if its crystal structure is clarified. The most convincing structure determinations are those based on single-crystal X-ray diffraction. In cases in which the X-ray scattering of different atoms is very similar, neutron diffraction may give important additional information. If various authors have assigned different unit cells to the same phase, this is not necessarily a contradiction, since there may exist a transformation between the two descriptions, although one of them may have lower symmetry. If just a unit cell is reported, this is less convincing because it may be due to a misinterpretation of the X-ray pattern of a two-phase sample. A good fit of X-ray intensities with a proposed crystal structure, however, is nearly impossible to create artificially from a not-single-phase sample, even though the crystal structure itself may be refined later. Of course, it must be checked whether the crystal structure is really proved by X-ray-intensity measurements. Some reported crystal structures are just concluded from the similarity of the X-ray pattern to that of a known structure.

The ratio of the numbers of positions of the different atoms belonging to the crystal structure of a stoichiometric phase can be taken as the best definition of its "ideal" composition. Often, however, more than one component can enter a sublattice or some positions may be vacant, then the actual composition is able to deviate significantly from the "ideal" composition.

Phases formed at lower temperatures are often detected only after longer annealing times and may be missed in most investigations. On the other hand, phases found only after rapid quenching are more likely to be metastable. Often that is proved definitively by transforming them into stable phases by annealing later. Taking all that into account, in most cases a consistent interpretation is possible.

After the first consideration on compatibility of the data, the optimization steps follow: the selection of the model for each phase and the decision regarding how many and which coefficients can be adjusted independently for each of the selected models.

At this level of the assessment it is very useful to produce a table indicating which kinds of experimental data are known for each phase. Table 6.1 presents a hypothetical set

Table 6.1 *A summary of experimental data on a hypothetical system*

Phase	Strukturbericht	Phase equilibria	μ, a	H	$H(T) - H(298\,\mathrm{K})$, C_p
Liquid		2-phase, 3-phase	$\mu_A(T)$, $\mu_B(T)$	$H(x)$	
$Phase_2$	A1	2-phase	$a(x)$		
$Phase_3$	$L1_2$	2-phase	$\mu_A(T)$, $\mu_B(T)$	$H(x)$	$H(T)$, $\partial H/\partial T$
$Phase_4$	C14	2-phase			$H_{Fusion}(T_0)$

of data organized by type. With this information, one can quickly answer some questions that are important for the model selection; see section 6.2.9. After the answers have been obtained, some simple decisions about the model can be made.

6.2 Modeling the Gibbs energy for each phase

As has already been mentioned, each phase in a system can be modeled independently, but, when two or more phases have the same or related crystal structure, they may be modeled as the same phase. Typical examples are when the terminal phases have the same crystal structure, or when an intermediate phase is an ordered form of a terminal phase.

6.2.1 *General considerations*

The appropriate description for the Gibbs energy of each phase should be selected from the models described in chapter 5. All the descriptions there are supported by physical models. Some of the polynomial excess terms, however, are only curve-fitting formulae to be used for description of small deviations from the model. This is necessary, because no model is able to take into account all possible physical effects. A model is well chosen if it has a physical background and experimental values fit the corresponding calculated quantities well using only a few model parameters.

All models contain simplifications, so it may be useful to add some terms with curve-fitting behavior summarizing the contributions of minor physical effects, which in detail might not be well known quantitatively. The contribution of this additional curve-fitting formula should be small compared with the contribution of the model formula itself. This problem had been addressed by Didier de Fontaine *et al.* (1995): "*Usually, it is impossible to treat the problem in a rigorous theoretical manner: even if magnetic effects would be described by an Ising model, a detailed mathematical treatment would be out of question as it is known that (a) the three-dimensional Ising model cannot be treated exactly, even in principle, and (b) finding accurate approximate solutions is one of the hardest problems in theoretical physics. We are forced, therefore, to adopt a purely phenomenological approach consisting of optimal fits of experimental data to an empirical analytical expression. The choice of this mathematical expression should not be completely arbitrary, and indeed its form should reflect as much as possible the essential physics of the problem. If that can be achieved, (a) at least some of the fitting parameters will have physical meaning and these derived values can be checked against known experimental information, (b) values of properties of metastable phases can be predicted with some chance of success."*

Some kinds of experimental data can give good hints about the selection of the model, in particular data on enthalpies and chemical potentials (activities) in the single-phase region.

After selecting the model for each phase, the parameters of the model should be analyzed. General models contain many parameters, and the ones to be used in the specific modeling of a given phase should be decided considering their physical meaning. Some models require parameters that cannot be provided by any experimental value; however, their use is necessary because they are intrinsically related to the model. Some parameters can be combined by constraints to give a single parameter or fixed arbitrarily to values that have no significant influence on G in the stability range of existence of the phase. Some parameters with curve-fitting behavior may be tolerated, but should be restricted to small deviations from a description based on physically meaningful parameters only.

Extrapolations provided by physically realistic models usually result in better descriptions of regions not covered by experimental data than do extrapolations done using purely curve-fitting formulae. Extrapolations deviate from reality only moderately for models based on sound theoretical physics. The more complicated a description, the more unpredictable the extrapolation if the model is misused as a purely curve-fitting formula.

The modeling of the phases may have to be compatible with existing multicomponent databases. That reduces the freedom in the selection of models and parameters. If the arguments in favor of selecting a better model for the phase are strong enough to motivate a change in the model used in the existing database, the recommended strategy is to make two assessments, using the different models. This will simplify a future change of the database.

For each phase one identifies the prototype crystal structure and other known structural characteristics. It is also important to find phases with the same prototype in other systems, in particular for intermediate phases. That will be discussed in more detail in section 6.2.5. The possibility of the extension of a binary to a ternary or higher-order system should always be taken into account.

6.2.1.1 Solubility and composition range

The liquid is often stable across the whole composition range. Even if there is a miscibility gap, the same model should be used on both sides of the miscibility gap for the liquid phase.

A terminal phase may also extend across the whole composition range if the pure elements have the same crystal-structure type. If the terminal phases have the same structure, they must be modeled as the same phase even if there is a miscibility gap and even if there are stable intermediate phases in between the terminal phases. See section 6.4.4.

All phases have some range of solubility, but, if it is small or no experimental data are available, one may approximate any phase by a stoichiometric compound. However, for example, in a semiconductor phase it may be important to model a composition range of mole fraction 10^{-6}, whereas for another type of phase even a solubility of mole fraction 10^{-3} may be insignificant. A stoichiometric compound is much simpler to model than a phase with a small composition range because in the former case the Gibbs energy is just a function of temperature and pressure. Fortunately, the phase boundaries of the other phases in equilibrium with the compound depend very little on the model selected for the compound, if the model selection is a choice between a stoichiometric phase and a phase with very small solubility. Thus this modeling can be changed later without significant influence on parameter values of neighboring phases.

In ternary systems a binary stoichiometric phase may dissolve considerable amounts of the third component. Then modeling as a line compound is the adequate procedure. Also terminal phases may be treated as stoichiometric phases, considering the same criteria. For example, graphite in many systems has no measurable solubility range and is consequently treated as stoichiometric.

For intermediate phases with small solubility ranges, constituents occupying the same sublattice are often distinguished as normal (major) constituents or as defects. There are several models to handle this case, which will be described in section 6.2.5.

Phases with wide solubility ranges may have ordering inside their composition ranges. This can be inferred from the structure type and determined by measuring the chemical potential or activity as a function of composition. For example, many intermediate phases have the B2 structure type and it is very likely that the properties of such a phase will change rather drastically at the ideal composition even if the phase diagram does not show any evidence of that. It must be kept in mind that a phase diagram shows only relations between phases; little can be deduced of the properties of the phases themselves.

In Fig. 6.2 two examples of phase diagrams and their Gibbs-energy curves at a fixed temperature are shown. In order to become a good assessor, it is important to develop a feeling for the relations between the Gibbs-energy curves and the phase diagram. Note that, even if a phase is stable only within a narrow composition range, the model often defines the Gibbs energy over a much wider range. In Figs. 6.2(c) and (d) these Gibbs energies are drawn for the whole range of compositions for the model of the phase. The assessment of the Fe–Mo system is from Fernández Guillermet (1982) and that of the Cu–Zn system from Kowalski and Spencer (1993).

6.2.1.2 Thermodynamic data

The thermodynamic information for the various phases will be reviewed here mainly in the context of the type of phase. A few general considerations can be made. If the enthalpy of mixing versus composition curve deviates from a parabolic shape, for example by having a sharp "V"-shape for a given composition, then strong LRO or SRO in the mixture should be expected.

If the enthalpy of formation is measured at several temperatures and no significant temperature dependence is observed, no excess heat capacity should be modeled, i.e.,

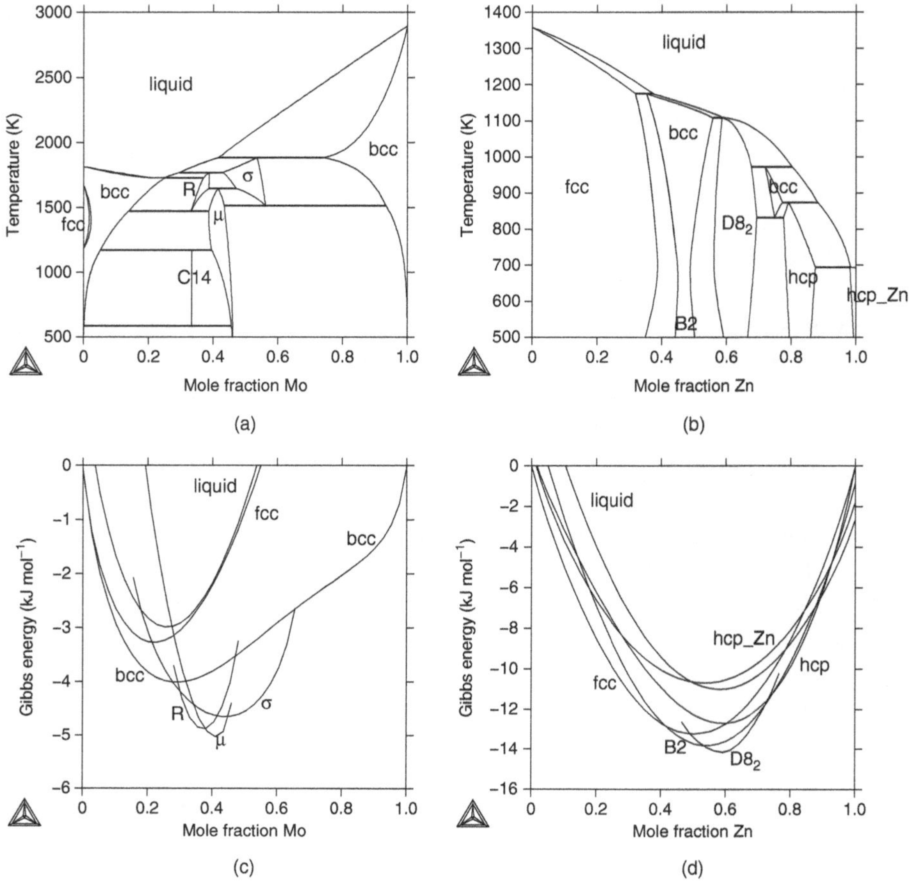

Figure 6.2 Examples of phase diagrams, (a) Fe–Mo and (b) Cu–Zn, and the Gibbs-energy curves of their stable phases: (c) Fe–Mo at 1673 K and (d) Cu–Zn at 673 K. Note that the second-order transition line between bcc and B2 in (c) is not shown.

the Kopp–Neumann rule should be used (section 5.2.3). Note that defects or associates contribute to the heat capacity because their fractions depend on temperature. In a few cases there exist enough heat-capacity data to enable evaluation of the heat capacity independently from those of the pure components for a solution phase.

A drastic change of the chemical potentials and the partial enthalpies over a small composition range indicates an ordering. Ordering can depend on various types of species and the most important type of species should be identified; it can be vacancies, anti-site atoms, interstitials, or some combination of all the foregoing. If the phase may disorder completely, like a B2 phase to A2 or an $L1_2$ phase to A1 structure type, one should use a model that describes both the ordered and the disordered state with the same description; see section 5.8.4.

In Fig. 6.3 four cases of integral and partial molar enthalpies versus mole fraction curves are shown together with the second derivatives of the excess Gibbs energy. The first three cases are related to the Redlich–Kister (RK) series described in section 5.6.2.1.

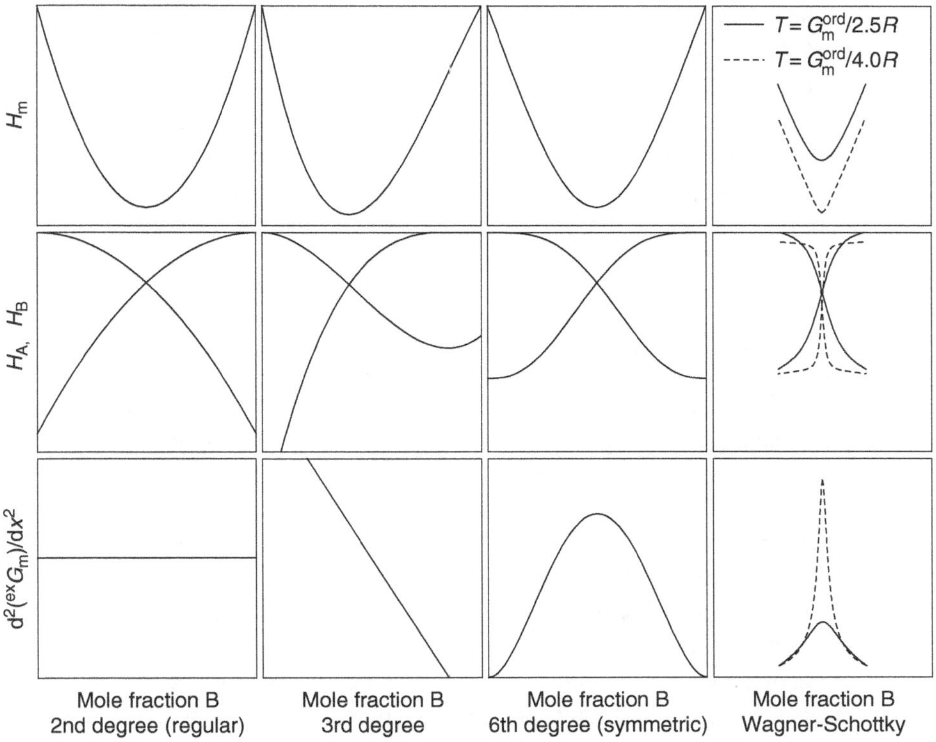

Figure 6.3 Enthalpies of mixing and d^2G_m/dx^2 calculated for a series of models.

For these figures the excess entropy is set to zero and thus the enthalpy of mixing, excess enthalpy, and excess Gibbs energy are identical. In the first case the enthalpy is described by a single RK coefficient; the second derivative is constant. In the second case two RK coefficients are used and the second derivative of the excess Gibbs energy changes linearly with composition. In the third case the first, third, and fifth RK coefficients are used (sixth degree, symmetrical). The fourth case shows typical curves of an ordered phase. For comparison with the RK cases also here d^2G_m/dx^2 of an ideal solution is subtracted from the calculated d^2G_m/dx^2 in order to get the second derivative of the "excess Gibbs energy." In this case the enthalpy and excess Gibbs energy are different. The quantities H_m, H_A, and H_B in this figure were chosen because they can be measured experimentally; d^2G_m/dx^2 is the "phase stability" introduced by Eq. (2.17).

The curves of integral and partial enthalpies related to the RK description in Fig. 6.3 are modeled as temperature-independent. With linearly temperature-dependent RK parameters they can also be modeled as being temperature-dependent, but the magnitude and shape change only moderately with temperature. For the ordered phase, however, magnitudes and shapes of all three curves in Fig. 6.3 strongly depend on temperature. This is shown by calculating them for two different temperatures, related to the ordering Gibbs energy of the phase. This drastic temperature dependence especially of d^2G_m/dx^2 cannot be well reproduced by a temperature-dependent RK description.

The conclusion from this figure is thus that, for phases for which the H_m versus x curve at lower temperatures exhibits a sharp kink leading to step-like curves of the partial enthalpy versus mole fraction plots, the RK formalism is not a good choice; instead, a model like the Wagner–Schottky model is needed. The associate-solution model and the partially ionic-liquid model behave similarly and may be alternatively chosen, if they are better suited for the phase in question.

6.2.1.3 Estimation methods

If insufficient data are available experimentally, some estimation methods may be useful. Theoretical approaches and semi-empirical estimates are important in selecting details of the models to be used for each phase and for reducing the number of adjustable model parameters. First-principles methods, total energies, and quantum and non-quantum-mechanical Monte Carlo and molecular-dynamics atomistic calculations can be used to compute thermodynamic properties useful for estimating model parameters.

Also trends of thermodynamic properties in series of related systems can successfully be used for estimates, for example along a sequence in the periodic table. Several such methods for estimating enthalpies of formation are reviewed by P. J. Spencer in a special issue of *Thermochimica Acta* dedicated to computational thermodynamics (Spencer 1998). Another useful source of data is the book by Bakker (1998).

An important estimate is given by the Kopp–Neumann rule for heat capacities (C_p) of compounds (Kopp 1864, Neumann 1865, Grimvall 1999). If there is not enough information to evaluate both an enthalpy and an entropy value, it is usually better to set the entropy value proportional to the enthalpy one, like $a_1 = a_0/T_0$, where T_0 has been estimated up to 3000 K for excess parameters by Lupis (1967) and Kubaschewski *et al.* (1967). This estimation was proposed to depend on the melting temperatures of the pure components and refined by Tanaka *et al.* (1996). Estimates for atomic volumes, the standard entropy, the Lindemann melting rule, and relations between C_p and thermal expansion can be found in the chapter "Estimations and correlations" in Grimvall (1991).

Typical estimates using the position in the periodic table are those of Bakker (1998) and Miedema *et al.* (de Boer *et al.* 1988) for the enthalpy of mixing in the liquid state and for enthalpies of formation of compounds.

First-principles total-energy calculations, using DFT methods (see chapter 3), even if done at 0 K, can also provide values for the formation enthalpies of real compounds as well as for fictitious end members in the CEF. Formation enthalpies of several compounds in different structures are available in databases compiled by Sluiter on the website (http://www.www-lab.imr.edu/~marcel/enthalpy/enthalp.htm) and by Colinet (2003). The alloy database prepared by the Widom group is available in a website (http://alloy.phys.cmu.edu/) and provides also auxiliary values. It is important to keep in mind that the values provided by first-principles calculations must be assessed critically because the values depend on the methods and approximations used and they scatter similarly to how experimental values do.

Systematic behavior is very useful information in modeling when no experimental data are known. A comparison with similar but well-known systems can be a good guide in certain cases (Shao 2001).

6.2.2 *The gas phase*

The gas phase is normally assessed by techniques different from those described in this book. Thermodynamic descriptions of gaseous molecules can often be obtained by *ab initio* calculations and from spectroscopic data. Several databases with multicomponent gas data are available, for example from the SGTE (*Landolt-Börnstein* 1999).

6.2.3 *Modeling the liquid phase*

The liquid phase usually exists across the whole composition range, but it may change its type of bonding with composition and temperature. This can be observed in measured enthalpies of mixing, for example. The type of bonding or the shape and magnitude of the enthalpy of mixing versus mole fraction curve, ${}^{\mathrm{mix}}H^{\mathrm{liq}}(x)$, should be considered in the modeling. For mainly metallic or van der Waals bonding, which is direction-independent, the regular solution is usually a good model. A variety of composition-dependent effects, which do not need to be identified individually, may influence the energies of the bonds and make them composition-dependent. To fit this behavior, usually a polynomial (RK) series is used as the curve-fitting formula.

In some phase diagrams one may find very deep eutectics between two intermediate phases, for example for the Mg–Zn system. It is unusual that the shape of the liquidus curve differs significantly on the two sides of a congruently melting compound. If such behavior is experimentally well determined to occur, one has to consider a model in which $\partial^2 G/\partial x^2$ of the liquid changes rapidly between the two sides, as can be deduced from the Gibbs–Konovalov rule, Eq. (2.50). Appropriate models are, for example, the associate solution (5.7.1) and the partially ionic-liquid model (5.9.4).

6.2.3.1 **Miscibility gaps**

Liquid oxides are often treated with a model different from that used for the liquid metal because many metal–oxygen systems have wide miscibility gaps, but it is always possible to find a continuous path from one side of the miscibility gap to the other in a binary system by adding a third component. For example, the Cu–S system has a miscibility gap between a Cu-rich liquid and a liquid with the average composition Cu_2S. In the ternary Cu–Fe–S system, however, the miscibility gap closes and thus a single model must be used to describe the liquid phase already in the binary Cu–S system.

In some binary phase diagrams the liquidus is almost horizontal for a significant composition range, for example for the Cu–Fe system. That is an indication of a metastable liquid miscibility gap just below the liquidus.

6.2.3.2 Short-range ordering in liquids

A "V"-shaped enthalpy of mixing versus mole fraction curve is experimental evidence for SRO and the apex of the "V" indicates the stoichiometry of an "entity" (sometimes called a molecule or associate, but actually an atomic configuration that appears statistically more frequently than others) or the ratio of the charges of the ions (compare this with Fig. 6.3).

In the phase diagram the "V"-shaped enthalpy of mixing curve is often matched by a congruently melting phase at the same composition. For example, see the Mg–Sn and Pb–Te phase diagrams in Figs. 6.4(a) and (b), respectively.

If the ${}^{\mathrm{mix}}H^{\mathrm{liq}}(x)$ curve is less pronouncedly "V"-shaped, several associates or several different cations or anions may exist (e.g., Fe^{2+}, Fe^{3+}, SiO_4^{4-}, $Al(OH)_4^-$, and $Al(OH)_6^{3-}$). However, if these cannot be well characterized, the RK series may still be sufficient as a curve-fitting formula.

If the bonding is less pronouncedly covalent or ionic, the "V"-shaped enthalpy may be an indication of SRO, which can be represented by the quasi-chemical description (section 5.7.2.1) which has a non-random entropy of mixing. Alternatively, the deviation of the entropy from that of random mixing may be represented by excess-entropy terms in the RK formalism. The latter method, however, is a curve-fitting method and does not describe the excess heat capacity connected with the decrease in enthalpy of formation due to the decrease of short-range ordering on heating. Usually very little is known about excess heat capacities for liquids, but some may be deduced from enthalpy-of-mixing measurements performed at various temperatures.

6.2.3.3 Freezing-point depression

Another useful item of information for modeling that can be extracted from the phase diagram is related to the "freezing-point depression."

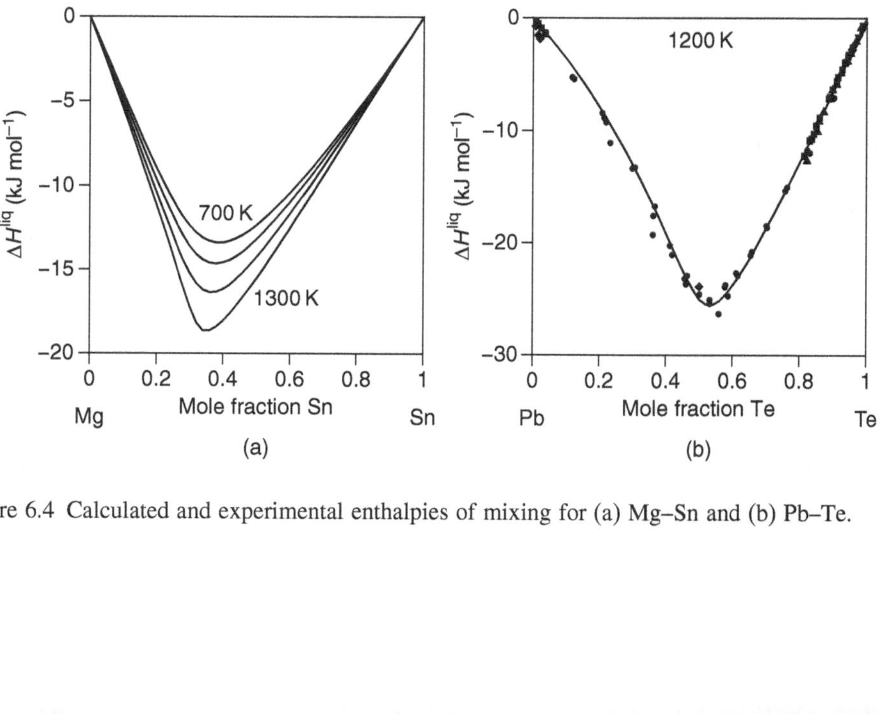

Figure 6.4 Calculated and experimental enthalpies of mixing for (a) Mg–Sn and (b) Pb–Te.

The width of the two-phase region close to the pure element is determined by the "van 't Hoff law," i.e., the difference in composition is related to the enthalpy of melting of the pure element. The depression of the melting temperature T_f of pure A on adding B when there is no solubility of B in pure A is

$$x_B^L = \frac{\Delta H^{\alpha_A/L_A}}{RT} \frac{T_f - T}{T_f} \tag{6.1}$$

This can easily be derived from the equilibrium conditions, see for example Hillert *et al.* (1998), p. 270. Equation (6.1) indicates that the freezing-point depression per mole fraction of addition is independent of the alloying element. If experimental information for the liquidus reveals a strong deviation from the slope given by Eq. (6.1), it is an indication that one mole of added element corresponds to a different number of moles of dissolved species (either one atom creates an additional species, e.g., an ion, or several dissolved atoms aggregate into a single molecule or associate).

6.2.4 Modeling terminal phases

By terminal phases we mean phases that exist with a pure component of the system. The solubility ranges of these phases may vary from very limited to extending across the whole composition range. In most cases it is useful to describe the Gibbs energy of the terminal phase for the whole composition range even if the real phase has limited solubility. If the two elements in their stable states have different crystal structures, then the Gibbs energy of a metastable state (lattice stability) of one of the elements must be related to that of its stable state. These relationships form a unary database. A first version of it is included in Dinsdale (1991); there is an update in *Landolt-Börnstein* (2002). Usually the unary dataset is provided without charge in the web pages of agencies producing databases.

Most of the pure elements have an fcc, bcc, or hcp lattice. Gibbs energies for these structures are collected in *Landolt-Börnstein* (2002) for almost all elements. For phases like graphite and diamond the solubility of other elements is usually very small and often can be neglected.

6.2.4.1 Substitutional solutions

Elements that dissolve substitutionally in a terminal phase can often be described with the RK formalism. If the two elements have the same crystal structure, complete solubility across the whole composition range may exist (e.g., Cr–V and Ag–Au). In other systems differences in atomic sizes (Ag–Cu) or in chemical properties create miscibility gaps or intermediate phases are formed and interrupt the solubility range, as in Cr–Mo and Fe–Mo. If there are many intermediate phases, one may check the modeling of the terminal phases during the assessment by calculating the metastable phase diagram with just the terminal phases (as discussed later in section 6.2.4.4).

In other systems the components may form ordered phases based on the same lattice as the terminal solution (see section 6.2.4.3).

6.2.4.2 Interstitial solutions

The typical case of an interstitial solution is carbon in steels. Iron can be ferritic, i.e., has a bcc lattice, or austenitic, i.e., has an fcc lattice. When carbon is dissolved in steel, it will not substitute iron on the lattice sites but occupies interstitial sites between the Fe atoms. In bcc there are octahedral interstitial sites in the middle of the edges and additionally at the centers of the planes of the unit cell and there are three times as many interstitial sites as matrix atom sites in the bcc lattice. Thus the sublattice model $(Fe, Cr, \ldots)_1$ $(Va, C, N)_3$, where Va denotes vacancies, is used.

In fcc there are two kinds of interstitial sites, tetrahedral ones surrounded by four matrix sites and octahedral ones surrounded by six matrix sites. In a unit cell of fcc containing four matrix sites, there are eight tetrahedral and four octahedral sites. Carbon and nitrogen dissolve in the octahedral sites only, so the model used for austenite is thus $(Fe, Cr, \ldots)_1(Va, C, N)_1$.

The hcp lattice is also very common for metallic phases. There is also one octahedral interstitial site per matrix atom. Nevertheless, an old modeling with only half the interstitial sites is still used. There is no significant difference when it is applied to dilute interstitial solutions, but it performs worse for the cases in which more-concentrated interstitial solutions are stable (O in Ti, Zr, and Hf). This old description was deduced from carbides like W_2C and Mo_2C that were assumed to be strongly ordered. However, for these carbides the crystal structure is described as 50% of all octahedral sites statistically occupied by C (Schubert (1964), p. 267).

There may exist intermediate phases that seem to correspond to "full" occupation of the interstitial sublattice; however, they may be ordered phases, wherein the symmetry is diminished and a single Wyckoff position of empty sites of the pure solvent phase splits into different sites, only one of which is occupied. In the dilute solution the symmetry cannot yet be diminished and the number of empty sites is the full number of sites in the Wyckoff position of the pure solvent structure. For dilute solutions, therefore, this has to be taken as the number of sites of the interstitial sublattice.

As an example the hexagonal-close-packed (hcp, A3, Mg-type) structure shall be considered. Its crystal structure can be described as

space group: $P6_3/mmc$

2 metal atoms (Me) in 2(c) $(\frac{1}{3}\frac{2}{3}\frac{1}{4})$ $(\frac{2}{3}\frac{1}{3}\frac{3}{4})$
2 octahedral voids (Va,X) in 2(a) (000) $(00\frac{1}{2})$

By reducing the symmetry, the two-fold site of the octahedral voids can split into two independent one-fold sites:

space group: $P\bar{3}m1$

2 metal atoms (Me) in 2(d) $(\frac{1}{3}\frac{2}{3}x)$ $(\frac{2}{3}\frac{1}{3}\bar{x})$
1 octahedral void (Va,X) in 1(a) (000)
1 octahedral void (Va,X) in 1(b) $(00\frac{1}{2})$

In the Cr_2C structure the 2(d) position is occupied by Cr and the 1(a) position by C; the 1(b) remains empty. From the Cr_2C structure it was deduced that in hcp only half of the octahedral voids can be occupied by interstitials and a ratio of 2 : 1 was set for the sites of the matrix and interstitial sublattice, respectively. In dilute solutions, however, that cannot be true, since the diminution of symmetry on going from space group $P6_3/mmc$ to $P\bar{3}m1$ can take place only if enough interstitials are present to interact and expand the lattice at the 1(a) positions, i.e., if chemical ordering can occur.

Fortunately, this difference is mainly theoretical and the two selections of the sublattice site ratio in dilute solutions give virtually the same Gibbs energy versus mole fraction ($G(x)$) curve, if the parameters $G^{\mathrm{hcp}}_{\mathrm{Me:X}}$ differ just by a term $RT\ \ln(2)$ in the two descriptions.

Similarly, in the bcc (A2, W-type) structure there are three octahedral voids per matrix atom. After tetragonal distortion, as happens in martensite, one third of these sites is enlarged, the other two thirds are shrunk. For dilute solutions (ferrite) the ratio 1 : 3 must be taken for matrix to interstitial sites. At higher solute concentrations, ordering may occur (martensite), splitting the interstitial sublattice into two, with one and two sites per matrix site, respectively. There are also tetrahedral interstitial sites in bcc and often it can be difficult to know which sites are occupied.

The CEF needs a parameter $G_{\mathrm{Me:Va}}$ for the pure solvent and a parameter $G_{\mathrm{Me:X}}$ for the very fictitious case of fully occupied interstitial sites. Interaction parameters ${}^{\nu}L_{\mathrm{Me:Va,X}}$ may be used to adjust the composition dependence of the Gibbs-energy description. Where the homogeneity range is restricted to dilute solutions, the sum $G_{\mathrm{Me:Va}} + \sum_{\nu} {}^{\nu}L_{\mathrm{Me:Va,X}}$ is the only independently adjustable parameter. If more than one of the terms of this sum are used, all except one can be chosen arbitrarily. Nevertheless, it may be necessary to use at least two parameters, since using $G_{\mathrm{Me:X}}$ alone may describe the Gibbs energy well in the range of homogeneity of the phase, but additionally describe another range of stability, where the phase is not stable in reality. In that case for $G_{\mathrm{Me:X}}$ an arbitrary and large enough positive value or linear function of T is chosen and ${}^{0}L_{\mathrm{Me:Va,X}}$ is used to adjust the Gibbs energy in the range of homogeneity.

6.2.4.3 Ordering phenomena

Adding elements to a terminal phase can change its crystal structure. For example, the site occupation of Al in an Fe bcc lattice can change from random A2 to ordered B2 with addition of Al. This ordering is of "second order" since it appears gradually and there is no two-phase region between the A2 and B2 phases, as shown in Fig. 6.5. In the same diagram the ferromagnetic transition is shown; this is also of second order. The low-temperature $D0_3$ phase was not modeled in the dataset used to calculate Fig. 6.5.

For other types of ordering like the γ' phase in the Ni–Al system, where the occupancy of the fcc lattice changes the structure from the disordered A1 to the ordered $L1_2$ type, the ordered phase appears as a separate intermediate phase in the phase diagram. However, as described in section 5.8.4, both the ordered and the disordered phase can be described by the same Gibbs-energy function.

There are many "families" of structures that are connected by reduction of symmetry and it is important to have that in mind when selecting models.

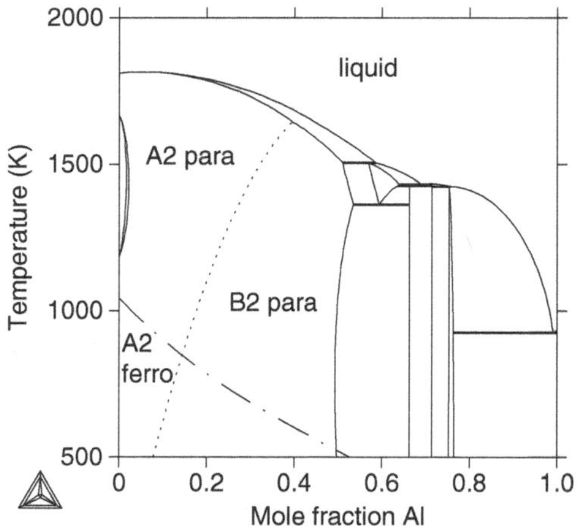

Figure 6.5 The Al–Fe phase diagram with the second-order transition between disordered A2 and ordered B2 marked by a dotted line. The second-order transition between paramagnetism and ferromagnetism is marked by a dot-dashed line. Where the two curves meet, there may be a change to a first-order transition, but that is not included in the dataset used for this diagram.

6.2.4.4 Metastable extrapolations of terminal phases

When the terminal phases do not extend across the system, either due to a miscibility gap or due to intermediate phases, it is still interesting to extrapolate the metastable two-phase field between the liquid and the terminal phases using the assessed parameters. There should not be any strange curvatures or both maxima and minima of this two-phase field. If such behavior shows up in the calculation, it is probably due to there being too many coefficients in the RK series for one or both phases. In Fig. 6.6 the metastable extrapolation of the terminal phases is shown as dashed curves for the Al–Cr and Fe–Mo phase diagrams.

These checks by use of metastable diagrams were proposed and done by J. J. van Laar as early as 1908 (van Laar 1908a, 1908b). His very pioneering works were reviewed recently by van Emmerik (2005).

These checks were used as standard procedure in the assessments performed by Larry Kaufman (2002) and Himo Ansara (2001) and are nowadays largely used by constructors of multicomponent databases. Some metastable extrapolations can be checked by experiments such as those done by Perepezko *et al.* (Perepezko and Wilde 1998). These results are especially useful for application in microstructure simulations such as of phase fields, since metastable regions are very often scanned, and the presence of model artifacts can be very detrimental to the calculations.

6.2.4.5 Terminal phases in quasibinary systems

A quasibinary system is a system with three components that behaves exactly like a binary system of two components. A typical case is the mixing of two oxides like MgO and CaO. The solid phases in this system have the same B1 structure type and the liquid

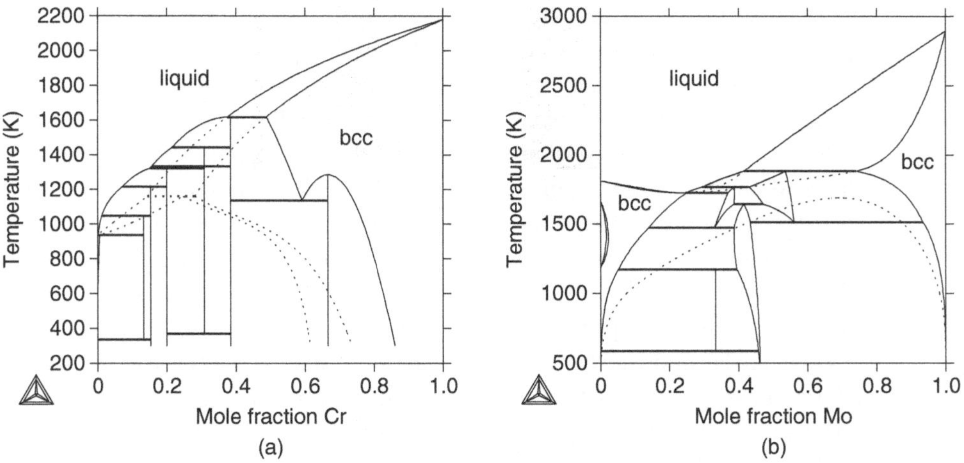

Figure 6.6 Metastable extrapolations of terminal phases in two systems: (a) Al–Cr and (b) Fe–Mo. For the Fe–Mo system the bcc phase should form a continuous solid solution at high temperature if there were no intermediate phases and there is a metastable miscibility gap at low temperatures. The latter is indicated by the shape of the liquid/bcc two-phase region.

phase is completely miscible between the oxides, but neither phase has any appreciable range of solubility in terms of the amount of oxygen. This makes it possible to treat the MgO–CaO system as a binary one and ignore the phases outside the composition range defined by the two oxides. For the thermodynamic definition of components, the condition $N_O = N_{Mg} + N_{Ca}$ reduces the number of independent components to two.

The essential concept for quasibinary systems is that the endpoints of the tie-lines between the phases must be inside the section defined by the compounds selected as quasibinary components. One of the phases of a quasibinary system may exist outside this section, but, if there exists another one also outside, tie-lines between these two phases usually are not exactly in the section. Then the system may be only approximately a quasibinary one.

There are many systems that are treated as quasibinaries although they do not have the tie-lines exactly in the plane. This is, for example, the case when an element may have several valencies, like Fe. The mixture of the two oxides CaO and FeO is thus not a quasibinary system since there are always some Fe^{3+} ions present. The CaO–FeO phase diagram is an isopleth but can be called a "pseudobinary" system since the endpoints of the tie-lines will be slightly outside the composition plane defined by the compounds.

Since it is not uncommon to assess quasibinary systems and quasiternary systems, some features of these are discussed below. It is possible that the liquid, or an intermediate phase, may be stable outside the quasibinary section, but there may be stoichiometry restrictions such that, in any two-phase equilibrium, only one of the phases can exist outside the quasibinary section. Such a constraint ensures that there is always one tie-line between the two phases inside the quasibinary section.

In a calculation using "double-precision" numbers (about 14 significant digits) a deviation of the mole fraction by 2×10^{-14} from the section is treated as falling outside the section. This may give confusing results. To avoid this problem, the quasibinary system may be calculated not by giving conditions between mole fractions to select the line of section in the ternary system, but either by defining the quasibinary components as the two components of a binary system or by selecting another condition in the ternary system. In many oxide systems defined by three elements, it is sufficient to fix the chemical potential of O to a moderately negative value. So long as it is not too negative, no reduction of the oxides will take place and at higher temperatures the calculation will remain in the quasibinary system until the gas phase appears. In a few cases this method might be not applicable, but, where it can be used, also systems that are not strictly quasibinary, but only approximately so, can be calculated.

One may be surprised by how some thermodynamic properties behave in a quasibinary solution phase. In the Al_2O_3–Y_2O_3 system the corundum phase is stable throughout the whole range between the quasibinary components. The physically reasonable modeling for this solid solution is $(Al^{3+},Y^{3+})_2(O^{2-})_3$. If Al_2O_3 and Y_2O_3 are defined as components, μ_O is kept constant (somewhere between $-10\,000$ and $-100\,000\,\mathrm{J\,mol^{-1}}$), and the activity of Y_2O_3 in the corundum phase is calculated as a function of the mole fraction of Y_2O_3, one obtains the curve in Fig. 6.7 even if there are no interaction parameters in the model. For an ideal substitutional model one would expect the activity to be equal to the mole fraction given by the diagonal, shown as a dashed line, in this figure.

The reason for this is that the activity of Y_2O_3 in phase Φ is defined as

$$a^{\Phi}_{Y_2O_3} = \exp[(\mu^{\Phi}_{Y_2O_3} - {}^{0}\mu^{\mathrm{ref}}_{Y_2O_3})/(RT)] \tag{6.2}$$

The μ-value of a species is the sum of the μ-values of the atoms present in the species. Consequently the activity of the species is the product of the activities of the atoms contained in the species. Thus the better choice for the quasibinary components

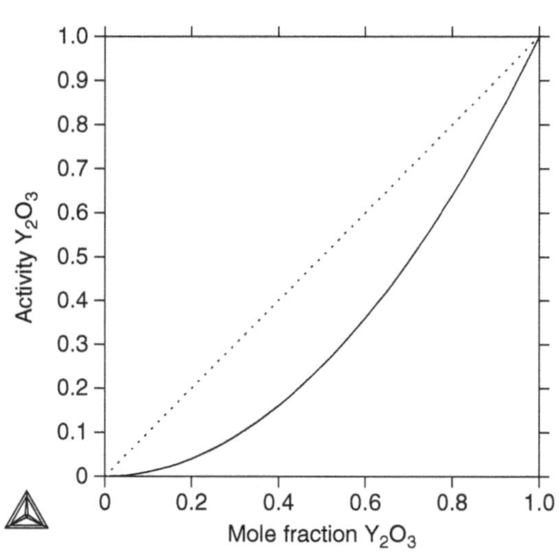

Figure 6.7 The activity of Y_2O_3 as a function of the mole fraction of Y_2O_3 for the sublattice model $(Al,Y)_2O_3$ without any interaction parameters.

is $AlO_{1.5}$ and $YO_{1.5}$. The chemical potential of Y_2O_3 is twice that of $YO_{1.5}$ and thus the activity of Y_2O_3 is the square of the activity of $YO_{1.5}$. A similar problem would arise if the species Cu_2 and Ni_2 in the Cu–Ni system were defined as components and the activity of Cu_2 were plotted against the mole fraction of Cu_2 for the fcc (Cu, Ni) solid solution. Of course, in the Cu–Ni system there is no reason for such a definition of components, but it illustrates the problem.

6.2.5 *Modeling intermediate phases*

The term "intermediate" phases means all phases that do not extend to the pure components of a system. It includes phases with no or little solubility and phases having very wide ranges of composition. Some of these phases may have the same lattice as is found also for terminal phases; for example, an intermediate phase with the bcc lattice appears in the Al (fcc)–Cu (fcc) system. But often the intermediate phases have more complicated crystal structures. As mentioned several times before, the identification of the structure type of a phase is the first step in the modeling. Crystallography gives information about the different sublattices and, by comparing with other phases with the same structure type, one can obtain important information about the modeling. If a phase with the same structure type has been modeled already, that is very useful information. However, phases with the same structure can also have mixing on different sublattices, depending on the relative sizes of the different atoms.

In some systems many intermetallic phases may have to be considered in the modeling, for example σ, μ, and Laves phases. A good review of these can be found in the proceedings of the 1996 Ringberg workshop (Ansara *et al.* 1997a).

$$(\mathrm{Fe, Ni, \ldots})_{10}(\mathrm{Cr, Mo, \ldots})_4(\mathrm{Cr, Fe, Ni, Mo, \ldots})_{16} \tag{6.3}$$

Even with this simplified model there is a large number of end members that must be determined from the scattered and insufficient experimental data. As an approximation, it has been proposed that one should consider the coordination number, i.e., the number of bonds at each site, and treat those in the first sublattice as being part of an fcc phase because the atoms there have 12 bonds, those in the second as a bcc phase since they have 14 bonds, and those in the third also as a bcc phase since the number of bonds is about 14, although they are of slightly varying length. (In reality a bcc structure has just eight nearest neighbors, but the six next-nearest neighbors are at almost the same distance, making it 14 bonds.) In fcc the six next-nearest neighbors are much further away than the 12 nearest neighbors. Thus an end member like $Fe_{10}Mo_4Ni_{16}$ could be estimated as

$$^{\circ}G^{\sigma}_{\mathrm{Fe:Mo:Ni}} = 10\,^{\circ}G^{\mathrm{fcc}}_{\mathrm{Fe}} + 4\,^{\circ}G^{\mathrm{bcc}}_{\mathrm{Mo}} + 16\,^{\circ}G^{\mathrm{bcc}}_{\mathrm{Ni}} \tag{6.4}$$

One should also consider systems other than the particular one to be modeled, in order to take into account what will happen when the model is used for higher-order systems. The same element may dissolve in different sublattices of the same phase when alloyed with different elements. For example, the σ phase can dissolve more V than considered according to the model above in the Ni–V system. This can be handled by considering V to dissolve also in the first sublattice, although that increases the number of end members.

Some intermediate phases with the structure types $L1_2$, $L1_0$, and $D0_{22}$ can all disorder to the A1 structure type; the structure types B2, $D0_3$, $L2_1$, etc. can disorder to A2 and $D0_{19}$; B_{19} etc. disorder to A3. When an intermediate phase can transform to a disordered state, it is important to use the same model both for the ordered and for the disordered state of this phase. The modeling for such cases is described in more detail in section 5.8.4. For ordered phases derived from bcc, fcc, or hcp but with a complicated structure and which never transform directly into a disordered state, it is recommended that one should model these phases as different from the disordered phase. A typical example is the $D0_{22}$ structure.

Phases with order–disorder transitions are commonly described as a sum of two Gibbs-energy expressions, one describing the disordered state and depending only on the mole fractions of the constituents, the other depending on the site fractions describing the ordering. This partitioning is described in section 5.8.4.2 and, in a slightly modified way, it can be applied also to phases that never undergo disordering.

6.2.5.1 Crystal-structure information

The structure information is sometimes too complex to be modeled in detail. If a phase has four or five sublattices, it may be necessary to reduce this to two or three because there might not be enough experimental information to determine the necessary model parameters. The number of parameters for the "end members" of the phase is obtained by multiplying together the numbers of constituents on each sublattice. There are seldom enough data to fit more than two end members for a binary phase.

In most compounds two (or more) different atoms occupy crystallographically different positions and this may give an enthalpy curve with a very sharp minimum at the ideal composition. In other cases the intermediate phases may be considered as substitutional-solid solutions, which could have a wide composition range if not limited by other phases. The crystal structure is again the primary information for the modeling.

The crystal-structure data should also be checked for information on whether a range of homogeneity is created by anti-site atoms, vacancies, or interstitials.

Good crystal-structure determinations by X-ray diffraction can even identify some crystallographic positions to be occupied randomly by different atoms and can even give values of the site fractions of the different constituents on these positions by Fourier synthesis from the intensities of the various reflections of the X-ray pattern. An example of such a determination is the paper of Bonhomme and Yvon (1996) on the $Mg_{24}(Y,Mg)_4Y$ phase. Also partially occupied sites can be detected and modeled as a random distribution of atoms and vacancies. Interstitial atoms are also a distribution of vacancies and atoms on the same sublattice, but here vacancies are the major constituent and in the ideal compound these sites are not counted (Joubert 2002).

Similarly to the distinction between substitutional and interstitial solid solutions, the site fractions of vacancies can be determined by pycnometric density measurements connected with X-ray lattice-parameter measurements. The volume of the unit cell times

Avogadro's number divided by the pycnometrically determined volume of one mole of atoms gives the number of atoms per unit cell. If, within the experimental error, this is less than the number of sites, the difference must be interpreted as vacancies. The presence of more atoms than lattice sites must be interpreted as the existence of interstitial atoms. Thus, if the range of homogeneity of the phase is not too small, it is possible by use of density measurements combined with X-ray-diffraction data to distinguish whether the deviation from ideal stoichiometry is due to the formation of anti-structure atoms, vacancies, or interstitial atoms.

In some cases, if the atomic volumes are very different, a solid solution may be formed by pairs of small atoms randomly replacing the large atoms. This happens in the $TbCu_7$ structure type (Buschow and van der Goot 1971), which is the $CaCu_5$ type with 85% of the Ca positions occupied by Tb atoms and 15% by pairs of Cu atoms.

If the composition dependence of density or lattice parameters is not available, the type of solution to be modeled has to be selected by estimation, considerations involving atomic volumes, or comparison with similar phases.

In modeling sublattices after the crystal structure has been determined, all sites of a set of equivalent positions have to be treated equally. This excludes a modeling in which only one site of a three- or four-fold position can be substituted. The only exception is when a group of atoms can be considered as a single molecule.

In any other case, when a description treats atoms of equivalent positions to be in different sublattices, it may facilitate reproduction of the experimentally observed range of homogeneity, but the crystallographic interpretation of the sublattice is totally lost. Therefore such a "modeling" would be a misuse of a formalism, using it as a mere curve-fitting formula with physically meaningless coefficients.

The crystal structure gives the ideal occupancy for a stoichiometric compound and might not describe the stable composition range, even if the crystallographic information indicates that there is mixing on several sublattices. If the stable composition range is wide, one may have to consider constituents in other sublattices rather than ideal occupation by this constituent. Information from the same phase in other systems may provide information about the mixing on the sublattices. See for example the modeling of the μ phase discussed in Ansara *et al.* (1997a) and Joubert and Feutelais (2002).

6.2.5.2 Compatibility with models used in databases

The model for the σ phase recommended by Ansara *et al.* (1997a) and adopted in this book distinguishes the five different Wyckoff positions as three different sublattices with 10 $(8+2)$, 4, and 16 $(8+8)$ sites, respectively. See also the case study in section 9.3. This recommendation is based on the coordinations of the five Wyckoff positions. In order to describe the whole composition range of σ without considering Cr and V on all sublattices, a model for the σ phase with 8, 4, and 18 $(8+8+2)$ sites is frequently used in databases. In new assessments one should use the recommended model, but it may be necessary to fit also the old model for the sake of backward compatibility. When a sufficient number of revised assessments is available, it may become possible to remove the old model of the σ phase.

One should always consider how the phase extrapolates into higher-order systems. This may demand the use of more sublattices or consideration of more constituents in a sublattice than needed for a particular binary system. For example, the Laves phase C15 can dissolve Cr in the second sublattice in $HfCr_2$ and Cr in the first sublattice in $CrTa_2$. In the ternary Cr–Hf–Ta system one thus has to model Cr on both sublattices and therefore it should be included in the binary assessments.

6.2.5.3 Thermodynamic information

The heat capacity of a compound for which there are no measurements can be estimated by applying the Kopp–Neumann rule, see section 5.2.3, that it is the stoichiometric average of the heat capacities of the pure elements.

The coefficients a_i in Eq. (5.2) describe the heat capacity. They can be adjusted only if the heat content $H(T) - H(298\,\mathrm{K})$ or the heat capacity C_p (section 4.1.1.2) has been measured. However, heat-content measurements allow maximal adjustment of a_2 and a_3. The coefficients a_4, a_5, etc. can be only determined by C_p measurements. The only alternative is to use semi-empirical estimates, for example after Kubaschewski and Ünal (1977). Chase *et al.* (1995) recently proposed expressing C_p by the adjustable linear function $a_2 - 2a_3T$ and the Debye function, characterized by the Debye temperature Θ_D. Instead of the Debye function, the Einstein function, characterized by the Einstein temperature Θ_E, may be used. Equation (5.2) does not enable one to calculate Debye or Einstein functions, but, since Calphad-type calculations are meaningful only at temperatures at which reactions toward equilibrium are possible, only the transition from the Debye or Einstein function to the Dulong–Petit function is of interest. This can be expressed by the term a_5T^{-1}. Low-temperature heat-capacity measurements are of interest when integrated to standard entropies.

In some cases there are peculiar shapes of the heat-capacity data. This can be due to magnetic transformations, see Fig. 5.3(a), a drastic change of constituent fractions, see Fig. 9.8(b), later, or some internal order–disorder transformation. Another common feature is that the heat capacity of a compound increases drastically just before it melts. This increase of heat capacity is often due to the formation of small amounts of liquid in the sample, so such data should not be included in the model of the compound.

Further thermodynamic data besides heat capacities are measurements of the enthalpy of formation at 298.15 K, H_{298}, and calculations of standard entropies S_{298} from low-temperature heat-capacity data. Entropies can be fitted independently of heat capacities also to phase-diagram data.

Even if there are no measured values for some thermodynamic quantities, one must check explicitly that the assessed values are reasonable, since it is possible to have good fit to all experimental data but still unphysical values of heat capacities, for example. Both the heat capacity and S_{298} of a compound must always be positive. It is possible to obtain a very good description of all experimental data in a system but still have the wrong value of S_{298}. For example, an assessment of all data on the Al–B system (Ansara 1998b) gave a nice phase diagram, but for the compound AlB_2 there resulted a negative value of $S_{298} = -1.79\,\mathrm{J\,mol^{-1}K^{-1}}$. This was corrected in a reassessment by Mirkovic *et al.* (2004).

Further thermodynamic information may be based on chemical potentials determined by emf or vapor-pressure measurements. This information is more similar to phase-diagram data. Thus its value as supplement to phase-diagram data is not as great as that of enthalpy data.

6.2.5.4 Stoichiometric phases

For stoichiometric phases a Gibbs-energy function like Eq. (5.2) can be adjusted to the available experimental data. If the phase is magnetic, additional parameters are needed. If the Gibbs energy of a phase is determined only in a narrow range of temperatures and no enthalpy values are measured, for the Gibbs energy $\Delta G = \Delta H - T \cdot \Delta S$ only a single coefficient or a single (linear) combination of coefficients can be adjusted.

One of the two coefficients ΔH and ΔS (a_0 and $-a_1$ in Eq. (5.2)) may be estimated, for example by setting ΔS proportional to ΔH, an empirical correlation first proposed by Lupis (1967) and Kubaschewski *et al.* (1967) and later refined by Tanaka *et al.* (1990). For a stoichiometric phase $\Phi = A_mB_n$, partial Gibbs-energy measurements in a special case are equivalent to a direct measurement of its integral molar Gibbs energy. If the phase is in equilibrium with one of the elements (A) in nearly pure state, the partial Gibbs energy $\mu_A = 0$ and, from the condition $G_m^{A_mB_n} = m \cdot \mu_A + n \cdot \mu_B$ it follows that $G_m^{A_mB_n} = n \cdot \mu_B$. The molar Gibbs energy here is defined for one mole of A_mB_n, i.e., $m+n$ moles of atoms of Φ.

6.2.5.5 The Wagner–Schottky model

The Wagner–Schottky model (Wagner and Schottky 1930, Wagner 1952) was the first model using the crystallographic positions of different atoms in sublattices. It was developed for binary intermediate phases with small homogeneity ranges. The "ideal phase" is defined to have on each sublattice only one occupant. In the real phase there are also defects on the sublattices. The *Wagner–Schottky model* contains the following simplifications.

1. The defects are always so dilute that interactions between them can be neglected and the Gibbs energy of formation of the defects inside the stability range can be treated as independent of composition.
2. On both sides of stoichiometry the defect with lowest Gibbs energy of formation is the only one considered from among the three types (anti-structure atoms, vacancies, and interstitials), since the others will exist only in smaller amounts at equilibrium and thus can be neglected.
3. Random mixing of the constituents, separately on each sublattice, is assumed.

As always, one should consider crystallographic information when using this model, as described in section 5.8.2.3.

The CEF may be interpreted as a generalization of the Wagner–Schottky model, dropping the first two of the three conditions above. Libowitz (1971) was the first to use interaction parameters in the Wagner–Schottky model.

6.2.5.6 Phases with order–disorder transformations

The structure type of an intermediate phase may be an ordered form of the bcc, fcc, or hcp lattice. In some cases the disordered phase is also stable in the system, but in other cases only the ordered form exists as a stable phase. Even if the phase does not undergo disordering inside the binary system, this may happen in a ternary or higher-order system, so one should use a model that can describe the order–disorder transition. See section 5.8.4 for more information.

6.2.5.7 Carbides and nitrides

Two important kinds of carbides and nitrides were described in section 6.2.4.2, the "MC" and "M2C" types, which can be modeled in the same way as an interstitial solution of carbon or nitrogen in fcc and hcp, respectively. There are many other phases having preferred sites for different kinds of atoms. If the atoms are very different, like metal and non-metal atoms, they usually keep strictly to their respective sublattices, but sometimes there is more than one type of sites for the metallic elements. For example, in the "M23C6 carbide," of which $Cr_{23}C_6$ is the prototype, two of the metal sites can accommodate bigger atoms more easily. These sites are preferred by W and Mo and it is almost impossible to dissolve more W or Mo than can be accommodated by these two sites. A model for this phase is thus

$$(\mathrm{Cr}, \mathrm{Fe}, \ldots)_{21}(\mathrm{Cr}, \mathrm{Fe}, \mathrm{Mo}, \mathrm{W}, \ldots)_2\mathrm{C}_6 \tag{6.5}$$

This gives a maximum solubility of W and Mo and the parameter ${}^\circ G^{\mathrm{M23}}_{\mathrm{Cr:W:C}}$ for the "end member" $Cr_{21}W_2C_6$ can be used to determine the actual solubility of W. In this model Fe is considered to enter all metallic sublattices. The phase diagram in Fig. 6.8 for the

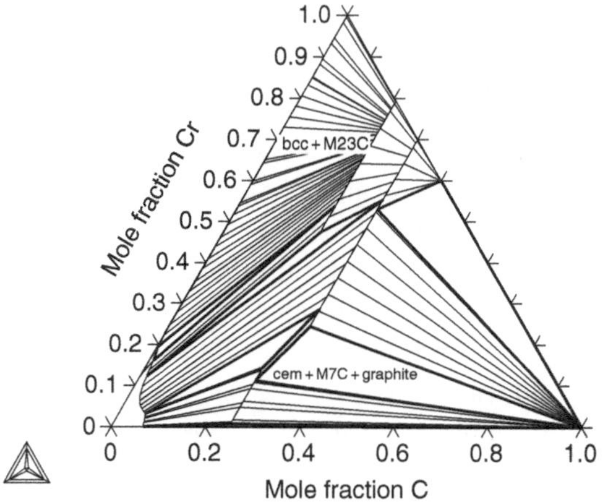

Figure 6.8 A ternary isothermal section of the Fe–Cr–C system at 1000 °C showing the large solubility of Fe in the M23 and M7 carbides. Note that cementite, which is metastable in the binary Fe–C system, is stable in the ternary system.

C–Cr–Fe system shows that the M23 carbide can dissolve considerable amounts of Fe, which makes it easy to determine the parameter $^{\circ}G^{\mathrm{M23}}_{\mathrm{Fe:Fe:C}}$ for the metastable "end member" $Fe_{23}C_6$. The model contains two additional end members, $Cr_{21}Fe_2C_6$ and $Fe_{21}Cr_2C_6$, which represent compositions inside the stability range of the phase and thus behave almost like interaction parameters in the model. The experimental information is usually not sufficient to determine individual values for all these parameters, so often their enthalpies of formation are set equal.

6.2.5.8 Ionic crystalline phases

For phases with strongly ionic behavior, mainly oxides, sulfides, chlorides, etc., the compounds often have no compositional variation. The reason for this is that the elements usually have fixed values of their valencies and the phase must remain electrically neutral. A variation in composition is usually due to vacancy formation and to the fact that some elements may have multiple valencies. One example of this is the wustite phase in the Fe–O system: Fe can have two valencies, +2 or +3. In the wustite phase the oxygen ions form a cubic-close-packed (fcc) lattice with the iron ions in the octahedral interstitial sites. The model used for wustite in an assessment by Sundman (1991a) is $(Fe^{2+}, Fe^{3+}, Va)_1(O^{2-})_1$. The vacancies are necessary in order to keep the phase electrically neutral.

Another more complicated case is the spinel phase, see Fig. 6.9(b). In the MgO–Al_2O_3 system the ideal stoichiometry is Al_2MgO_4. In this structure the O^{2-} ions form a cubic-close-packed (fcc) lattice; the Al^{3+} ions occupy half of the octahedral interstitial sites and the Mg^{2+} ions occupy one eighth of the available tetrahedral interstitial sites. However, both Al and Mg can be at the "wrong" sites. If some Al^{3+} ions according to the ideal formula are replaced by Mg^{2+}, the electro-neutrality condition requires additionally

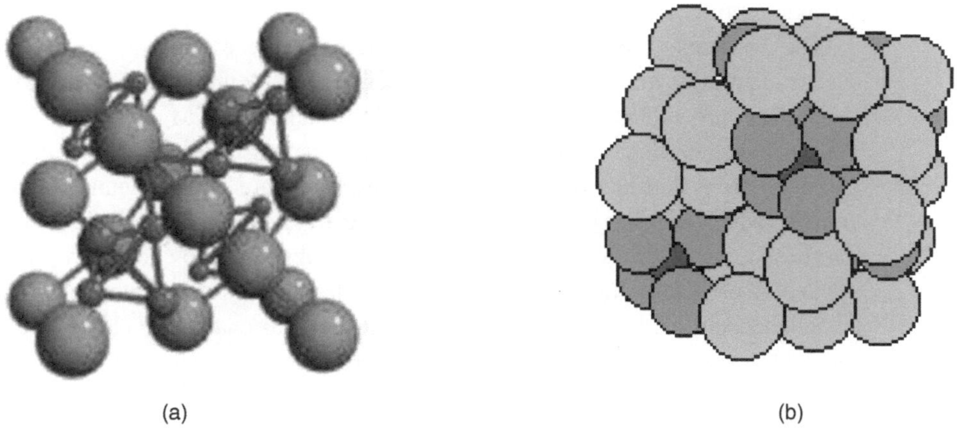

Figure 6.9 The C15 Laves phase (a) and $H1_1$ spinel (b) structures. From the Crystal Lattice Structures web page (http://cst-www.nrl.navy.mil/lattice).

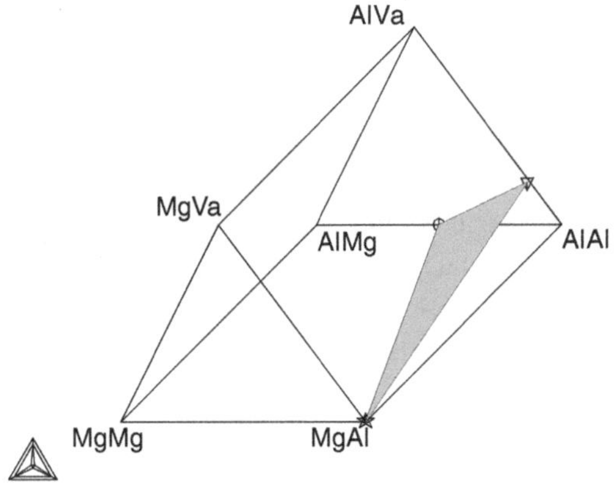

Figure 6.10 A prism representing the end members of the spinel phase. The small triangle, shaded gray, inside the prism represents all possible neutral combinations of constituents.

the occupation of some of the remaining octahedral interstitial sites. A model used by Hallstedt (1992) is

$$(Mg^{2+}, Al^{3+})_1(Al^{3+}, Mg^{2+}, Va)_2(Va, Mg^{2+})_2(O^{2-})_4 \tag{6.6}$$

The model for the spinel, without the third interstitial sublattice, can be visualized as the prism in Fig. 6.10. The front triangle represents all combinations of Mg^{2+} in the first sublattice with all three constituents of the second sublattice; the rear triangle represents all combinations of Al^{3+} in the first sublattice with all three constituents in the second. The small triangle inside the prism marked with symbols at the endpoints represents the neutral combination of the constituents. The front corner of this triangle is the "normal" spinel. The point on the bottom line of the rear triangle represents the "inverse" spinel, which is found e.g., in magnetite (Fe_3O_4). The point on the sloping side of the rear triangle represents the structure known as γ-alumina.

The vacancies on the second sublattice are necessary in order to maintain electro-neutrality when the spinel is off stoichiometry by virtue of the presence of excess Al^{3+}. The third sublattice is mainly vacant. In principle both Mg^{2+} and Al^{3+} ions should be considered on this sublattice. Since their site fractions remain very small, a valid simplification is to consider only one kind of ions. There are 12 end members in this model, most of which are charged. Only six electro-neutral combinations of these end members can be assessed. Thus six of the charged end members can be chosen arbitrarily and the six others used to define the six independent electro-neutral combinations. The number of independent parameters shall be further reduced to enable their evaluation from the available experimental information. If the third sublattice is ignored, three independent reciprocal relations can be written as below:

$$Al' : Al'' + Mg' : Mg'' \rightleftharpoons Mg' : Al'' + Al' : Mg''$$

$$Al' : Mg'' + Mg' : Va'' \rightleftharpoons Mg' : Mg'' + Al' : Va''$$

$$Al' : Va'' + Mg' : Al'' \rightleftharpoons Mg' : Va'' + Al' : Al''$$

A reasonable estimate is to assume that the reciprocal reactions should have zero ΔG because the mean occupation on each sublattice remains unchanged during these reactions. This reduces the number of parameters to be optimized by three. The reciprocal reactions are represented by the square surfaces of the prism in Fig. 6.10. There are three such surfaces characterized by the triangle sides Al″–Mg″, Mg″–Va″, and Va″–Al″. The remaining three parameters can now be attributed to the normal spinel (Mg′ : Al″), the inverse spinel (Al′ : Mg″ + Al′ : Al″)/2 and the γ-alumina (5 Al′ : Al″ + Al′ : Va″)/6. The γ-alumina parameter serves also as end member for deviations from stoichiometry caused by excess Al_2O_3.

If the third sublattice is assumed to be occupied by Mg^{2+}, one can draw a similar prism. A reasonable estimate for simplification may be to relate all six end members of this prism to the corresponding ones of the prism for Va‴ (vacancies on the third sublattice) by adding the same value, which must be large and positive in order to model the difficulty of occupation of this sublattice due to the narrowing caused by the atoms of the first sublattice. This prism itself does not contain any electro-neutral combination of end members, but the combination (Mg′ : Mg″ : Mg‴ + Mg′ : Mg″ : Va‴)/2 is electro-neutral and independent of the already-defined three electro-neutral end-member combinations. It serves well as end member for deviations from stoichiometry caused by excess Mg^{+2}. There are many papers on modeling ionic phases, for example Liang *et al.* (2001), Gueneau *et al.* (2002), and Grundy *et al.* (2006).

6.2.6 *Semiconductor compounds*

There is no doubt that a thermodynamic database for semiconductor materials will be very useful for the future development of electronic devices. Simulating the combination of different elements to optimize the properties on the computer is much simpler than the normal trial-and-error method.

The most commonly used semiconducting phases have the diamond (A4) or the zincblende (B3) structure type. These have the same atomic positions, but A4 belongs to the higher-symmetry space group Fd3m, for which all sites belong to the same Wyckoff position, (8a), whereas in B3, space group F$\bar{4}$3m, these sites form two different Wyckoff positions, (4a) and (4c). The B3 compounds consist of nearly perfectly ordered compounds from the third and fifth groups of the periodic table or from the second and sixth groups. Deviations from the ordering of magnitude 10^{-7} to 10^{-5} can be treated as defects by the Wagner–Schottky model. If the number of electrons of the defect species deviates from that of the matrix atom, the defect works as doping. To compensate for its electric charge either a nearly free electron or an electron hole (a missing electron in an ideally completely filled electron band) is formed. These electrons or holes obey the Fermi–Dirac statistics and do not contribute to a mixing enthalpy as in Eq. (2.11). Thus, in the CEF, they must not be treated with extra sublattices; see section 5.10.2.

To find the best routes for the production of B3-phase semiconductors, phase-diagram calculations regarding solid solutions of group-3 elements on one sublattice or of group-5 atoms on the other one are very important. Owing to the strongly directed bonding of these phases, the diffusion rate is extremely slow and it is nearly impossible to homogenize the material by annealing once it has crystallized. Thus, for growing single crystals or epitaxial films from liquid, the composition of the liquid has to be chosen to be in equilibrium with the desired composition of the material to be grown. Several databases for the group-3 and group-5 semiconductors have been published, of which the most complete is that of Ansara *et al.* (1994). A model for treating the variation in stoichiometry as well as the holes and electrons has been used by Chen *et al.* (1998).

6.2.7 *Phases with miscibility gaps*

The simplest kind of miscibility gap is when a phase appears on both sides of a two-phase region in a binary system. Normally the miscibility gap closes at a high enough temperature, where the configurational entropy will dominate, but that may be higher than the temperature at which one can obtain any measurements. The term "composition set" with a number will be used to identify the two different instances of the same phase. The concept of a composition set can be extended to cases that normally are not considered as miscibility gaps, but still have the same phase appearing with different compositions at the same equilibrium. This happens in many cases of ordering if the disordered and ordered forms are described as a single phase. For example, the structure types A1, $L1_2$, and $L1_0$ can be described with a single Gibbs-energy function and treated as a single phase. The different ordered forms can be treated as "composition sets" and can appear in equilibrium with each other, and this is a kind of miscibility gap.

In many cases one may have the same terminal phases in a system, but no continuous solution because there are stable compounds in between. However, if the compounds are excluded from the calculation, the terminal phase may have a metastable miscibility gap.

A miscibility gap can be described with the parameters of a single phase. At the beginning of an assessment it may be useful to try to fit the parameters that describe this gap together with the thermodynamic information before involving parameters from more phases.

A miscibility gap results from a positive interaction between the components, but can also occur in an ideal reciprocal system; see section 5.8.1.2.

During an optimization it may happen that miscibility gaps appear where there should not be any; a common feature is that there are "inverted" miscibility gaps at high temperature in the liquid. This may sometimes be difficult to detect, but, at the end of an assessment, one should calculate across the whole compositon range at a sufficiently high temperature, at least 4000 K, to ensure that no solid phases reappear and that there are no miscibility gaps.

It is possible to include a check on miscibility gaps during the assessment in the PARROT software; see section 7.3.10. The stability function for a phase, called QF, can be checked after calculating an equilibrium at some temperatures and compositions. If QF returns a negative value, the equilibrium is inside the spinodal.

6.2.8 *Hume-Rothery phases*

There is a class of phases with which the Calphad method still has some difficulties, namely the Hume-Rothery phases. For such a phase, the Gibbs energy has a significant contribution from the energy of the electron gas, which is part of the metallic bonding. The interpretation of this contribution by Hume-Rothery (Hume-Rothery *et al.* 1969) is described here, following a model proposed by Mott and Jones, but it should be remembered that this work was done before the establishment of the DFT and can be reinterpreted after systematic calculations of the density of states using the quantum-mechanical methods available nowadays. For a recent discussion on Hume-Rothery phases, see Paxton *et al.* (1997) and Mizutani *et al.* (2006).

Following Hume-Rothery, this electron gas is modeled as nearly free electrons, obeying Fermi–Dirac statistics, so that two electrons can never have identical quantum numbers simultaneously for all states. The electrons form stationary waves in the metal, for which the quantum numbers can be represented by a lattice in reciprocal space. The energy of the electrons increases with the quantum numbers. The lattice in reciprocal space is filled starting from the states with lowest energies. The highest energy of occupied states is called the Fermi energy. The corresponding quantum states in reciprocal space form the Fermi surface. The wavelength of the stationary waves belonging to states near the Fermi energy is of the same magnitude as the period of the crystal-structure lattice. This is visualized in reciprocal space by the "Brillouin zones." The number of quantum states versus the energy of the electrons, called the density of states, exhibits discontinuities where the Fermi surface touches Brillouin zones.

If in a phase for example a Cu atom is replaced by a Zn atom, one additional electron is added to the electron gas, which contributes proportionally to the Fermi energy to the derivative $\partial H/\partial x$. The change of the Fermi energy versus number of electrons added is the reciprocal of the density of states. Thus the contribution of electrons added to the enthalpy H is proportional to $\iint \mathrm{DOS}^{-1}\,\mathrm{d}x^2$, where DOS is the density of states and x is the valence-electron concentration (VEC), the number of electrons per atom in the electron gas. The VEC is a linear combination of the mole fractions of the elements constituting the phase.

The simplest model, called the rigid-band model, assumes the density of states to be independent of composition. This simplification is too crude, but it may be taken as a first approximation. The most important feature of this model is that the discontinuities of the DOS versus VEC prevent this contribution to H being represented well by a polynomial. Thus the RK formalism necessarily has limited accuracy for this class of phases. Furthermore, for the extrapolation to ternary and higher-order systems one has to use the VEC as a single composition variable rather than Muggianu's formalism. This statement is valid only for the electron-gas contribution to the enthalpy H. Contributions from other sources may be described well by the Redlich–Kister–Muggianu formalism.

For successful use of this consideration in modeling G, first-principles calculations of the density of states are desirable and some ideas regarding how to simplify the results into a formalism that can be handled by the Calphad method without significant loss of accuracy have been proposed. The contribution of the electron gas to G may be identified with that to H since, due to the Fermi–Dirac statistics, the contribution of $-TS$ is very small. A manageable simplified description is needed in order to describe how the density of states depends on the phase composition (Mizutani *et al.* 2006).

6.2.9 *A summary of phase-model selection*

For Table 6.1, even without knowing details of the shape of the measured quantities, one can already make some decisions about the modeling, for example the following.

- $Phase_2$ and $Phase_3$ should be modeled by a single Gibbs energy.
- There are no enthalpy-of-formation data for $Phase_4$; therefore, a measurement, an estimate, or a first-principles calculation for that quantity should be sought.
- For $Phase_2$ a constraint relating H and S should be imposed because there are not enough data for modeling H and S independently for a solution phase.
- $Phase_3$ is well studied and C_p can be modeled; however, since the heat content was also measured, one can test for conflicts between these two sets of data.

To restrict the kind of model and the number of independent coefficients allowed by the amount of experimental data further, one should consider Table 6.2.

More details of the parameter selection follow in the next section.

6.3 Determining adjustable parameters

The CEF formalism described in chapter 5 is very versatile and contains many adjustable parameters, but only some of them can be used in an assessment, depending on the available experimental data. Certain considerations and estimates should be used in order to decrease the number of independently adjusted parameters either by fixing some parameters to constant values or by setting constraints relating two or more parameters, thus expressing them all by a single parameter.

6.3.1 *Selecting parameters*

In adjusting a thermodynamic description to experimental values, the final value of each adjustable coefficient depends on many of the diverse measurements and each measured value contributes to many of the coefficients. The advantage of the least-squares method is that these influences need not be known quantitatively, since the strategy of the method is to select the best possible agreement of all the coefficients and all the experimental values as described in section 2.4. Many of the coefficients of the descriptions, however, are not able to improve the fit between measurements and descriptions significantly. Using them may lead the calculation to follow just the scatter of the experimental values, creating maxima and minima where a smooth line is physically more plausible.

To judge whether a certain coefficient is well defined by the available set of measured values, the effect of each coefficient on the shapes of calculated curves should be known at least qualitatively. It must be discussed for each coefficient, if its influence on the shapes of calculated functions really is necessary to improve the fit between calculation results and the experimental dataset. In the following paragraphs some general considerations for this check are given. Among the examples given in chapter 9, this check is described in detail for some systems.

Table 6.2 *A quick guide to model selection for a phase*

1	Is the phase stoichiometric? **If no, skip to 2**		
	1.1	Is the phase non-stoichiometric in another system? If yes, skip to 2 and consider also data from that other system.	
	1.2	Is the heat capacity known? Use Eq. (5.2) in section 5.2.1 and check how many of the coefficients a_2 to a_5 can be adjusted.	
	1.3	Not the heat capacity, but the heat content is known. Do not use more than coefficients a_0 to a_3 of Eq. (5.2) in section 5.2.1.	
	1.4	Neither the heat capacity nor the heat content is known. Use the Kopp–Neumann rule and only coefficients a_0 and a_1, possibly with a constraint.	
2	Is the composition range narrow? **If no, skip to 3**		
	2.1	The crystal structure is known.	
		2.1.1	If it is B2, $D0_3$, or $L2_1$, model it as ordered bcc; if it is $L1_2$ or $L1_0$, model it as ordered fcc; if it is $D0_{19}$ or B19, model it as ordered hcp. See section 5.8.4.
		2.1.2	Use crystallographic information to decide on sublattices. Simplifications may be needed; see section 6.2.5.4 also about using defects.
	2.2	The crystal structure is not known.	
		2.2.1	Is the solubility really well known? **If no**, model the phase as stoichiometric, select a formula reproducing the composition, and skip to 1.
		2.2.2	Try the Wagner–Schottky model with two sublattices. If the atomic sizes are not extremely different, use anti-site defects on both sites. Choose the end members (pure elements) arbitrarily to have at all temperatures more positive G values than those of the stable elements. Use only ${}^0L$ parameters.
3	The phase should be modeled as a solution phase, although not necessarily across the whole range.		
	3.1	Is the enthalpy known as a function of temperature?	
		3.1.1	If the heat capacity is known, a $T\ln(T)$ coefficient may be used in the first regular excess parameter, but check whether it converges sufficiently well during the optimization.
		3.1.2	In almost all other cases, use only a linear temperature dependence (a_0 and a_1 in Eq. (5.2)).
	3.2	Is the enthalpy of mixing known as a function of composition?	
		3.2.1	Is it "V"-shaped or presenting kinks? For the liquid phase, see section 6.2.3.2. For solids, use sublattices depending on the crystal-structure information; see section 6.2.5 or section 6.2.4.3.
		3.2.2	Is it asymmetrical in the composition dependence? Use at least one odd RK coefficient, normally both odd and even ones; see section 5.6.2.1.
	3.3	Read carefully about models for solution phases in this chapter and chapter 5.	

6.3.2 *Reducing the number of parameters*

The analysis of experimental data does not always make evident whether a certain parameter is related to the available experimental data. This can be found out by applying a least-squares method twice, with and without the coefficient. Comparison of the two results makes a decision possible in most cases. It is recommended that one start the assessment with as few coefficients as possible and then include additional coefficients as necessary. A systematic misfit between some series of experimental points and the corresponding calculated curve usually gives enough hints to clarify which coefficients should be added. If a coefficient is not defined well enough, it usually does not show up in the comparison of measured values and calculated curves. It may, however, have a bad influence on the behavior of the extrapolation of the calculation into areas not covered by experiments.

6.3.3 *Constraining parameters*

For phases with several sublattices and several constituents on each, there are many reciprocal relations, as described in section 5.8.1. If there is not enough information, taking into account also crystallographic symmetries, to determine each end member of a reciprocal relation independently, a possible method is to assume that the reciprocal energy is zero. If three end members of a reciprocal relation are known, this assumption makes it possible to fix the fourth from Eq. (5.100). For cases in which there are several possible reciprocal relations, the order in which one should select these has been suggested by Hillert (1997b).

For intermediate phases with a homogeneity range, there often exists an "ideal" composition for which each sublattice is occupied by a single constituent only, equivalent to an "end member." In order to model the homogeneity range, one needs information about the kind of defects causing the deviation from ideality. Possible types of defects are anti-site atoms, interstitials, vacancies, and several kinds of defects simultaneously. When optimizing the properties of the phase, one should first optimize the parameter that describes the ideal composition, i.e., a single end member. If the homogeneity range is small, modeling this end member first as a stoichiometric phase often facilitates the optimization. If this description converges, it is fixed and the parameters that determine the deviation from the ideal stoichiometry are added and assessed. Finally, all parameters are made adjustable again and assessed simultaneously in the final calculation step. This procedure in most cases prevents large jumps of the parameters in the first calculation steps, which may lead to a parameter set in which the opposite occupation of sublattices appears to be the more stable one. When a reasonable fit has been obtained with the available data, but more than one independent defect structure is defined for the phase in question, one may start varying also the parameters that determine the relative fractions of these defects.

When there are many end members of a phase, little or no thermodynamic information, and no data on the actual occupancy, one may reduce the number of optimizing variables by assuming the same entropy and heat capacity for all the end members. In some cases, it may suffice to set an end-member parameter equal to its "Kopp–Neumann" value like in Eq. (5.39).

Setting an end-member parameter equal to zero is a very bad estimate.

6.3.4 Coefficients in the Redlich–Kister series

In solutions for which the RK formalism gives a good description of the Gibbs energy (sections 6.2.3 and 6.2.4), the question of how many coefficients ${}^{\nu}L$ of Eq. (5.65) can be independently adjusted arises, or, what is the maximal power ν of the series? The composition variations of G_m due to the various coefficients are shown in Fig. 5.10.

The simplest case is a dilute solution starting from one of the pure components. Here Henry's law describing an ideal solution (Eq. (5.42)) between the pure solvent and a fictive state of the solute expressed by ${}^{\circ}G_B - H_B^{SER}$ may be applied. However, ${}^{\circ}G_B - H_B^{SER}$ is formally a unary parameter, which must be the same in several binary systems and cannot be fitted to a particular binary system. Then the Henrian solution is described as a regular or quasiregular solution with a single coefficient ${}^{0}L_{A,B}$ in Eq. (5.65). If the solution is dilute, only the sum ${}^{\circ}G_B - H_B^{SER} + {}^{0}L_{A,B}$ is significant. If the coefficient ${}^{0}L_{A,B}$ has to be adjusted then whether it should be treated as a constant (regular solution) or as a linear function of temperature (quasiregular) depends on the experimental information.

Measured solubilities in an extended temperature range or independent measurements of a solubility at one temperature and an enthalpy of solution allow the independent adjustment of two coefficients, a_0, and a_1, of ${}^{0}L_{A,B} = a_0 + a_1 \cdot T$. If, however, the solubility is known in a narrow temperature range only and no enthalpy of solution (mixing) is available, only one coefficient should be adjusted independently. Then either a_1 is set zero or it is set proportional to a_0: $a_1 = a_0/T_0$, where T_0 is estimated to be 3000 K by Lupis (1967) and Kubaschewski *et al.* (1967). This estimation was proposed to depend on the melting temperatures of the pure components and refined by Tanaka *et al.* (1996).

For some systems there may be enough experimental information to adjust two or three RK coefficients for one of the solution phases in a system, for example the liquid. Even if there is little or no information except phase-diagram data on the other solution phases in the system, one may find it necessary to use the same number of coefficients for the other solution phases as for the liquid in order to obtain a good description.

6.4 Decisions to be made during the assessment

The use of software to optimize the system can be divided into two steps. The first is to get a set of parameters that can roughly reproduce the main features of the data. The second step is fine-tuning the parameters to the selected data.

6.4.1 Steps to obtain a first set of parameters

An experienced assessor may be able to guess parameter values for the models that can be used to calculate a phase diagram in reasonable agreement with the critical dataset. For beginners, however, guessing the first set of parameters is usually a very big problem.

Most types of software require that one can calculate the experimental equilibrium in order to compare the measured quantity with the corresponding quantity calculated from the model. If all parameters are initially zero, that is usually not possible (the reason for this is explained in section 7.1.1).

In the BINGSS software there is an option IVERS=3 that does not require a full equilibrium calculation to compare the model value with the experimental one; see section 7.2.4. In the PARROT software the same feature is called ALTERNATE mode and is described in section 7.3.7.3.

6.4.2 *Contradiction between experiments*

6.4.2.1 Experimental data of the same kind

When one has several measurements of the same quantity by various authors, one may sometimes find that their results are so scattered that one cannot fit them all. If all possibilities of systematic errors, impure samples, bad calibration, etc. have been checked but the differences remain, the assessor must decide which data are not to be used. It is not advisable to include several contradicting values of the same quantity, since the least-squares method will just fit the mean value. This initial selection may have to be reconsidered later during the assessment when the fit to other kinds of data may have clarified that the originally selected value is less compatible with these other data than the rejected one.

6.4.2.2 Experimental data of different kinds

After the selection of the models and obtaining a first set of parameters by running the optimizing program, it may be found that some sets of experimental data cannot be fitted simultaneously. This can be detected by excluding the suspected datasets one by one and checking whether the fit of the non-excluded datasets improves. Which dataset finally should be excluded is up to the judgment of the assessor, who should consider the estimated accuracy of the datasets and the overall fit to other data.

Before excluding a set of data, one should analyze carefully the original paper and look for a possible reason why the experimental result does not describe the physical reality. The reason behind the contradiction between different types of data is seldom evident. In order to understand the problem, one can use some rules like the Gibbs–Konovalov rule, Eq. (2.50).

There are cases in which different kinds of data provide the same information, for example when the liquidus, chemical potentials, and enthalpy of mixing of the liquid phase are known. A conflict may be found in that the value of the entropy obtained by combining the enthalpy and the liquidus data and that obtained by combining the chemical potential and enthapy data may be contradictory. In this case, the entropy is overdefined. Some action should be taken to use the combinations separately, observing the results obtained.

To determine that datasets are really in contradiction one should question the assumed model. The main question is the following: is there a model that will cope with the different sets of data? This question is usually difficult to answer.

6.4.3 *Weighting of experiments*

The least-squares minimization in Eq. (2.53) can be affected by selecting different weightings for the experimental data. The uncertainty assigned to each piece of experimental data is the first level of weighting, since an experimental result with a very small uncertainty will force the optimizer to try to fit this result well. The uncertainty should never be adjusted during an optimization; one should use the value provided by the experimentalist, with a reasonable amount of skepticism.

If some important data are not fitted well during the optimization, one must first reconsider the model and the parameters selected for it. For example, an asymmetrical miscibility gap requires at least two RK coefficients. Increasing the weighting of the experimental data will not improve the fit if only a regular parameter is used.

It is also possible that the parameter set has become stuck in a local minimum, such that it may help to change the parameters drastically, even change signs, in order to try to reach a global mimimum. This may make it difficult to calculate some, or all, of the experimental points and one may have to use again the technique provided for bad initial values; see section 7.1.1.

However, if the model and model parameters are reasonable, one may increase the weighting of data from the most important experiments. That will necessarily make the fit to data from other experiments less good and it is not reasonable to increase all weightings. In the PARROT software, see section 7.3.10, one may also adjust the weightings to give equal importance to a few phase-diagram points if one has much thermochemical data.

6.4.4 *Phases appearing where they should not or missing from where they should appear*

Terminal phases with small composition ranges or intermediate phases may sometimes appear to be stable in parts of the phase diagram where they are not stable. This can be checked only by calculating the whole diagram with all phases for the current set of parameters. Such calculations should be done regularly during the assessment in order to discover such problems as early as possible.

Some such cases can easily be avoided. If the artificial appearance of a phase is at temperatures far above its real stability range, it may be sufficient to set a breakpoint in the description of the temperature dependence of one or more of the end members of the phase and continue with constant C_p, i.e., without terms containing T^2, T^3, or even higher powers of T. Outside the stability range of the phase, high powers of T may lead to extremely unphysical C_p descriptions. In a few cases it has also been found that a constant C_p is not sufficient to suppress the reappearance of the phase; one should then, after the breakpoint, allow C_p to decrease to the Kopp–Neumann value.

It is normally not possible to relate the appearance of a phase at the wrong place in a phase diagram to a particular optimizing parameter. Sometimes it can be the Gibbs energy of a particular end member becoming too negative or the coefficients in the RK series adding up to too negative a value where the phase should not appear to be stable. This problem can be solved by finding an expression stating that the phase should be less stable than the stable equilibrium in the region, which means that the phase appearing at the wrong place must have a negative driving force at the stable equilibrium at this place. The driving force of a phase is explained in section 2.3.6.

In the PARROT software, see section 7.3.10, this difference can be prescribed as a negative driving force, and the problem is easily solved. This is done by prescribing that the driving force of the phase should be smaller than some prescribed negative number. This information can usually be added to an existing experimental tie-line or similar data that have already been optimized. This is described in section 7.3.7.9.

It is possible to use the same feature of PARROT to force a phase to become stable in a region where it is missing for the current set of optimizing parameters. One can then require that the phase has zero driving force at the desired composition and temperature.

6.4.5 *Reasonable values of parameters*

There are certain limits of the model parameters that should be considered. In most cases the coefficients a_0 of temperature-independent and a_1 of linearly temperature-dependent terms are the only ones to be optimized and these can be related to enthalpies and entropies, respectively. If heat-capacity data have been assessed, the enthalpy and entropy parts must be calculated from the full Gibbs-energy expression and one cannot look at just the a_0 and a_1 coefficients, for example. This check must be performed before the assessment is finished, but, now and again during the assessment, one may decide to reset or discard parameter values that are unreasonable for the reasons discussed below. For a sensible discussion of parameter values, one should have a reasonable fit to all experimental thermochemical data.

There are limits of enthalpies and entropies that can be checked even when experimental data are lacking, as described in section 6.2.5.3, for example that the total entropy for a given composition must always be positive. For enthalpy values one should expect that their absolute value is less than a few times $100\,\mathrm{kJ\,mol^{-1}}$ (moles of atoms). For entropies they should be less than $100\,\mathrm{J\,mol^{-1}\,K^{-1}}$. This should apply to each coefficient in the RK series, even if the sum of all coefficients gives a reasonable enthalpy.

6.5 Checking results of an optimization

It is not easy to know when the best possible set of parameters has been reached. The answer will depend on the use for which one wants the description obtained. Supposing that one wants to have the best possible descriptions using the full potentiality of the method, for which this book is intended, the following procedures should be implemented.

A well-optimized set of parameters for the Gibbs energies of the system should be able to reproduce the available experimental set in the best possible way.

New experimental evidence may indicate that the description should be modified. What are the criteria for deciding that a description is well established for the known experimental evidence?

1. The visual check of the agreement between experimental data and calculated values is an important tool. Tendencies shown by the experimental data but not reproduced by the calculated curves should be carefully analyzed. They can express physical behavior and may require a change of the modeling. This fact is illustrated by the assessment of the Mg–Zn system (Agarwal *et al.* 1992). There, the steep slope of the liquidus line found experimentally was reproduced only after the introduction of a temperature-dependent enthalpy of mixing into the modeling. Further experimental enthalpy results specially provided for the optimization confirmed the existence of a large excess C_p in the liquid phase. The background of this problem can be explained by the Gibbs–Konovalov rule, Eq. (2.50).
2. The sum of squares of errors obtained from the least-squares fit is another important tool, although one should not take it as the unique criterion.
3. Extrapolate to higher-order systems. Sometimes it can happen that a binary system is quite well described, but the extrapolation of the description to a ternary system does not succeed in reproducing the ternary experimental information. In this case the ternary information is used for improving the binary description.
4. Analyze the plausibility of the values of the parameter found by the least-squares fit. For example, for a stoichiometric phase for which no C_p is known and the Kopp–Neumann rule is used to estimate C_p, just two coefficients are left to be optimized. With such considerations of the selection of the number of coefficients to be adjusted and with a reasonable set of experimental data used for the optimization without contradicting data, the values found by the fitting also should be reasonable. Nonetheless, one should check whether the signs and values of the coefficients are plausible. Usually, if the value of the coefficient of the a_1T-term is very large, that should be taken as an indication that a_0 and a_1 cannot be optimized independently and the constraint proposed by Tanaka *et al.* (1990) should be considered an adequate estimate.
5. Removing non-significant digits. Parameters with many digits may give the impression that they are very well determined, although actually only the first few digits are significant. However, one cannot arbitrarily round off parameters without the danger of losing agreement with some experimental data. Even if an enthalpy and an entropy of melting are known with at most two significant digits, their quotient, the melting temperature, may be known to within four digits. A method of safe rounding might be to select the parameter with the highest relative standard deviation, set it to a rounded value, and reoptimize the remaining parameters. This can be repeated until all parameters have been rounded to the appropriate number of significant digits; see section 7.4.4.
6. Check that S_{298} and C_p of all phases are within reasonable limits.

6.6 Publishing an assessed system

Before your paper is ready to be published, you should check that the following items are reported. These checks may also be used by reviewers.

- A list of experimental data used and of those that were not used.
- A complete report of the model parameters. If that is not possible, state explicitly the reason. If the parameters are just partially reported, indicate which parameters are missing and where to find them (for example a commercial database). Provide parameters in a well-documented format on a computer file for the reviewer. This facilitates the work of the reviewer, who will wish to check that parameters and text agree.
- A table of the invariant equilibria, including azeotropic maxima or minima.
- Crystal-structure information and lattice parameters.
- Standard enthalpies of formation (units)
- A report about metastable phases.
- Diagrams with calculated and experimental data together. All experimental data, even datasets excluded from the assessment, should be plotted, maybe in separate diagrams if it would otherwise be confusing.
- The range of validity of the description.

These reports can be done systematically. It would be interesting to create a format as for reporting crystal structures. Such a standard format would facilitate further improvements of the modeling of the system in question. Instead of a completely new publication, just a short note providing an update would be sufficient.

6.7 How the experts do assessment

During the preparation of this book and after so many years of cooperation, it was found that each author had his or her own way of doing assessments. For further opinions on assessments, read also Kumar and Wollants (2001) and Schmid-Fetzer *et al.* (2007), and read several published assessments.

When modeling the Gibbs energy for phases in a binary system, one must take into account that they will extend into multicomponent systems when added to a database. This means that one should look at least at some ternary systems including this binary in order to find out which phases are important. Phases stable within a small range in a binary system may extend far into higher-order systems and the parameters assessed for the binary system will affect the stability of the phase in the multicomponent system. The experience with the Calphad technique is that phases assessed in binary systems taking ternary information into account can be extrapolated to higher-order systems with a high degree of confidence.

6.7.1 Expert type 1

This kind of expert will look at the general outline of a system and, if there are compounds, he will estimate metastable extrapolations of the terminal phases and possible metastable three-phase equilibria with the liquid. After obtaining a preliminary fit to this, he will add the intermetallic phases one by one and, if necessary, estimate congruent melting points or metastable invariant equilibria with the already-assessed phases. When all phases have finally been incorporated, he will optimize with just the experimental data, but often it may be necessary to include some estimated equilibria in order to avoid the appearance of phases at "wrong" compositions or temperatures. This technique facilitates the use of *ab initio* data and comparisons with other systems. It emphasizes the use of the assessment in higher-order systems but it requires long experience in estimating metastable extrapolations. Often such extrapolations should be smoothed without any maxima or minima of the liquid–solid two-phase equilibria. It is called the Kaufman–Ansara method.

6.7.2 *Expert type 2*

This expert may be more interested in the thermodynamic models of the system than in the specific system being assessed. He will try models and test model parameters that generate prototype diagrams similar to the system and try to find relations between the parameters and the various features of the system. He will also estimate or even invent experimental data that should be representative of the system and try to determine which model parameter can be used to describe them. Like expert type 1, he will usually work with one phase at a time and pay great attention to the metastable extrapolations of the system. This is a very powerful feature of the "Stockholm" school.

6.7.3 *Expert type 3*

This expert starts by analyzing all the experimental data available in order to avoid discrepancies between the assessment and any experimentally determined phase diagram as well as thermodynamic data. He will use all experimental data from the beginning and may be less focused on metastable extrapolations insofar as they are not meaningful for higher-order descriptions in experimentally accessible ranges. His objective is to model the present system as well as possible, without limitations demanded by merely formal problems of the existing database. He may prefer to simplify, if there are not enough experimental data really enabling him to determine parameters of a more-sophisticated model. For example, intermediate phases will be treated as stoichiometric, if the solubilities are not well defined by experimental data. This is a method mainly envisaging an academic use of the final description. That procedure reflects the "Stuttgart-PML" (Pulvermetallurgisches Laboratorium) school.

6.7.4 *The ultimate expert*

It is more or less easy to identify the ways to do assessments by searching the literature. The experts of type 2 will start their publications by describing the models; experiments come later, when the discussion has been completed. Experts of type 3 start with the experimental data, followed by a discussion of the selection of the model using arguments based on the existing experimental data. One does not find many papers with the first approach, but this is the approach of people thinking in terms of multicomponent databases for real materials.

Which is really the best method for an assessment? There are good features in all methods and, if you can make a combination of them with a lot of "good sense," you will certainly obtain a reasonable result.

7 Optimization tools

In this chapter, two of the most commonly used types of software for optimization, BINGSS and PARROT, are described.

7.1 Common features

7.1.1 *Handling bad starting coefficients*

The definition of the "error" (v_i in Eq. (2.52)) is based on the "calculated value" ($F_i(C_j, x_{ki})$), which is often defined by an equilibrium calculation with two or more phases. The initial set of adjustable coefficients may result in improper Gibbs-energy functions, with which this equilibrium cannot be calculated. As an example, in Fig. 7.1 such a situation is shown for a two-phase equilibrium, liquid + bcc. There G^{bcc} at all compositions is larger than G^{liquid} and the construction of a common tangent is impossible, so also no equilibrium can be calculated numerically.

The experimental information is either "at temperature T_1 there is a two-phase equilibrium, liquid + bcc, for which the composition of the bcc phase was measured as ($x^{\text{bcc}} = x'$)" or "in a single-phase bcc sample of composition x' on heating the first liquid appears at temperature T_1" (see Fig. 4.4).

In the least-squares calculation for Eq. (2.52) no "calculated value" ($F_i(C_j, x_{ki})$) can be provided as long as the starting values for the Gibbs-energy descriptions of liquid and bcc phases behave as in Fig. 7.1. To find better starting values by trial and error is not easy and is time-consuming. Therefore it is desirable to have a method whereby this problem is avoided and that can start even with a very bad initial set of adjustable coefficients.

Zimmermann (Zimmermann 1976, Lukas *et al.* 1977) defined the error of a two-phase equilibrium by a linear expression relating the Gibbs energies and their derivatives with respect to the composition x as shown in Fig. 7.2

$$G' + (x'' - x') \cdot \frac{\mathrm{d}G'}{\mathrm{d}x'} - G'' = \text{error} \tag{7.1}$$

The first two terms together give the point at x'' on the tangent to the curve $G'(x)$ at x'. All three terms together thus give the distance between this point on the tangent

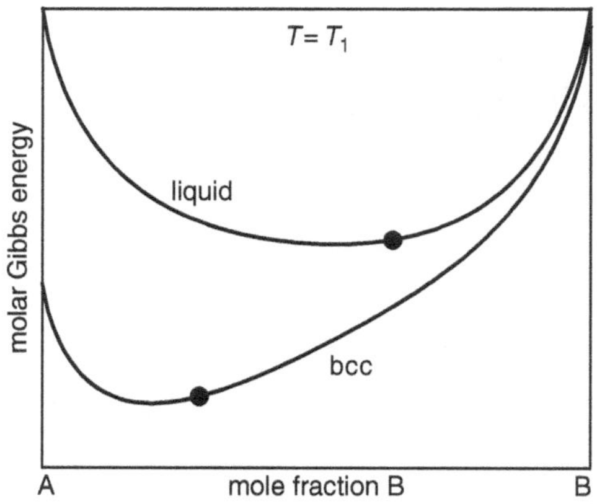

Figure 7.1 At fixed temperature T_1 no equilibrium between the phases can be calculated with the Gibbs-energy curves in the diagram because the bcc phase is more stable than the liquid at all compositions. The experimental information is a tie-line between the compositions marked by symbols.

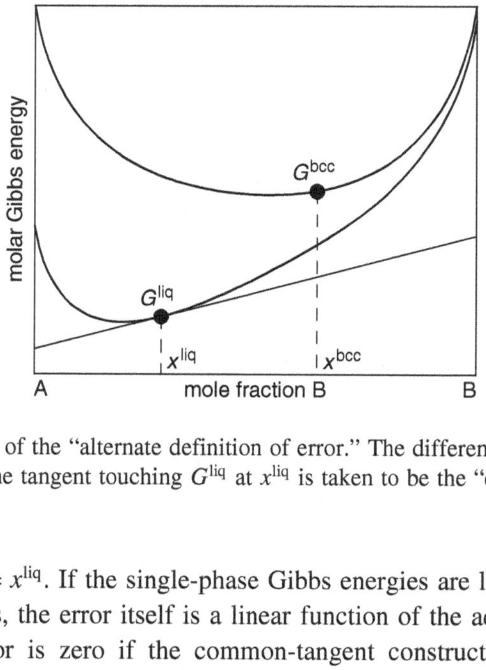

Figure 7.2 Construction of the "alternate definition of error." The difference taken at composition x^{bcc} between G^{bcc} and the tangent touching G^{liq} at x^{liq} is taken to be the "error."

and $G'' = G^{\text{liq}}$ at $x'' = x^{\text{liq}}$. If the single-phase Gibbs energies are linear functions of the adjustable coefficients, the error itself is a linear function of the adjustable coefficients. The thus-defined error is zero if the common-tangent construction fits exactly with the measured concentration x'. In the following Eq. (7.1) will be called the "alternate definition of error." In contrast to the "normal" definition of the error, using x' calculated by solving the equilibrium conditions, the alternate definition always has a result, even if the two-phase equilibrium liquid + bcc cannot be calculated. There is, however, also a disadvantage: the alternate error depends on x'', which is either just estimated or measured by other, independent, experiments. This may give an incorrect behavior of the least-squares optimization.

Therefore the alternate definition of error should be used only at the beginning. Mostly then the result is good enough to serve as starting values using the "normal" error, i.e., the error obtained from a normal equilibrium calculation.

An alternate error can be defined also for other types of measurements. Generally we define it as a linear expression relating Gibbs energies and their derivatives with respect to independent variables that becomes zero if an equilibrium calculation using the current set of adjustable coefficients exactly fits the experimental values. Auxiliary state variables, like x'' in the above example, may be used, but should have little influence on the value of the alternate error in the vicinity of the final solution.

In BINGSS, besides measured T or x' of a two-phase tie-line, alternate definitions of error are used for measured chemical potentials μ_i in two-phase equilibria and for measured temperatures of invariant three-phase equilibria. In

$$\text{alternate error} = \frac{G' \cdot x'' - G'' \cdot x'}{x'' - x'} - \mu_1(\text{measured}) \tag{7.2}$$

$$\text{alternate error} = \frac{G'' \cdot (1 - x') - G' \cdot (1 - x'')}{x'' - x'} - \mu_2(\text{measured})$$

the common tangent is replaced by a line connecting G' and G'' calculated for the compositions x' and x'', respectively, without specifying it as a tangent. If the optimized Gibbs-energy descriptions of the two phases exactly fit the measurement, this line becomes the common tangent, if also x' and x'' are correctly given. Even if x' and x'' slightly deviate, the line is expected to be very near to the common tangent. For a three-phase equilibrium the common tangent touches all three Gibbs energy versus mole fraction curves. The alternate definition of error in BINGSS demands that "the three calculated Gibbs energies must be on the same straight line" without specifying this line to be a tangent:

$$\text{alternate error} = \begin{vmatrix} 1 & 1 & 1 \\ G' & G'' & G''' \\ x' & x'' & x''' \end{vmatrix} = G' \cdot (x'' - x''') + G'' \cdot (x''' - x') + G''' \cdot (x' - x'') \tag{7.3}$$

In PARROT there is no limitation in definition of the error for each experiment, thus in principle the alternate definition can be chosen freely, too. The PARROT module of Thermo-Calc provides a setting called "set alternate mode," by use of which alternate errors can be calculated from a "normal" experimental data file as differences of Gibbs energies rather than differences between the measured and calculated quantity. More about that is given in section 7.3 and in the user manual of Thermo-Calc.

7.1.2 Usage of calculated phase-diagram information during optimization

In optimization by least squares, only single pieces of information are used as experimental data, but it is important to calculate the complete phase diagram several times during the assessment. Such calculations show how the thermodynamic properties interpolate between the experimental points and such calculations may show that some phases appear in regions where they should not be stable. In PARROT it is possible to add as experimental data that a phase should not be stable, as described in section 7.3.7.9.

7.1.2.1 Using calculations of metastable phase diagrams and other diagrams

The temperature–composition phase diagram, T–x, is the normal presentation of metallic phase diagrams and it is common to overlay the calculated diagram with the experimental data.

It is also useful to calculate metastable phase diagrams if, e.g., all phases except the liquid and the terminal solution phases have been suspended. It may also be interesting to calculate the metastable solubility lines between the liquid and each solid phase including the terminal phases.

When there are measured values of the chemical potential in two-phase regions, one may calculate T–μ diagrams, and add the experimental data to them. With Thermo-Calc, many different diagrams can be plotted using various types of axes in the post-processing of the same phase-diagram calculation. In BINGSS, this type of diagram is particularly useful to elucidate regions where the $T-x$ diagram contains metastable parts, which are not automatically recognized by BINGSS. Calculating the $T-\mu$ diagram by BINFKT enables one to distinguish between stable and metastable regions of the two-phase fields.

7.2 How to use BINGSS

7.2.1 General features of BINGSS

BINGSS is restricted to binary systems. For ternary systems there is a similar program called TERGSS. Quasibinary systems can often be treated like binary systems by BINGSS.

BINGSS was initially constructed for phases described by the RK formalism, with the assumption that the molar functions of state of a phase are completely determined by T, p, and the mole fractions of the components. Site fractions as used in the CEF and similar variables initially were not used. Later, most of the models defined for binary phases in the CEF were implemented. However, there is now a distinction between cases in which the site fractions are already defined by the mole fractions of the components (for example austenite, the interstitial solution of C in fcc Fe) and cases in which one or more degrees of freedom remain when the site fractions are related to the mole fractions. Models in which the number of degrees of freedom is more than one are not implemented in BINGSS. Also pressure dependence is not implemented in BINGSS. Since these cases are not often modeled, most of the binary assessments can be done by BINGSS.

The equilibrium conditions are used in BINGSS as formulated in Eq. (2.22), with all functions of state formulated as functions of T, p, and x_j^α (not $y_j^{\alpha,s}$). The site fractions are related to the x_j^α and a single site fraction is selected as the independent one and called y. They are calculated for internal equilibrium of the α phase by solving the condition $\partial G_{\rm m}^\alpha / \partial y = 0$. With the resulting site fractions, all molar functions of state of phase α and all their derivatives can be calculated using the concept of implicitly defined functions (see section 2.3.1).

The data necessary to run the program are collected in two input files, which must be prepared by a text editor before running the program. One of these files contains the analytical descriptions of all the phases, namely the model selected for each of the phases and the selection of which coefficients shall be adjusted and which kept fixed. It has FORTRAN file code 01 and is called "xxxx.coe." The second file contains the experimental data; it has FORTRAN file code 03 and is called "xxxx.dat." The name "xxxx" is arbitrary, but must be the same for both files; usually the symbols of the two elements are taken for it.

Some decisions regarding how to start the calculation and how to continue are either provided in an extra file, FORTRAN file code 04, "xxxx.bgl," or given interactively from the terminal, if file xxxx.bgl does not exist or is empty. Mostly the program is started with file xxxx.bgl and continued interactively.

For the data output also two files are used, "store.coe" and "output.lst." The file store.coe has the same structure as xxxx.coe, but contains the final numerical values of the adjusted coefficients. If the calculation was successful, it is copied on request to xxxx.coe, overwriting the former file. To avoid overwriting with unsatisfactory results, it is not automatically copied.

The file output.lst contains information about how the calculation progressed in the various steps. Its content is determined by a variable "LAUF" in file xxxx.bgl or given interactively. By use of the selection LAUF=4 a table can be prepared, giving for each measurement the difference between the measured value and the value calculated with the current set of adjustable coefficients.

7.2.2 *The coefficient file xxxx.coe*

This file is created and modified by a text editor. It is usually prepared as a formatted file using FORTRAN format specifications. Free format (integers, fixed-point real values, floating-point real values, or strings ("abc123xyz") separated by blanks or commas) is also supported by BINGSS. Lines starting with the symbol "$" are comments and are not used by the computer. Such lines may appear everywhere in the file.

If the unary systems are elements and their descriptions are taken from Dinsdale (1991), these may be copied from a file called "unary0x.atd," where 0x denotes the year of the last update. Please check the SGTE website for the latest update. This file contains also templates for all formats of file xxxx.coe. To use these templates most effectively, the text editor should be switched into overwrite (or replace) mode rather than insert mode.

For quasibinary systems the "quasiunary" descriptions may be taken from the SGTE substance database (Ansara 1999), or from another appropriate source.

In the first line of the file some dimensions are given. This line is usually edited successively after the corresponding parts are finished. The line looks like

```
4 5 CaMg 3 2 9 1 3000.00
```

The first integer, NP = 4, is the number of phase descriptions in the "excess-term part" of the file. The second integer, NT = 5, is the number of terms used in the "excess-term part" to describe the temperature dependence of parameters according to Eq. (5.2). It is restricted to the six terms explicitly given by Eq. (5.2). After one blank there follow two-character names of the two components. These names are used only for table headings in the file output.lst, not for the identification of the components. These are identified as components 1 and 2 in the sequence given here. Therefore this sequence must be kept throughout all the files; it also determines the sign of odd RK terms. The following integers, 3, 2 and 9 are the variables NPREF(1), NPREF(2) and NTU of the following "reference states part." The integer 1 specifies that Eq. (5.2) is used to describe the temperature dependence of G. Initially another formulation of this equation was used, which was derived from the description of C_p by a polynomial in T:

$$G = b_1 - b_2 \cdot T + b_3 \cdot T \cdot [1 - \ln(T)] - b_4 \cdot T^2/2 - b_5 \cdot T^{-1}/2 - b_6 \cdot T^3/6 \tag{7.4}$$

This description of temperature dependence can be selected by integer 0. Finally, the real number 3000.00 is the maximum temperature for which it is recommended that the dataset should be used.

The following "reference-state part" contains Gibbs-energy descriptions of the stable phases of the unary systems, called reference phases in the following text. They are functions of temperature after Eq. (5.2), here not restricted to six terms. Also here several different sets of coefficients a_i may be used for different temperature ranges. The number of different reference phases is given in the first line for both elements separately as integers NPREF(1) and NPREF(2). Also the maximum number of different coefficients a_i is given in the first line as the integer NTU. The description for each reference phase starts with a line giving the number of different temperature ranges and a name. This name is not used by the computer and may be considered as a comment. Then, for each temperature range, depending on NTU, there follow one (NTU $\leq$ 4), two (5 $\leq$ NTU $\leq$ 9), or three (10 $\leq$ NTU) lines giving the lower limit of the temperature range and the coefficients a_i. An example for NTU=9 is

```
3Al_(fcc)
298.00 -7.97615000D+03 1.37093038D+02 -2.43671976D+01 -1.88466200D-03
7.4092000D+04 -8.7766400D-07 0.000000D+00 0.000000D+00 0.000000D+00
700.00 -1.12762400D+04 2.23048446D+02 -3.85844296D+01 1.85319820D-02
7.4092000D+04 -5.7642270D-06 0.000000D+00 0.000000D+00 0.000000D+00
933.47 -1.12783780D+04 1.88684153D+02 -3.17481920D+01 0.00000000D+00
0.0000000D+00 0.0000000D+00 0.000000D+00 0.000000D+00 -1.231000D+28
3 Al liquid
298.00 3.02887900D+03 1.25251171D+02 -2.43671976D+01 -1.88466200D-03
7.4092000D+04 -8.7766400D-07 0.000000D+00 7.933700D-20 0.000000D+00
700.00 -2.71210000D+02 2.11206579D+02 -3.85844296D+01 1.85319820D-02
7.4092000D+04 -5.7642270D-06 0.000000D+00 7.933700D-20 0.000000D+00
933.47 -7.95996000D+02 1.77430178D+02 -3.17481920D+01 0.00000000D+00
0.0000000D+00 0.0000000D+00 0.000000D+00 0.000000D+00 0.000000D+00
```

The first line specifies three temperature ranges and the name "Al_(fcc)," the next pair of lines gives the lower temperature limit and the nine coefficients a_1 to a_9, the following pair of lines is for the temperature range starting at 700 K, and the last pair is for the range above 933.47 K. The last range is the range of the metastable solid above the melting temperature containing the term $a_9 \cdot T^{-9}$ (Andersson 1985). The following reference phase is liquid Al, containing the term $a_8 \cdot T^7$ (Andersson 1985) in the ranges below the melting temperature.

The various reference phase descriptions are identified by integer numbers indicating their sequence in the file. They are counted starting with No. 1 first for the first element and starting again with No. 1 for the second element.

The following part of the file is the "excess-term part." It consists of descriptions of each phase of the binary system. For each phase in one line the model is chosen and a name is given. Several models need some stoichiometric coefficients for complete description, which are also given in this line. Finally, the reference states of both elements used in the ${}^{\rm srf}G^{\theta}$ part of the description of $G^{\theta}_{\rm m}$ in Eq. (5.1) are selected by two integers from the "reference phases" of the previous part of this file. The following lines give the parameters for this phase as functions of T in accord with Eq. (5.2), using the coefficients a_1 to a_6. The parameters are identified by their sequence, differently for each type of description. For each of these coefficients a one-digit integer is used as a logical variable to decide whether the coefficient will be adjusted during the optimization or kept constant. For a phase described by the RK formalism with terms ${}^0L^{\phi}_{\rm A,B}$ to ${}^2L^{\phi}_{\rm A,B}$ the description of the excess part may be as follows:

```
 5phi     12 0 0 0 0 0 0 0 2 1
000000    0.00  0.00000  0.000000  0.000000     0.  0.00000
000000    0.00  0.00000  0.000000  0.000000     0.  0.00000
110000    0.00  0.00000  0.000000  0.000000     0.  0.00000
110000    0.00  0.00000  0.000000  0.000000     0.  0.00000
100000    0.00  0.00000  0.000000  0.000000     0.  0.00000
```

In the first line the number "5" means that there are five parameters for this phase; "phi" is the name; and "12" is the code for Redlich–Kister. By this code also the parameters are defined as ${}^0G^{\phi}_{\rm A} - {}^0G^{\rm ref}_{\rm A}$, ${}^0G^{\phi}_{\rm B} - {}^0G^{\rm ref}_{\rm B}$, ${}^0L^{\phi}_{\rm A,B}$, ${}^1L^{\phi}_{\rm A,B}$, and ${}^2L^{\phi}_{\rm A,B}$. The following four zeros are reserved for stoichiometric constants and are not used for the RK formalism. The next two zeros are for counting the parameters of a magnetic description (here, 0) and the parameter p (given as integer $1000 \cdot p$). The next two numbers indicate that for element A the second reference phase is chosen and for element B the first reference phase is chosen. The reference phases correspond to ${}^0G^{\rm ref}_i - H^{\rm SER}_i$ for element i.

The following lines give the six one-digit integers and six coefficients of the five parameters. Both ${}^0G^{\phi}_i - {}^0G^{\rm ref}_i$ are zero here. For ${}^0L^{\phi}_{\rm A,B}$ and ${}^1L^{\phi}_{\rm A,B}$ two coefficients are adjusted and for ${}^2L^{\phi}_{\rm A,B}$ only one. The number of coefficients to be adjusted must be less than or equal to the maximum number of coefficients used per line, which is given by the integer NT in the first line of the file.

For referencing in file xxxx.dat and everywhere else during the calculations, the phases are identified by integer numbers denoting their sequence in the file. There may appear several different descriptions for the same phase, which in calculations by BINFKT may be used to compare these different descriptions on the same plot using different line styles (dashings or colors).

In the last part linear constraints relating the coefficients being adjusted can be introduced. These constraints are of the form $\Delta\text{coeff} = \text{factor} \cdot \Delta V_i$, where V_i are the "independent coefficients" to be adjusted and "factor" means the factor by which an independent coefficient contributes to the coefficient appearing in the formalism of the description. Each "independent coefficient" is identified by an integer and each "coefficient of the formalism" by its line and column number in the "excess-term part" of this file. As an example the correlation of the two coefficients a_1 and a_2 of parameter ${}^0L$ (line 3) according to the estimate of Tanaka *et al.* (1990) with a factor of $T_0 = 8400\,\text{K}$ is shown below:

```
1 3 1 8.40000000D+03
1 3 2-1.00000000D+00
```

The file is finished by a line containing only zeros and blanks. Behind that line comments may be added. A set of items useful for complete documentation is given in the file "unary.atd." It is strongly recommended that all these comments be added, as long as they are available without needing to search for them again in old protocols.

7.2.3 *The experimental data file xxxx.dat*

The file containing the experimental data, xxxx.dat, must be created in parallel to the xxxx.coe file. The sequence of the components and the numbering of the phases must be common to xxxx.coe and xxxx.dat. That must be decided before starting to construct the files.

Each experiment must be completely described by the FORTRAN variables NTYP, NPHA, ITU(1) to ITU(4), W±DW, T±DT, TT, DTT, and X(1) to X(6), some of which are not used for some types of experiments. The experiments are classified into types by NTYP and NPHA, where NTYP distinguishes between partial Gibbs energies of element A (1) or B (2), enthalpy differences between single-phase samples at the same temperature (3) or different temperatures (4, 12), enthalpy differences between two-phase equilibrium samples (5), entropy differences (9, 10, 11), phase-diagram mole fractions (6) or temperatures (8), equilibrium phase amounts (7), equilibrium mole fractions at isopiestically given chemical potentials (14, 15), or site fractions of a species on a sublattice (13). NPHA gives the number of phases involved in the experiment. Reference states need not be counted in NPHA, although they must be specified as phases. The array ITU(I) identifies the phases including the reference states. W±DW is the measured value in SI units. T±DT and, if necessary, TT are temperatures in kelvins. If TT is used, DTT describes the accuracy of (T−TT), not of TT itself. The X(I) are mole fractions.

Experiments are classified into standard types, for which the equations of error (Eq. (2.52)) are summarized in Table 7.1. This classification covers nearly all possible experiments giving quantitative information connected with thermodynamic functions, except pressure dependence. Which of the above variables are used and their specific meanings depend on NTYP and NPHA, as summarized in Table 7.2.

Table 7.1 *Equations of error related to binary experimental data*

Type		IVERS	Calculated value – measured value
1	1	All	$\mu'_1 - {}^\circ G_1^{\mathrm{ref}} - \underline{\mu_1}$
1	2	1, 3, 5, 6, 7	$G' \cdot x''/(x''-x') + G'' \cdot x'/(x'-x'') - {}^\circ G_1^{\mathrm{ref}} - \underline{\mu_1}$
1	2	2, 4, 8, 9	$\mu_{1,\mathrm{calc}}^{\mathrm{Eq}} - {}^\circ G_1^{\mathrm{ref}} - \underline{\mu_1}$
2	1	All	$\mu'_2 - {}^\circ G_2^{\mathrm{ref}} - \underline{\mu_2}$
2	2	1, 3, 5, 6, 7	$G'' \cdot (x'-1)/(x'-x'') + G' \cdot (x''-1)/(x''-x') - {}^\circ G_2^{\mathrm{ref}} - \underline{\mu_2}$
2	2	2, 4, 8, 9	$\mu_{2,\mathrm{calc}}^{\mathrm{Eq}} - {}^\circ G_2^{\mathrm{ref}} - \underline{\mu_2}$
3	1	All	$H'(T) - (1-x') \cdot {}^\circ H_{x=0}^{\mathrm{ref1}}(T) - x \cdot {}^\circ H_{x=1}^{\mathrm{ref2}}(T) - \underline{H}$
3	2	All	$H'(T) - H''(T) - \underline{H}$
3	3	All	$H'(T) - (1-m''') \cdot H''(T) - m''' \cdot H'''(T) - \underline{H}$
4	2	All	$H'(T) - H''(T_2) - \underline{H}$
4	3	All	$H'(T) - (1-m''') \cdot H''(T_2) - m''' \cdot H'''(T_2) - \underline{H}$
4	4	All	$H'(T) - [(1-m'''-m'''') \cdot H'' + m''' \cdot H''' + m'''' \cdot H''''](T_2) - \underline{H}$
5	3	2–9	$H'(T) - \left(\dfrac{x'''-x^0}{x'''-x''} \cdot H'' + \dfrac{x^0-x''}{x'''-x''} \cdot H'''\right)(T_2)_{\mathrm{calc}} - \underline{H}$
5	4	2–9	$\left(\dfrac{x''-x^0}{x''-x'} \cdot H' + \dfrac{x^0-x'}{x''-x'} \cdot H''\right)(T)_{\mathrm{calc}}$ $-\left(\dfrac{x''''-x^0}{x''''-x'''} \cdot H''' + \dfrac{x^0-x'''}{x''''-x'''} \cdot H''''\right)(T_2)_{\mathrm{calc}} - \underline{H}$
6 or 8	2	1, 3, 5–9	$G'(T) + (x''-x') \cdot \partial G'/\partial x(T) - G''(T)$
6	2	2, 4	$x'_{\mathrm{calc}}(T) - \underline{x'}$
6 or 8	2	2, 4	if $x' = x''$, extremum, $T_{\mathrm{calc}} - \underline{T}$
6 or 8	3	1, 3, 5, 6, 8	$[G' \cdot (x'''-x'') + G'' \cdot (x'-x''') + G''' \cdot (x''-x')](T)$
6 or 8	3	2, 4, 7, 9	$T_{\mathrm{calc}} - \underline{T}$
7	2	2–9	$m'_{\mathrm{calc}}(T, x^0) - \underline{m'}$
8	2	2, 4	$T_{\mathrm{calc}} - \underline{T}$
9	1	All	$S'(T) - (1-x') \cdot {}^\circ S_{x=0}^{\mathrm{ref1}}(T) - x \cdot {}^\circ S_{x=1}^{\mathrm{ref2}}(T) - \underline{S}$
9	2	All	$S'(T) - S''(T) - \underline{S}$
9	3	All	$S'(T) - (1-m''') \cdot S''(T) - m''' \cdot S'''(T) - \underline{S}$
10	2	All	$S'(T) - S''(T_2) - \underline{S}$
10	3	All	$S'(T) - (1-m''') \cdot S''(T_2) - m''' \cdot S'''(T_2) - \underline{S}$
11	3	All	$S'(T) - (1-m''') \cdot S''(T) - m''' \cdot S'''(T_2) - \underline{S}$
12	3	All	$H'(T) - (1-m''') \cdot H''(T) - m''' \cdot H'''(T_2) - \underline{H}$
13	1	2–9	$(T_2 = 0)$ $y'_{\mathrm{Va,calc}}(T) - \underline{y'_{\mathrm{Va}}}$
13	1	2–9	$(T_2 \neq 0)$ $y'_{\mathrm{Va,calc}}(T) - y'_{\mathrm{Va,calc}}(T_2) - \underline{(y'_{\mathrm{Va}}(T) - y'_{\mathrm{Va}}(T_2))}$
14, 15	1	1, 3, 5, 6, 7	$\mu'_i - {}^\circ G_i^{\mathrm{ref}} - \underline{\mu'_i}$ $(i = \mathrm{NTYP} - 13)$
14, 15	1	2, 4, 8, 9	$x'_{\mathrm{calc}}(\mu_i) - \underline{x'(\mu_i)}$

The underlined symbols are the measured values. All other symbols are either independent variables, given in file xxxx.dat as specified by Table 7.2, or calculated from the current set of adjustable coefficients. The subscript "calc" indicates the result of a Newton–Raphson calculation of a two- or three-phase equilibrium. Va denotes vacancies.

Table 7.2 *A summary of binary experimental values used in the file xxxx.dat*

TYP		W	DW	T	DT	TT	DTT	X(1)	X(2)	X(3)	X(4)	X(5)	X(6)	n
1	1	$\underline{\mu_1}$	$\Delta\mu_1$	T	ΔT	-	-	x'	$\Delta x'$					1
1	2	$\underline{\mu_1}$	$\Delta\mu_1$	T	ΔT	-	-	x'	$\Delta x'$	x''	$\Delta x''$	-	-	2
2	1	$\underline{\mu_2}$	$\Delta\mu_2$	T	ΔT	-	-	x'	$\Delta x'$					1
2	2	$\underline{\mu_2}$	$\Delta\mu_2$	T	ΔT	-	-	x'	$\Delta x'$	x''	$\Delta x''$	-	-	2
3	1	$\underline{H}$	ΔH	T	ΔT	-	-	x'	$\Delta x'$					1
3	2	$\underline{H}$	ΔH	T	ΔT	-	-	x'	$\Delta x'$			$(x' = x'')$		1
3	3	$\underline{H}$	ΔH	T	ΔT	-	-	m'''	$\Delta m'''$	x''	$\Delta x''$	x'''	$\Delta x'''$	2
4	2	$\underline{H}$	ΔH	T	ΔT	T_2	$\Delta\Delta T$	x'	$\Delta x'$			$(x' = x'')$		1
4	3	$\underline{H}$	ΔH	T	ΔT	T_2	$\Delta\Delta T$	m'''	$\Delta m'''$	x''	$\Delta x''$	x'''	$\Delta x'''$	2
4	4	$\underline{H}$	ΔH	T	ΔT	T_2	$\Delta\Delta T$	m'''	Δm	m''''	x''	x'''	x''''	2
5	3	$\underline{H}$	ΔH	T	ΔT	T_2	$\Delta\Delta T$	x^0	Δx^0	x''	x'''	$(x^0 = x')$		2
5	4	$\underline{H}$	ΔH	T	ΔT	T_2	$\Delta\Delta T$	x^0	Δx^0	x'	x''	x'''	x''''	2
6	2	-	-	T	ΔT	-	-	$\underline{x'}$	$\Delta x'$	x''	-	$(x' \neq x'')$		2
6	2	-	-	$\underline{T}$	ΔT	-	-	x'	$\Delta x'$	x''	-	$(x' = x'')$		2
6	3	-	-	$\underline{T}$	ΔT	-	-	x'	-	x''	-	x'''	-	2
7	2	$\underline{m'}$	$\Delta m'$	T	ΔT	-	-	x^0	Δx^0	x'	-	x''	-	2
8	2	-	-	$\underline{T}$	ΔT	-	-	x'	$\Delta x'$	x''	-	-	-	2
8	3	-	-	$\underline{T}$	ΔT	-	-	x'	-	x''	-	x'''	-	2
9	1	$\underline{S}$	ΔS	T	ΔT	-	-	x'	$\Delta x'$					1
9	2	$\underline{S}$	ΔS	T	ΔT	-	-	x'	$\Delta x'$			$(x' = x'')$		1
9	3	$\underline{S}$	ΔS	T	ΔT	-	-	m'''	$\Delta m'''$	x''	$\Delta x''$	x'''	$\Delta x'''$	2
10	2	$\underline{S}$	ΔS	T	ΔT	T_2	$\Delta\Delta T$	x'	$\Delta x'$			$(x' = x'')$		1
10	3	$\underline{S}$	ΔS	T	ΔT	T_2	$\Delta\Delta T$	m'''	$\Delta m'''$	x''	$\Delta x''$	x'''	$\Delta x'''$	2
11	3	$\underline{S}$	ΔS	T	ΔT	T_2	$\Delta\Delta T$	m'''	$\Delta m'''$	x''	$\Delta x''$	x'''	$\Delta x'''$	2
12	3	$\underline{H}$	ΔH	T	ΔT	T_2	$\Delta\Delta T$	m'''	$\Delta m'''$	x''	$\Delta x''$	x'''	$\Delta x'''$	2
13	1	$\underline{y_{Va}}$	Δy_{Va}	T	ΔT	$(T_2$	$\Delta\Delta T)$	x'	$\Delta x'$					1
14	1	μ_1	$\Delta\mu_1$	T	ΔT	-	-	$\underline{x'}$	$\Delta x'$					1
15	1	μ_2	$\Delta\mu_2$	T	ΔT	-	-	$\underline{x'}$	$\Delta x'$					1

The measured quantities are underlined. The other values are independent variables or used as starting values of a Newton–Raphson calculation.

$H \pm \Delta H$, $\mu \pm \Delta\mu$ = value in J mol^{-1} of atoms
$T \pm \Delta T$, T_2 = temperatures (K)
$\Delta\Delta T$ = accuracy of $T - T_2$
$x' \pm \Delta x'$ = mole fraction of phase ITU(1)
$x^0 \pm \Delta x^0$ = overall mole fraction
$m' \pm \Delta m'$ = amount of phase ITU(1) in moles of atoms per total amount
- = value is not used
TYP = two integers, NTYP and NPHA
n = number of lines used in file xxxx.dat for this value

7.2.3.1 Compiling experimental values using the program BINDAT

The file xxxx.dat can, in principle, be written as an ASCII file by any text editor. Usually several calculations are necessary in order to get the values for the variables in the file from the data reported in a publication, such as transformations from mass fractions to mole fractions or changing units to SI units. To facilitate this, a program called BINDAT was written.

This program is constructed so that one need enter only once data that usually are equal throughout a set of measured values, thus minimizing the input for a single value to the few numbers characteristic for this value. Values that are equal only for a subseries may be prompted intermediately. Because of the manifold different ways to report data of the same kind, this idea may turn out not to be used 100% consequently. At each level of (sub)series a condition is prompted, by which the program switches to the previous level. This condition is an input that is impossible for the variables prompted, namely a "phase number" zero, a "mole fraction" outside the range 0–1, or a "temperature" below 0 K.

For the accuracies of the measurements a standard estimation procedure is used. The estimated accuracy of each value is additively composed of a part proportional to the value and of a constant part, $v = v \pm \Delta v$, with $\Delta v = v \cdot \mathsf{DWP} + \mathsf{DWA}$. The constant part reflects the minimum quantity detectable by the method applied and the proportional part reflects the usual trend that the inaccuracy of large measured quantities increases proportionally to the quantity itself. The factor of proportionality as well as the constant part are given once for a series of measurements. The accuracy of mole fractions is estimated similarly. Since a mole fraction near unity corresponds to a small mole fraction, the proportional part is taken as a fraction of $(1-x) \cdot x$ rather than as a fraction of the mole fraction x itself. Since the maximum of $(1-x) \cdot x$ is 1/4 in this context, this fraction is multiplied by 4 in order to make the maximum estimated inaccuracy equal to the factor plus the constant value given by $x = x \pm \Delta x$ with $\Delta x = x \cdot (1-x) \cdot \mathsf{DXP} + \mathsf{DXA}$. For the accuracy of the temperature only a constant value $\Delta T = \mathsf{DT}$ is estimated. For the variables DWP, DWA, DT, DXP, and DXA default values of 0.05, 0.0, 5.0, 0.004, and 0.001 are proposed. It is necessary to estimate these accuracies to be equal for experimental data of similar quality, even if the accuracies given in the papers are different. If all the values of the inaccuracies were changed by the same factor, this would not change the result of the least-squares calculation, since it is just equivalent to a change of meaning of the inaccuracies, for example from a mean error to a limit of the 99% confidence range.

Since equivalent values may be reported differently in the literature, besides the variables NTYP and NPHA a third variable ISELEC is used to select the proper type for a series of measured values. For each series a two-character label is requested. This label in BINGSS is used to change the weighting by multiplying it by a dimensionless factor (including zero for omitting this series). In BINFKT the label is used to select different symbols for plotting the measured points. Therefore, if data of the same kind and the same origin might need to be distinguished later, they should be compiled as different series with different labels.

During the interactive input to BINDAT all the data input is collected into a file called yyyy.out. This file may be renamed to yyyy.bin and BINDAT can be executed again, using this file as input rather than interactive input. Typing errors may be corrected and new values added to yyyy.bin by a text editor. The formatted experimental data file appears as file yyyy.app. Several of these files with different names "yyyy" may be merged by a text editor to give a single file "xxxx.dat."

7.2.3.2 An example of running BINDAT

Starting BINDAT gives the following messages on the screen (the first two questions are answered by "0" and "MgY", respectively):

```
=================================================================
 Give: from terminal                    (0)
          from file 01 = "sys".bin      (1)
0

=================================================================
 Give: 2 element names
 (2 characters each without blank between and before)
MgY

=================================================================
 Give: LABEL, NTYP, NPHA, ISELEC
                     ========== NTYP = 0: Stop ==========

 LABEL: 2 characters without blank before

 NTYP = 1 or 2,     NPHA = 1 or 2,    ISELEC = 1: my (emf, RTlna,..)
        1 or 2             1 or 2              2: my as A + BT + CTlnT
        1 or 2             1 or 2              3: vapor pressure by dew point
        1 or 2             1 or 2         4 or 5: like 1...2, but excess my
        1 or 2             1 or 2              6: activity
 NTYP = 3,          NPHA = 1,         ISELEC = 1: H(mix) or H(form)
 NTYP = 3 or 4,     NPHA = 2,         ISELEC = 1: Delta-H of two phases
        4                  2                   2: Cp-values
 NTYP = 3,4 or 12,NPHA = 3 or 4,      ISELEC = 1: Enthalpies of mixing
                                                  (series)
        3                  3                   2: partial enthalpies
        3                  3 or 4              3: Series, reported as H(mix)
 NTYP = 6 or 8,     NPHA = 2 or 3,    ISELEC = 1: phase diagram
        6 or 8             1                   1: for plot by BINFKT only

  LABEL = "$ ": Comment
 0  0 0 0
```

This is a short list of all the types of experiment implemented in BINDAT and their usual descriptions in the literature.

The most effective way of running BINDAT is to give interactively some few values and interrupt the program as done here by giving four items "0 0 0 0" with 0 for NTYP. BINDAT creates a file MgY.out, which can be used as a template for adding more input data by a text editor. The file is then renamed MgY.bin. Starting BINDAT again and answering the first two questions by "1" and "MgY" compiles the data. The file MgY.out contains at the end of each line, after the relevant numbers, the names of the data given in that line.

Part of the file MgY.bin is listed here containing two series of calorimetric measurements, for which cold samples (298 K) were thrown into a calorimetric bath at $T = 955$ or 974 K containing initially Y1 moles of Mg and Y2 moles of Y. Each sample contained Y3 moles of pure Y (X3 = 1, ITU(3) = 3), the measured heat effects were HM J. (ITU(3) = 3 is defined by the file MgY.coe to be the phase hcp_A3.) This type of measurements is explained in Fig. 4.2.

```
MgY   Element names
$ FA [95Aga] R. Agarwal, H. Feufel, F. Sommer, J. Alloys Comp. 217
$    (1995) 59-64
FA  12  3  1 LABEL,NTYP,NPHA,ISELEC
 1    2   1  1 ICONC,IELEM,ITEMP,IVAL
 .050000       .00    5.00    .004000 .001000 DWP,DWA,DT,DXP,DXA
 2   2   955.00   298.00    5.00  ITU(1),ITU(2),T,TT,DTT
   .240880    .000000  Y1, Y2
   .003580   -123.3000 1.000000   3  Y3,HM,X3,ITU(3)
   .003930   -115.4000 1.000000   3  Y3,HM,X3,ITU(3)
   .005200   -141.3000 1.000000   3  Y3,HM,X3,ITU(3)
   .005200   -118.6000 1.000000   3  Y3,HM,X3,ITU(3)
   .006270   -112.5000 1.000000   3  Y3,HM,X3,ITU(3)
   .005660   -42.8000  1.000000   3  Y3,HM,X3,ITU(3)
   .007330    35.8000  1.000000   3  Y3,HM,X3,ITU(3)
 0. 0. 0. 0    new series
 -1. -1.    new phases,  new temp.
  2 2   974.00   298.00    5.00  ITU(1),ITU(2),T,TT,DTT
   .142330     .000000   Y1, Y2
   .001260   -65.0000   1.000000   3  Y3,HM,X3,ITU(3)
   .001490   -48.4000   1.000000   3  Y3,HM,X3,ITU(3)
   .001150   -48.1000   1.000000   3  Y3,HM,X3,ITU(3)
   .001610   -63.6000   1.000000   3  Y3,HM,X3,ITU(3)
   .001990   -19.7000   1.000000   3  Y3,HM,X3,ITU(3)
   .001910   -54.1000   1.000000   3  Y3,HM,X3,ITU(3)
   .001920   -47.7000   1.000000   3  Y3,HM,X3,ITU(3)
   .001940   -44.1000   1.000000   3  Y3,HM,X3,ITU(3)
   .001870   -30.4000   1.000000   3  Y3,HM,X3,ITU(3)
 0.  0. 0.  0    new series
 -1. -1.       new phases, new temp.
 0 0 0. 0. 0.      new LABEL, NTYP, NPHA
   0  0  0  0      End of input, stop
```

Now running BINDAT, giving the two answers "1" and "MgY," results in

```
Give: from terminal                        (0)
          from file 01 = "sys".bin         (1)
```

```
1
 Please give file (+ path) name (without apostrophes)
MgY
    16 values
```

Sixteen experimental items were created and compiled into a newly created file, MgY.app, listed here:

```
$ FA [95Aga] R. Agarwal, H. Feufel, F. Sommer, J. Alloys Comp. 217
$     (1995) 59-64
FA 2 2 3 012 3  -504.38  25.22  95500  500  29800  500  1464452   123088
          0  100000100000000   100000
FA 2 2 3 012 3  -464.59  23.23  95500  500  29800  500  1582189   124914
   1464452  123088100000000   100000
FA 2 2 3 012 3  -557.20  27.86  95500  500  29800  500  2050554   132136
   3023471  146913100000000   100000
FA 2 2 3 012 3  -458.29  22.91  95500  500  29800  500  2009351   131504
   5012027  176173100000000   100000
FA 2 2 3 012 3  -424.43  21.22  95500  500  29800  500  2365502   136953
   6920669  203067100000000   100000
FA 2 2 3 012 3  -158.10   7.90  95500  500  29800  500  2090721   132752
   9122463  232644100000000   100000
FA 2 2 3 012 3   128.75   6.44  95500  500  29800  500  2636217   141068
  11022459  256920100000000   100000
FA 2 2 3 012 3  -452.68  22.63  97400  500  29800  500   877498   113917
          0  100000100000000   100000
FA 2 2 3 012 3  -333.61  16.68  97400  500  29800  500  1027020   116264
    877498  113917100000000   100000
FA 2 2 3 012 3  -328.93  16.45  97400  500  29800  500   786432   112484
   1895506  129753100000000   100000
FA 2 2 3 012 3  -430.19  21.51  97400  500  29800  500  1089015   117234
   2667031  141534100000000   100000
FA 2 2 3 012 3  -131.48   6.57  97400  500  29800  500  1328172   120969
   3727002  157410100000000   100000
FA 2 2 3 012 3  -356.53  17.83  97400  500  29800  500  1258732   119886
   5005673  176082100000000   100000
FA 2 2 3 012 3  -310.43  15.52  97400  500  29800  500  1249512   119742
   6201397  193069100000000   100000
FA 2 2 3 012 3  -283.42  14.17  97400  500  29800  500  1246787   119700
   7373422  209276100000000   100000
FA 2 2 3 012 3  -193.05   9.65  97400  500  29800  500  1187528   118775
   8528278  224815100000000   100000
 000000000000000000000000000000000000000000000000000000000000000000000000
```

Each measurement is represented by two lines, which, by the FORTRAN format specification,

```
FORMAT (A2,6I2,F9.1,F7.1,2(F7.2,F5.2),8P,2F9.0/T2,8P,4F9.0)
```

are related to the variables given in Table 7.2. A list giving these variables in a more-readable form can be created by BINGSS by specifying the variable LAUF = 3.

7.2.4 *Running BINGSS*

BINGSS is open-source code; it is available from the website of this book. There BINGSS can be used and tested following the instructions on line.

If the files xxxx.coe and xxxx.dat are ready, BINGSS can be run interactively by following the prompt commands. The first standard answers on these prompt commands preferably are given by a file xxxx.bgl looking like

```
1 0 0
1 3 2 9 1 0 1.000E-11 1.000E+00
 0Sv
9 3 2 1 0 0 1.000E-11 1.000E+07
```

The first integer in the first line is non-zero and specifies that the file xxxx.bgl exists. The two following integers, 0 0, specify that files xxxx.coe and xxxx.dat are to be formatted. Non-zero integers specify free-format input from the corresponding file.

The second line contains six integers named LAUF, IVERS, IALGOR, NMAX, NOTAUT, and NOTPHA, and two real numbers, EPS and PMARQ.

- LAUF determines the amount of output given in file output.lst: "1" gives the minimal output, the mean square of error after each complete run of the optimization algorithm; "7" allows one to copy the current set of coefficients into the file "store.coe"; and "9" switches to interactive input.
- IVERS defines the use of the "alternate mode" (section 7.1.1): "1" selects only alternate mode (data for which no alternate mode is defined are skipped); "2" selects normal mode for all values; and "3" is like "1," but values without definition of alternate mode are kept and used in normal mode. Other integers define a mixture of alternate and normal mode, which sometimes is useful in intermediate steps. The corresponding equations of error are given in Table 7.1.
- IALGOR selects between Gauss (1) and Marquardt (2) algorithms. Usually the Marquardt algorithm is selected.
- NMAX gives the number of complete runs of the optimization algorithm before the next prompt asks for continuation.
- NOTAUT gives the number of different "labels" for which the weighting defined by Eq. (2.60) is multiplied by a dimensionless factor.
- NOTPHA: to improve convergence, some phases may be suspended during the first optimization steps, this integer declares how many.
- EPS is a convergence criterion. If the corrections of all coefficients are smaller than the absolute value of the coefficient times EPS, the calculation is defined to have "converged" and stops.
- PMARQ is the Marquardt parameter (see section 2.4.3). It is decreased by a factor of 10 after each successful step and increased by a factor of 10 after each diverging step.

The following line is formatted. It contains NOTAUT groups of a four-digit integer (0 to 9999) and a two-character label. The weighting of the values in xxxx.dat marked by this label is multiplied by this integer taken as percent. There may be eight such groups in the line, up to three lines. If NOTPHA is not zero, there follows a line defining the phases to be suspended by three-digit integers (numbers defined by the sequence of phase descriptions in xxxx.coe).

In the above example the following line gives LAUF = 9 and thus switches to interactive input.

7.2.5 *Plotting of diagrams for comparison with experimental data*

The result of an optimization must be checked by comparing all experimental data with the corresponding values calculated using the optimized dataset. This is usually done by plotting diagrams.

For this purpose the program BINFKT is provided in the BINGSS package. It enables one to calculate tables and to produce plots of molar functions of state of single phases versus temperature or mole fraction, of invariant equilibria (three-phase equilibria or azeotropic points of two-phase equilibria), or of two-phase equilibria between two selected temperatures. "Mapping" of phase diagrams is not provided. All equilibria are calculated between specified phases irrespective of whether they are the most stable ones. This allows plotting of metastable two-phase equilibria. BINFKT also enables one to introduce experimental data compiled in the file xxxx.dat into various types of suitable diagrams.

To check the stability ranges by use of BINFKT, the calculation of T versus μ_i plots is recommended (two-phase equilibrium chemical potentials).

BINFKT uses the same files as BINGSS, namely xxxx.coe and xxxx.dat, as well as an extra file called xxxx.bfl defining all the tables and curves to be calculated. This file may be created by running BINFKT interactively, but usually it is easier to take an existing such file as template and create xxxx.bfl by use of a text editor.

7.2.6 *Some differences from PARROT*

1. BINGSS cannot be used for ternary or higher-order systems.
2. BINGSS has some restrictions regarding the usage of less-common models.
3. BINGSS has built-in equations of error for most types of experimental data.
4. BINGSS can start an optimization with all parameters set to ZERO without any extra preparation of the experimental data used (start with IVERS = 1). If only models in which the number of independent site fractions does not exceed the number of independent mole fractions are selected, this guarantees convergence in very few steps. This procedure creates a set of starting values that will usually give the final optimization in a few steps.
5. BINGSS outputs a complete list of final errors (differences between calculated and measured values) on request (LAUF = 4).

7.3 The PARROT module of Thermo-Calc

With the program PARROT, which is part of the Thermo-Calc software system, one can fit thermodynamic model parameters to all kinds of experimental information that can be measured at an equilibrium, stable or metastable. Any kind of model parameter can be optimized, including magnetic and pressure-dependent parameters, in all models that have been implemented in the Gibbs Energy System (GES), which is also a part of Thermo-Calc. A special version of PARROT, not described here, can also optimize mobilities and activation energies for diffusion.

PARROT is not limited to binary systems but can handle experimental information for systems with up to twenty components. In practice systems with more than five components have never been used in an assessment and in most of those cases binary and ternary parameters only were adjusted or added to fit the multicomponent information.

PARROT is a fully interactive program. It is possible to give all information about the system to be assessed directly from the keyboard to the program, but it is recommended that the user start by creating a number of text files for data and commands, of which the most important are the *setup file* and the *experimental data file*. PARROT has a main module for manipulating the optimizing conditions and a special EDIT-EXPERIMENT module for manipulating each individual experimental equilibrium.

This chapter does not replace the user's guide, which is available together with the software, but it contains practical advice based on experience and will convey the flavor of the system. The user must be aware of the fact that it is his own judgment that is critical. The software is only a tool for trying out various options of model selection and weighting of experimental data. It requires a user with a good understanding of thermodynamics and phase diagrams and some experience of modeling.

7.3.1 The optimization method

For equilibria for which all independent state variables are determined with negligible inaccuracies, the criteria for the best fit, derived from the maximum-likelihood principle, will be minimum in the sum of squares of weighted residuals. Inaccuracies in experimental conditions can be taken into account in two ways in PARROT.

1. The inaccuracies in conditions, i.e., independent state variables, can be prescribed in the POLY-3 interface. In this case an equilibrium will be calculated with the experimental values of independent state variables. The standard deviations of the dependent state variables will be calculated by use of the error-propagation law, presuming linear dependences of the dependent state variables on the independent state variables.
2. The "true" value of the condition can be optimized by using one of the defined variables as the condition. This can be obtained by the IMPORT command in the experimental data file. In this case the experimental observations of the independent state variable should be specified in the EXPERIMENT command in the experimental

data file. The commands that can be used in the experimental data file are a subset of the commands available in the POLY-3 module, with a few extensions.

Both methods can be transformed to the problem of finding the minimum of the sum of squares. Method 2 can be used when several experiments have been performed under the same, badly determined, conditions. The two methods can be mixed in the same optimization run.

The set of adjustable coefficients, in PARROT called variables, that give a minimum of the sum of squares is found by a numerical subroutine called VA05A from the Harwell Subroutine Library.

7.3.2 *The use of PARROT*

The assessor should prepare a number of files that will be used during the assessment. These will be briefly described here. They are

POP file with experimental data,
SETUP file with models and known and unknown parameters,
EXP file with experimental data to be plotted, and
MACRO files for quick calculation of various diagrams.

These files will be described below in more detail. A simple description of the flow of the assessment work would be as follows.

1. Prepare the SETUP and POP and EXP files with a text editor (not a word processor!).
2. Start PARROT and run the SETUP file once to create the work file, usually called the PAR file since its extension is ".PAR." The PAR file is machine-dependent and cannot be read by a text editor. It can be manipulated only through the PARROT module. The PAR file will always contain the last results and is automatically updated whenever it is used in PARROT. Whenever a user wants to "freeze" a reasonable set of model parameters but perhaps continue trying to change the weightings or set of model parameters, it is advisable to make a copy of the PAR file.
3. COMPILE the POP file inside the PARROT module. The experimental data will be stored on the PAR file.
4. Select which variables to optimize.
5. SET-ALTERNATE-MODE ON and optimize all equilibria until they have converged.
6. RESCALE the variables to set the start values to the final values.
7. Optimize and rescale until no more changes occur.
8. SET-ALTERNATE-MODE OFF.
9. Calculate diagrams and compare the results with experimental data. This should be done whenever needed during the steps below also. The sum of errors is not a sufficient measure of the overall fit. You may find it convenient to make MACRO files to calculate several diagrams.

10. Use the EDIT-EXPERIMENT module to COMPUTE-ALL equilibria. Some equilibria might not converge or may converge to results far away from the experimental data. Some hints on how to handle that are given below. Experimental data that cannot be calculated should have SET-WEIGHT zero. SAVE when finished with the EDIT module.
11. Optimize the variables zero times and check the errors carefully, using the LIST-RESULT command. This output gives an overview of the current fit to all experimental data. You may have to use the EDIT module again to correct or remove (SET-WEIGHT zero) some equilibria.
12. Optimize and rescale the variables until the calculation has converged. You may find that some variable becomes very large or very small. That may be due to a lack of experimental information. You may have to increase or decrease the number of optimized variables and also use the EDIT module to select the weightings of the various experimental equilibria until you get reasonable results.
13. You may have to optimize "in parts," keeping the variables for some phases fixed and optimizing others with respect to selected sets of experimental data. The selections of experiments are made in the EDIT module.
14. You may have to iterate several times through all points above, even editing the POP and SETUP files, before you are satisfied. You may try various models for the phases and various numbers of variables for each phase.
15. A final optimization with all variables and all experiments with their selected weightings should be carried out.
16. Write the report. When you do this, you may find that you cannot explain some decisions made during the optimization; and you may have to go back to optimize and try various new options.

7.3.3 The experimental data file, the POP file

The experimental data on a system, taken from the literature or measured by the assessor, should be written onto a file called a "POP" file (because the default extension is .POP). The experimental equilibria and measurements are described with POLY commands, with some additional features. The commands that are legal in a POP file are described in a special section of the POLY manual. It is very important to understand the state-flexible variable notation used in POLY and PARROT.

The POP file is a very important form of documentation because it describes the known experimental data for a system. The POP file is intended to be self-documenting and readable both to a human and to the computer. The experimental data are described independently of the models selected for the phases. It is thus possible to use the same POP file to assess a system using different models for the phases. It is not uncommon that a system must be reassessed some years later when new information is available, or if a model for a phase should be changed. Since the reassessment may be done by someone other than the person who created the POP file, it is important that the information in the POP file is well organized and documented.

The recommended way to specify an experimental equilibrium is to document as closely as possible the actual experimental conditions. Usually the set of stable phases is known and also the temperature, pressure, and some or all compositions. We will now describe how to enter some typical experimental data.

7.3.3.1 Single-phase experiments

Experiments with a single stable phase most often concern enthalpies of mixing or chemical potentials. As an example, the following describes an experiment on the Au–Cu system:

```
CREATE-NEW-EQUILIBRIUM 1 1
CHANGE-STATUS LIQUID=FIX 1
SET-CONDITION T=1379 P=1E5 X(LIQUID,AU)=0.0563
SET-REFERENCE-STATE AU LIQ * 1E5
SET-REFERENCE-STATE CU LIQ * 1E5
COMMENT Measurement by Kopor and Teppa, Met, Ann. Phys. 1927 p 123
LABEL ALH
EXPERIMENT HMR=-1520:200
GRAPHICS 5 1379 -1520 1
```

The first five commands in this example are standard POLY commands described in the user's guide, but the first command, CREATE, is rarely used in POLY and may deserve some comments. Each experimental equilibrium must start with the CREATE command and the first integer given after the command is a unique identifier that is later used interactively to set weightings, for example. The second integer is an initialization code; 0 means that all components and phases are initially suspended, 1 means that all components are entered but all phases suspended, and 2 means that all components and phases are entered initially.

The last four commands, COMMENT, LABEL, EXPERIMENT, and GRAPHICS are available only for the POP file and in the EDIT-EXPERIMENT module. EXPERIMENT specifies the quantity to be fitted by the optimization. The syntax of this command is similar to that of the command SET-CONDITION. It is followed by a state variable or a function name and a value and an uncertainty. The EXPERIMENT command is described in detail in the section of the POLY command manual about the EDIT-EXPERIMENT module.

The command COMMENT is followed by a text that will be saved in the work file of the optimization. One may also give comments after a dollar sign, "$", but these comments are lost when the experimental data file is compiled; see section 7.3.7.1.

The command LABEL provides a way to specify a set of equilibria that the user wants to treat as one entity when setting weightings. A label can have at most four characters and must start with the letter A.

The command GRAPHICS can be used to create automatically an experimental data file, the EXP file, when the POP file is compiled. This is a new feature from version P of Thermo-Calc. Read the documentation in the Thermo-Calc manual to understand this file format. The first integer after the command is the dataset number; the next two numbers are the horizontal and vertical coordinates, respectively. The final integer is the symbol to plot.

7.3.3.2 Two-phase experiments

Most experimental information, for example from the phase diagram, involves two or more phases. For example, the melting temperature of an Au–Cu alloy can be described as follows:

```
CREATE-NEW-EQUILIBRIUM 1 1
CHANGE-STATUS PHASE FCC=FIX 1
CHANGE-STATUS PHASE LIQUID=FIX 0
SET-CONDITION X(FCC,CU)=0.14 P=1E5
EXPERIMENT T=970:2
COMMENT H E Bens, J Inst of Metals 1962 p 123
LABEL ALS
GRAPHICS 1 0.14 970 2
SET-ALTERNATE-CONDITION X(LIQUID,CU)=0.16
```

All commands except the last have been described above. The last command specifies an estimated value of the liquid composition at the equilibrium. This command is not necessary except during the alternate-mode calculation described in section 7.3.7.3.

Note that the information in this case was the temperature. One could equally well describe the same melting point with the temperature as condition and the composition as experimental information because both are measured quantities. The selection of quantities used as conditions should be based on the experimental technique. Those known with the least accuracy should be used as experimental data.

7.3.3.3 Experiments on invariant equilibria

It is a peculiarity of PARROT that invariant equilibria are the most important experimental information to be provided for an assessment. It is thus advisable to have all invariant equilibria for a system on the POP file, even if some of them have not been measured explicitly. A reasonable estimate from the available experimental data can often be sufficient, but one should be careful using phase diagrams drawn when there is little data to limit the imagination of the artist. At the end of the assessment such estimated equilibria should be excluded, but they are very useful for obtaining a set of initial values for the model parameters.

An example of a three-phase equilibrium in a binary system follows:

```
CREATE-NEW-EQUILIBRIUM 1 1
CHANGE-STATUS PHASE FCC BCC LIQUID=FIX 1
SET-COND P=1E5
EXPERIMENT T=912:5
```

```
SET-ALTERNATE-CONDITION X(FCC,B)=0.1 X(BCC,B)=0.4 X(LIQ,B)=0.2
LABEL AINV
COMMENT Estimated compositions
GRAPHICS 1 0.1 912 3
GRAPHICS 1 0.2 912 3
GRAPHICS 1 0.4 912 3
```

The alternate conditions are needed only for alternate-mode calculations. Another example of an invariant equilibrium is a congruent transformation when the composition of both phases is the same (but may be unknown):

```
CREATE-NEW-EQUILIBRIUM 1 1
CHANGE-STATUS PHASE BCC LIQUID=FIX 1
SET-COND P=1E5 X(BCC,B)-X(LIQ,B)=0
EXPERIMENT T=1213:10
SET-ALTERNATE-CONDITION X(B)=0.52
LABEL AINV
COMMENT Estimated compositions
GRAPHICS 1 0.52 1213 2
```

With some experience from phase-diagram evaluation, it is possible to estimate metastable invariant equilibria. In particular, such estimated metastable equilibria are useful to reduce the number of phases to be assessed simultaneously. One may, for example, assume that a certain intermediate phase does not form and extrapolate the liquidus curves below the stable three-phase equilibrium and estimate temperatures and compositions of metastable three-phase equilibria between two other phases and the liquid.

Another useful technique is to extrapolate a liquidus line from a peritectic equilibrium to estimate the congruent melting temperature of the compound that melts peritectically.

7.3.3.4 Ternary and higher-order experiments

PARROT can handle optimization of ternary or higher-order information in the same way as binary. The only thing to note is that one more condition is needed for each component added. In practice quaternary and higher-order information is used mainly to optimize binary or ternary parameters. In ternary systems it may be more important to use the feature that one may have uncertainties also on conditions. A tie-line in a binary system is determined if the two phases, the temperature, and the pressure are known and the composition of one of the phases has been measured. For a tie-line in a ternary system, one must have measured at least two compositions, which often have the same uncertainty. One may then assign an uncertainty to the composition selected as condition. For example,

```
CREATE-NEW-EQUILIBRIUM 1 1
CHANGE-STATUS FCC BCC=FIX 1
SET-CONDITION T=1273 P=1E5 X(FCC,B)=0.1:0.02
```

```
EXPERIMENT X(FCC,C)=0.12:.02
LABEL AFB
SET-ALTERNATE-COND X(BCC,B)=0.17 X(BCC,C)=0.07
GRAPHICS 1 0.1 0.12 2
GRAPHICS 1 0.17 0.07 3
```

The alternate conditions will be explained in section 7.3.7.3.

One problem with binary assessments is that the experimental information can often be described almost equally well by very different sets of model parameters. It is often the extrapolation of these assessments into ternary systems that gives decisive information about which set of model parameters is the best. Sometimes information from several ternary systems may be needed in order to decide on the best description of a binary system.

7.3.3.5 Simultaneous use of binary and ternary experiments

PARROT allows simultaneous optimization of binary and ternary (and higher-order) information. By using the CHANGE-STATUS COMPONENT command, one may have experimental data from binary and ternary systems on the same POP file. Note that it is not good technique to set the fraction of a third component to zero for binary experimental information. The CHANGE-STATUS COMPONENT C=SUS is fragile, however, and may need manual setting for it to work properly. An example of a binary three-phase equilibrium in a ternary system is

```
CREATE-NEW-EQUILIBRIUM 1 0
CHANGE-STATUS COMPONENT A B = ENTERED
CHANGE-STATUS PHASE FCC BCC LIQ=FIX 1
SET-COND P=1E5
EXPERIMENT T=1177:10
LABEL AAB
COMMENT from A-B system
```

Note the use of initialization code 0 in the CREATE-NEW-EQUILIBRIUM command. This means that all components must be entered.

7.3.3.6 Experimental data given as inequalities

For some experimental data one does not know the actual value, just that it must be lower or higher than a certain value. It is possible to use such information in the POP file by using ">" or "<" instead of the equals sign in EXPERIMENT. The number after the colon in a factor for the "penalty function" specifies how big the error should be if the value is on the wrong side of the inequality. The smaller the number the quicker the error will rise. The inequality experiment is useful, for example, when a phase appears where it should not be stable, see section 7.3.7.9, but also in other cases, for example when a temperature or composition is not known, but it is known that it should be below or above a certain value.

7.3.4 *The graphical experimental file*

For plotting calculated diagrams together with experimental data, it is recommended that the experimental data be written into the POP file so that they can be used to create a file according to the “dataplot” format automatically. This can be done with the GRAPHICS command shown above. The graphical format is described in the Thermo-Calc manual. The default extension of a dataplot file is .exp. More elaborate graphics can be edited directly in this file by use of a text editor. The command looks like

```
GRAPHICS <dataset> <x-coordinate> <y-coordnate> <GOC>
```

Use different datasets for different types of data, i.e., phase diagrams, enthalpies, activities, etc. Each dataset can be selected separately to be overlayed on the calculated data in the post-processor with the APPEND-EXPERIMENTAL-DATA command. The Graphical Operation Code (GOC) can be just a number representing the symbol plotted at the coordinate or a D, meaning draw a line from the previous coordinate to this point (which is useful to indicate tie-lines, for example). See the Thermo-Calc manual for more examples.

7.3.5 *The model setup file*

The second step is to create a “setup” file with the names of the elements and phases, the models for the phases, and all known information such as “lattice stabilities” and Gibbs energies of formation. Most values for the pure elements can be found in the collection amassed by the Scientific Group Thermodata Europe (SGTE) and published by Alan Dinsdale (Dinsdale 1991) or from the SGTE website (http://www.sgte.org). These parameters can also be extracted from the PURE database, which is distributed free with Thermo-Calc. In the Gibbs-energy module there is a command LIST-DATA with option P that can be used to create a template setup file after extracting the data from the PURE database. This template must be edited and new phases and parameters must be added. The default extension of a setup file is .TCM; this is short for “Thermo-Calc MACRO” file. At the end of the setup file, the work file is created with the PARROT command CREATE-NEW-STORE-FILE. The work file cannot be edited directly and it is hardware-dependent. The default extension of the work file is .PAR.

7.3.5.1 Models for phases

The literature with experimental data collected for the POP file usually contain information useful for modeling the phases. If an assessment should be compatible with an existing database, the models for most solution phases often have to be taken from the database. For intermetallic phases it may be important to determine whether the phase has the same structure as a phase in another system.

In the chapter on methodology, chapter 6, the criteria for model selection are discussed in detail. In summary, one requires

- physical soundness,
- as few parameters as possible that must be optimized,
- reasonable extrapolations of the model, and
- consistency with previous assessments.

7.3.5.2 Model parameters

PARROT has 99 variables, also called coefficients or parameters in this book, for optimization and can handle up to 1000 experimental measurements, but there are limits to the simultaneous number of variables and experiments. At each optimization the program will list these limits. The variables are called V1 to V99 and they are used when entering functions and parameters to be optimized. For example,

```
ENTER-PARAM L(LIQUID,AU,CU;0) 298.15 V1+V2*T; 6000 N
ENTER-PARAM L(LIQUID,AU,CU;1) 298.15 V3+V4*T; 6000 N
ENTER-PARAM L(LIQUID,AU,CU;2) 298.15 V5+V6*T; 6000 N
```

makes it possible to use three RK parameters, each linearly temperature-dependent, to describe the excess Gibbs energy for the liquid phase in the Au–Cu system. The variables V1 to V6 can be optimized to describe the experimental information. In the setup file one often introduces more variables than will be needed, since it is convenient to have them in sequential order for each phase. In some cases the model requires that several thermodynamic parameters are related, which can conveniently be handled by using the same variables for several parameters. For example, the ordering parameters for a B2 ordered phase can be described by

```
ENTER-FUNCTION GAB 298.15 V10+V11*T+GHSERAA+GHSERBB; 6000 N
ENTER-PARAMETER G(B2,A:B) 298.15 GAB; 6000 N
ENTER-PARAMETER G(B2,B:A) 298.15 GAB; 6000 N
```

A stoichiometric compound with measured heat-capacity data may require several variables to describe its temperature dependence, for example

```
ENTER-PARAMETER G(MG2SI,MG:SI) 298.15 V20+V21*T+V22*T*LN(T)+
V23*T**(-1)+V24*T**2+V25*T**3; 6000 N
```

It is possible to optimize all kinds of parameters in the Gibbs-energy systems that can be described as functions of temperature and pressure. Examples of this are Curie temperatures for magnetic transformations, molar volumes, and thermal expansivities,

```
ENTER-PARAMETER TC(SPINEL,FE+2:FE+3:O-2) 298.15 V50+V51*P; 6000 N
```

7.3.6 *File names and relations*

Since all files have different extensions, it is possible to use the same name for all files for an assessment. Thus one may have aucu.POP, aucu.TCM, aucu.exp, and aucu.PAR. In particular, the work file (PAR) may exist in several copies during the assessment; but it is advisable to update the text files POP, TCM, and exp to reflect all changes made interactively in the PAR file.

The text files, i.e., the POP file, the TCM files, and the exp file, are important documentation of the assessment. At the end these should be updated in such a way that it is possible to run the setup file, compile the experiments, and optimize to get the final result directly. This requires that the final weightings be entered into the POP file and the final set of variables used as initial values. With such a set of files, it is easier to reassess the system using new experimental data or new models.

Some care must also be taken with the work (PAR) file. This always contains the last optimized set of variables and the weightings of the selected experimental data. The work file contains a workspace for POLY and for GES. When a diagram is calculated from the current work file, a POLY file is created. This POLY file will have a copy of the current set of variables. If some further optimization is done and the user then by accident tries to READ the old POLY file, he may destroy the new set of variables and overwrite them with the old set. Thus one *must never READ* a POLY or GES file while running PARROT, but one may, of course, SAVE new POLY or GES files, for example when calculating diagrams from the current set of variables.

7.3.7 *Interactive running of PARROT*

With the three files POP, setup, and dataplot, the user can start running PARROT interactively. This can be divided into some initial separate steps. Usually these steps have to be repeated cyclically, modifying weightings, modifying models, adding new information, etc. It is actually difficult to decide when an assessment is finished. Quite often the deadline for the publication sets the limit for the work.

The commands on the setup file are executed by the MACRO command. There is usually a number of error messages and the setup file must be corrected and re-run until there are no errors. In the PARROT module one can list interactively the description of the phases, the parameters' expressions, and the values of the optimizing variables. After a successful run of the setup file, this should be done to check that all models and parameters have been entered correctly.

7.3.7.1 Compilation of experiments

The next step is to "compile" the experimental data file. The command to do this is COMPILE. This compilation will usually also result in a number of error messages due to syntax errors. The compilation normally stops when it detects an error and gives an understandable message. These must be corrected and the file compiled again. It is convenient to use several windows for this, one for editing and one for compiling.

Sometimes an error message is less understandable and the error may have occurred some lines before the program actually discovers it. Consultation with an expert is usually the best way to correct these problems quickly, since it can be difficult to find the right place to look in the manual. Since the setup file and the POP file are text files, they can easily be E-mailed to experts anywhere in the world.

7.3.7.2 Global equilibrium calculations

The global equilibrium calculation is a new feature in Thermo-Calc version R. By default it is turned off in PARROT and, when using POLY from PARROT, one must explicitly turn it on if one wants to use it. Be very careful doing that, since the reason it is turned off is that global equilibrium calculations may automatically create new composition sets and that will destroy the experimental equilibria in the PAR file.

7.3.7.3 Setting alternate mode

When the experimental data file has been compiled correctly, the first really big problem arises. This is that one must try to calculate the experimental information from the models of the phases. Initially all model parameters are zero and in many cases it might not be possible to calculate a measured value from the model unless the parameters have some reasonable values. In PARROT the *alternate* mode makes it possible to calculate most of the experimental equilibria even with zero values of the variables. The command SET-ALTERNATE-MODE ON means that experimental equilibria involving two or more phases are calculated with an approximate technique described in section 7.3.9 below. Some extra information may be needed in the alternate mode, as described in section 7.3.9.2. The user may exclude some of the experimental information to be used by setting weights in the EDIT module.

The OPTIMIZE command in PARROT is then used with the alternate mode until it has converged. Several OPTIMIZE commands are usually needed and the user may have to change the selection of experimental data. This is again done in the EDIT module. The result of an optimization is obtained in a form readable by a human with the command LIST-RESULT. The workfile is continuously updated and always contains the last set of optimized variables and calculated results. Sometimes the user may want to save a copy of the current workfile when trying various selections of experimental data or models. This is done by making a copy of the workfile and giving the copy another name.

7.3.7.4 Plotting intermediate results

Reading the output from LIST-RESULT is usually not enough to understand how good (or bad) a fit one has obtained. Since PARROT is part of Thermo-Calc, it is possible to use POLY directly to calculate the phase diagram and other diagrams with thermodynamic properties and plot them together with the experimental data from the dataplot file. It is important to have MACRO files with the command sequences for such calculations because this should be done frequently. Calculating and plotting the phase diagram may give many surprising results when the variables to be optimized are far from their final values! Also the fit to activities, chemical potentials, and enthalpy data can be checked with MACRO files by comparing calculated values with the data on the EXP files.

7.3.7.5 Selection of experimental data

Initially it can be difficult to know which experimental data are good and which are problematic; see section 6.1.2 for advice. It is also important to determine a reasonable number of variables to be optimized; see section 6.3.

Optimization with the alternate mode using all data may give a set of initial values for the variables that reproduces the main features of the system. If not, one may have to exclude data and maybe phases; see section 7.3.10 for some hints.

When a first set of variables reproducing the main features of the system has been obtained using the ALTERNATE mode, the user should SET-ALTERNATE OFF and thenceforth calculate all equilibria using the normal mode. After turning off ALTERNATE and entering the EDIT module, calculate all experiments with the COMPUTE-ALL command. Several experimental equilibria may still fail to converge and the user may have to provide initial values manually or even remove some equilibria (by setting the weighting to zero). At a later stage in the optimization, when the optimizing variables are closer to their final values, the user may be able to restore and calculate all experimental equilibria.

When the selected set of equilibria can be calculated in the EDIT module, the user should calculate all experiments once again in the PARROT module using the OPTIMIZE command with only 0 iterations. The output from LIST-RESULT should be examined carefully.

In this output the experimental information that is badly fitted will be marked with an asterisk "*" in the rightmost column. It is not a problem that many experimental data are badly fitted at this stage, but one should be careful with errors that the optimizer may not be able to solve by itself. A typical case of such a problem is when a phase undergoes congruent melting and there is experimental information on both sides of the congruent transformation. If the composition of the phase is the experimental information, it may happen that the calculated equilibrium is on the "wrong" side of the congruent point and thus gives a large error. The user must correct such problems manually in the EDIT module. A similar error may occur for miscibility gaps when the experimental information is from one side and the calculation gives the composition of the other side.

7.3.7.6 The listing of results

The LIST-RESULT command gives a list of the current set of optimizing variables and their values and the fit to each selected experimental point. This may look like Table 7.3.

In the first section the variables are listed. The first column is the variable symbol as used in the model parameters; one may use variables V1 to V99. The second column is the current optimized value and the third column is the initial value of the variable. If they are

Table 7.3 *A variable and result listing produced by the LIST-RESULT command*

VAR.	VALUE	START VALUE	SCALING FACTOR	REL.STAND.DEV
V1	-3.51671367E+04	-3.51671367E+04	-3.51671367E+04	1.78212695E-02
V2	4.53814089E+00	4.53814089E+00	4.53814089E+00	1.91804677E-01
V3	-5.56268264E+03	-5.56101433E+03	-5.56101433E+03	2.54804276E-01
V4	-5.35543298E+00	-5.35382683E+00	-5.35382683E+00	2.90371977E-01
V20	-2.70811726E+04	-2.70811726E+04	-2.70811726E+04	1.06182097E-02
V22	-2.77312502E+03	-2.77312502E+03	-2.77312502E+03	9.25739632E-01
V23	-1.61636716E+01	-1.61636716E+01	-1.61636716E+01	1.98209960E-01
V32	-4.76078712E+04	-4.76078712E+04	-4.76078712E+04	1.63612167E-02
V33	3.58242251E+02	3.58242251E+02	3.58242251E+02	2.04819326E-03
V34	-6.92764170E+01			
V35	-5.19246000E-04			
V36	1.43502000E+05			
V37	-5.65953000E-06			

NUMBER OF OPTIMIZING VARIABLES : 9
ALL OTHER VARIABLES NOT LISTED ARE FIX WITH THE VALUE ZERO
THE SUM OF SQUARES HAS CHANGED FROM 2.95795193E+02 TO 2.95795094E+02
DEGREES OF FREEDOM 106. REDUCED SUM OF SQUARES 2.79051976E+00

...				
16 MUR(MG)=-6368	-6988.	4.18E+02	-619.8	-1.483
17 MUR(MG)=-4052	-4161.	3.03E+02	-109.3	-0.3606
18 MUR(MG)=-2316	-2268.	2.16E+02	48.39	0.2240
100 HMR=-3700	-3566.	6.00E+02	134.4	0.2241
101 HMR=-5610	-6161.	6.00E+02	-550.8	-0.9179
102 HMR=-6800	-7852.	6.00E+02	-1052.	-1.754
...				
316 T=1278	1273.	5.0	-5.370	-1.074
317 T=1303	1306.	5.0	2.567	0.5134
370 X(LAVES_C15,MG)=0.313	0.3169	8.00E-03	3.8732E-03	0.4841
371 X(LAVES_C15,MG)=0.346	0.3411	8.00E-03	-4.8968E-03	-0.6121
...				

almost the same, the optimization may be almost finished. The scaling factor is usually the same as the initial value and it is needed because the variables can be very different in magnitude and, by dividing the values by the scaling factor, the variables' values will, at least initially, all be set to unity. The final column is the relative standard deviation (RSD), which is an indication of the significance of the variable for the assessment. If the RSD is larger than 0.5, it may be possible to set this variable to zero and still have about the same reduced sum of squares (the last line of the second section in Table 7.3). It is also important to know the number of significant digits in the RSD; see section 7.4.4. Some of the variables have been kept constant during the last optimization.

After the list of variables, there follows some general information, the number of variables last optimized, that all variables not listed are zero, and how much the sum of squares was changed by the last optimization. The number of degrees of freedom here means the number of experimental points minus the number of optimizing variables. The reduced sum of squares is the sum of squares from the line above divided by the number of degrees of freedom. If this is less than unity, it means that, on average, all experimental data have been fitted to within their estimated uncertainties.

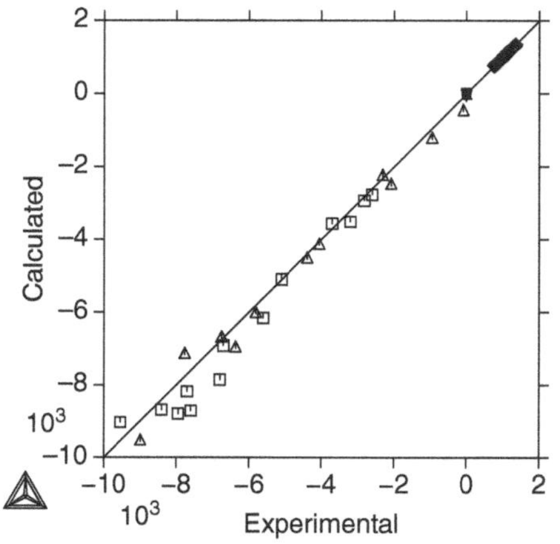

Figure 7.3 Calculated versus experimental values for all experiments from a LIST-RESULT command. If all points are on the diagonal, one has a perfect fit. Since the experimental values have a very wide range, one may have to magnify parts in order to see details. This kind of graph gives a clear indication of whether some experimental datasets are inconsistent with each other.

Finally, there follows one line for each selected experimental point. Only a few are listed. The first column in this list is the number assigned to the experiment with the CREATE command. Then comes the experimental quantity and the measured value, like MUR(MG) = −6368. The exact meaning of the experimental value is not self-evident from the listing; one must use the EDIT module in order to see all the conditions. The third column shows the calculated values for the current set of variables, the fourth column shows the estimated uncertainties, divided by the weightings assigned, and the fifth column presents the differences between the calculated and experimental values. The final column gives the quotient relating the fifth column and the fourth column; if this is less than unity, it means that one has fitted the experimental value to within the estimated uncertainty. It is the sum of the values in this column squared that is minimized.

It is sometimes useful to plot the experimental data versus the calculated data and there is a graphical option of the LIST-RESULT command that achieves this. Such an output is shown in Fig. 7.3.

7.3.7.7 Critical sets of experimental data and of parameters

Following the advice in section 6.1.2, a weighting for each experiment that specifies its importance relative to all the other experiments must be determined. The following points should be taken into account when determining this set.

- The reliability of the experimental technique.
- The agreement between independent measurements of the same quantity.
- The agreement between data obtained with different experimental methods.
- One should use only experimentally determined properties, not quantities that have been converted.

- One should be careful about the estimated accuracy of the experiments.
- One must correct systematic errors (in the temperature scale, for example).
- One should bring experience from previous assessments to bear.
- One should make use of "negative" information, for example that a phase should not be stable within a certain composition or temperature region.

When a sensible weighting of the experimental data has been found, this is called the "critical set."

The assessor may at any time change the set of model parameters to be optimized. On using more adjustable variables, the sum of errors usually decreases, but at the same time the variables become less well determined. A measure of the significance of the variable is given by the column RSD, which is listed for each variable optimized. The RSD is important only at the end of an assessment. The meaning of the RSD is that the variable can be changed by plus or minus this factor without changing the reduced sum of error by more than one unit. A large RSD thus means that the variable is badly determined. The RSD for all variables should be less than 0.5 in an acceptable assessment.

The value of the RSD is significant only when the user has run the optimization after a RESCALE and it has converged almost immediately, when the "scaling factor" is the same as the "value." If there are variables with RSDs larger than unity, it means that too many variables have been used. It is not necessarily the variable with the largest RSD that should be removed, although that is quite likely.

The value of the RSD depends also on the weighting of the experiments. It may be possible to reduce the RSD by changing the weightings.

If one has one or more variables with RSDs larger than unity, one should try to remove one or more optimizing variables by setting them to zero, or to some value that can be determined from other information, for example by semi-empirical estimation methods.

7.3.7.8 Optimize and continue to optimize

Using the critical set of experiments, which may be modified now and again, and trying various numbers of model parameters to be optimized, the user must use his skills to get the best possible result. The smaller the sum of errors the better and, by giving the command OPTIMIZE, CONTINUE, RESCALE, and OPTIMIZE again, the user may finally reach a point at which PARROT states that it cannot improve the set of optimizing variables. This should not be trusted, however, so a few more OPTIMIZE runs should be done. If PARROT converges with the same number of iterations as there are variables to optimize, one has to accept this set.

A problem with using OPTIMIZE several times to make the optimization converge is that sometimes a variable may suddenly start to change by several orders of magnitude. This behavior may lead to impossible values of variables and requires careful reconsideration of the weighting of the critical set of experimental data and of the model parameters that are being optimized. There is a reasonable range to which the values of variables should be restricted. Finding interaction parameters of the order of 10^7 for the temperature-independent part and 10^3 for the temperature dependence is a clear indication of a bad weighting of the experimental data or the use of too many parameters.

If the user is still not satisfied with the overall fit, he has to change the weights or add more information to force the optimization in the right direction. The success of such manipulations depends on the skill of the assessor.

7.3.7.9 Wrong phases at wrong places

During the optimization it may happen that a phase appears at the wrong temperature or composition in the phase diagram. Typically a phase may be stable only on one side of the diagram, but on calculating the diagram using the optimized variables, it may appear also on the other side. This can be handled most easily by the DGM state variable.

For example, if fcc becomes stable at high Mg content in an assessment of the Ag–Mg system, one may modify an experimental equilibrium on the high-Mg side by adding as experimental data the criterion that the driving force for fcc should be negative. For example,

```
CREATE-NEW-EQUILIBRIUM 900 1
CHANGE-STATUS PHASE HCP LIQ=FIX 1
SET-COND P=1E5 X(LIQ,AG)=0.12
EXPERIMENT T=900:10
LABEL AHL
```

can be modified by adding two lines as below:

```
CREATE-NEW-EQUILIBRIUM 900 1
CHANGE-STATUS PHASE HCP LIQ=FIX 1
CHANGE-STATUS PHASE FCC=DORMANT
SET-COND P=1E5 X(LIQ,AG)=0.12
EXPERIMENT T=900:10
EXPERIMENT DGM(FCC)<-.001:.001
LABEL AHL
```

The DGM experiment will give a contribution to the sum of errors if fcc is stable at this equilibrium and this will make the optimizer try to decrease the stability of fcc. One should give a small negative number rather than zero, since fcc is stable also when the driving force is zero. Thus fcc is set as dormant and will not affect the calculated equilibrium, but the driving force for forming fcc will be calculated.

This type of error is common when phases are modeled for the whole composition range but may be stable only on one side. If many RK coefficients, see section 5.6.2.1, are used to describe the phase, then these may have unexpected and unwanted effects on the other side.

The DGM experiment makes use of the feature in PARROT that one may specify a limit rather than a fixed value as experimental data, as described in section 7.3.3.6.

One may use a similar method if a phase is not stable over the whole temperature range within which it should exist. For example, when there are several intermediate phases in a system it may be difficult to make stable those that should be stable at room temperature. One may then use fictitious experimental data with just the DGM experiment to enforce the correct set of stable phases at room temperature. In many cases there are no experimental data for the enthalpy of formation of an intermediate phase, so one should be careful not to trust blindly hand-drawn extrapolations of stability lines.

7.3.7.10 Unwanted miscibility gaps

During an optimization it is possible that the model parameters for some phases create unwanted miscibility gaps in the phase. A typical example is that the liquid sometimes has a miscibility gap at high temperatures. It is possible to keep control of this by adding some "experimental" data that should be outside the spinodal, see section 2.1.10, at a certain temperature and composition like in this example:

```
CREATE-NEW-EQUILIBRIUM 910 1
CHANGE-STATUS PHASE LIQ=FIX 1
SET-COND T=4000 P=1E5 X(LIQ,AG)=0.4
EXPERIMENT QF(LIQUID)>1:.001
LABEL AQL
```

This equilibrium will calculate the single liquid at 4000 K and X(AG) = 0.4 and, if the liquid is inside the spinodal, the value of QF(LIQUID) will be negative. Here it is required that QF(LIQUID) be larger than unity in order to be on the safe side. QF is a "state variable" in PARROT and POLY that has the value of the smallest eigenvalue of the determinant of all second derivatives of the Gibbs energy with respect to all constituents.

One should be careful with added "experiments" like DGM and QF since they may restrict the optimization severely. Toward the end of the optimization, when one should be close to having found a global minimum, one should try to remove them.

7.3.7.11 Phases with order–disorder transitions

A model for a phase that can be ordered like the B2 and $L1_2$ phase will have several sublattices with the same set of constituents; see section 5.8.4. For an experimental equilibrium involving an ordered phase, the disordered state may become more stable than the ordered any time during the optimization. To have control of the state of order in a phase, one can add a function that calculates the difference between the site fractions and returns an error if the state of order is wrong. For example, the congruent transformation of $L1_0$ to A1 in the Au–Cu system can be written

```
ENTER FUNCTION DL10=Y(FCC#3,CU)-Y(FCC#3,CU#3);
ENTER FUNCTION DHTR=HMR(FCC)-HMR(FCC#3);
CREATE 320 1
CHANGE-STATUS PHASE FCC FCC#3=FIX 1
SET-CONDITION P=1 X(FCC,CU)-X(FCC#3,CU)=0
EXPERIMENT T=683:2
```

```
EXPERIMENT DHTR=4000:1000
EXPERIMENT X(CU)=.5:.01
EXPERIMENT DL10>0.1:0.001
```

The model for the ordered (and disordered) FCC phase is

$$(\mathrm{Au, Cu})_{0.25}(\mathrm{Au, Cu})_{0.25}(\mathrm{Au, Cu})_{0.25}(\mathrm{Au, Cu})_{0.25}$$

and the $L1_0$ ordering (indicated as FCC#3) has a high site fraction of Au on two sublattices and a high site fraction of Cu on the other two. The disordered phase (indicated as FCC) has all site fractions equal in all sublattices. The congruent transformation is specified by giving as a condition the stipulation that the two phases should have the same composition. As experimental data the temperature, enthalpy of transformation (FUNCTION DHTR), and overall composition are used. Additionally, a function called DL10 is used to calculate the difference between the site fractions of Cu on the first and third sublattices. If these are equal, the phase is disordered, which is wrong, and the value will be zero, which will give a large error in the EXPERIMENT DL10>0.1:0.001.

One should not use the "partition model," see section 5.8.4.2, for ordering when assessing a phase with an order–disorder transformation unless one has extensive information both on the ordered and on the disordered phase throughout the whole composition range, as in the Au–Cu system, for example. In other cases the partitioning will create problems because there is an ambiguity regarding how much enthalpy should be in the disordered state and how much in the ordered. The split can be made without any ambiguity when the assessment is added to a database.

7.3.7.12 Some more hints

Many problems and errors may occur and it is not possible to give simple explanations of how to handle them. The main recommendation is that one should exclude all experimental data that give strange results but make sure that all important invariant equilibria are reasonably calculated. If some invariant equilibria cannot be calculated, it may be better to exclude the phases that are involved in these, if they are intermediate phases, and just optimize the liquid and the most important solution phases in a first step with a full equilibrium calculation. When reasonable results have been obtained with the most important phases, the intermediate phases may be put back and optimized, keeping the model parameters for the already-optimized liquid and solid phases fixed. Consult section 7.3.10 for more hints, since it is impossible to give any more general advice at this stage.

7.3.8 Changes made interactively that require recompilation

It is possible to change almost everything from the initial setup and POP file interactively. For example, one may add more variables to be optimized, and modify, or add more, experimental information, but there are some changes that will destroy the data structure and thus require that the experimental data file be recompiled. An example of such a change is adding more composition sets to a phase. This actually changes the number of phases, which will destroy the links between the experimental data and the thermodynamic models. Thus the POP file must be recompiled and it is then important that the POP file reflect the changes made interactively in the EDIT-EXPERIMENT module.

One practical problem with recompiling the POP file may also be that the experimental equilibria will be calculated with the default initial values of all compositions. In some systems the equilibria require manual input of initial values and recompilation of the POP file may then require careful massaging of the experimental equilibria to make them converge again. To simplify the recompilation, it is recommended that one use the command AMEND-PHASE in the GES module to set the major constituents for the phases, since these are used as the default initial values (see the GES user's guide for details). It is possible to provide initial values of the constitutions of the phases in the POP file. It is possible to make a new POP file from the current PAR file by giving the command MAKE-POP-FILE in the EDIT module. However, this file does not contain the initial tables, etc.

7.3.9 *The alternate mode*

The information that PARROT uses to optimize the thermodynamic model parameters represents measurements at equilibrium in the system. The measurements can be made in a single-phase region, for example activities or enthalpies, in two-phase regions, for example solubilities or transformation temperatures, or with more than two phases involved. At each equilibrium, at least one quantity must have been measured in addition to those necessary to determine the equilibrium state. For a binary system in a single-phase region, one may have measured the temperaure, pressure, composition, and chemical potential. Three of these quantities are necessary in order to specify the equilibrium state and the fourth can be used as experimental information to model the phase. In a two-phase region at given temperature and pressure, one may have determined the stable phases and the composition of one or both phases. The temperature, pressure, and set of phases are sufficient to determine the equilibrium and the compositions can then be used as experimental data.

7.3.9.1 Common-tangent construction

It is easy to understand that it may be difficult to calculate an equilibrium between two or more phases when the model parameters for the phases are badly determined. The equilibrium calculation requires that one can find a "common tangent," i.e., that the chemical potentials for all components are the same in all phases. Such a common tangent might not exist or may be at a completely wrong composition or temperature for the initial set of model parameters; see section 7.1.1.

Instead of requiring that an equilibrium should be calculated between two or more phases, PARROT supports an "alternate" technique to use such experimental information. The alternate technique calculates the thermodynamic properties for each phase separately and the program uses as "experimental information" the difference in chemical potential for the components in each phase. The model parameters are then adjusted to make the chemical potentials of all components the same. This is not a new technique; it was possible to describe equilibria in this way in PARROT earlier, but it was cumbersome and difficult. In the BINGSS software the options IVERS=1 or IVERS=3 have been available from the beginning, and these make use of the same technique as the alternate mode to calculate chemical potentials for each phase separately rather than the full equilibrium.

7.3.9.2 Preparation of the POP file for the alternate mode

Alternate calculation of the experimental equilibria with two or more stable phases is set by the SET-ALTERNATE-MODE Y command in the PARROT module. The POP file will usually require additional information to handle this option because there must be enough information to calculate each phase separately. For example, if both compositions of a binary tie-line have been measured, this may be given in the POP file as

```
CREATE 1 1
CH-ST PH FCC BCC=FIX 1
SET-COND P=1E5 T=1000
EXPERIMENT X(BCC,B)=.2:.01 X(FCC,B)=.3:.01
```

The equilibrium above can be calculated with the alternate mode without any modification. PARROT will use the values given as EXPERIMENT as conditions when calculating the thermodynamic properties of each phase, but, if just one side of the tie-line has been measured, one must provide an estimate of the composition of the other phase. This can be added by the SET-ALTERNATE-CONDITION command in the POP file. A command SET-ALT-COND is ignored unless the alternate mode is set. The same example as above would then be

```
CREATE 1 1
CH-ST PH FCC BCC=FIX 1
SET-COND P=1E5 T=1000
EXPERIMENT X(BCC,B)=.2:.01
SET-ALT-COND X(FCC,B)=.3
```

When PARROT calculates the thermodynamic properties of the bcc phase, it will use the composition provided as EXPERIMENT. When calculating for the fcc, it will use the composition provided with SET-ALT-COND.

A three-phase equilibrium may have some compositions determined experimentally, with others provided as alternate conditions,

```
CREATE 1 1
CH-ST PH FCC BCC LIQ=FIX 1
SET-COND P=1E5
EXPERIMENT T=912:5 X(LIQ,B)=0.2:.02
SET-ALT-COND X(FCC,B)=0.1 X(BCC,B)=.4
```

The experiment below cannot be converted to alternate mode:

```
CREATE 1 1
CH-ST PH BCC LIQ=FIX 1
SET-COND P=1E5 T=1111
SET-REF-STATE B LIQ * 1E5
EXPERIMENT MUR(B)=-4300:500
```

The reason is that there is no information about the compositions of the two phases and, even if these were added as SET-ALT-COND, the alternate mode will end up with five conditions for each phase instead of the correct four because the alternate mode will keep the condition MUR(B) for both phases. The alternate mode may be able to handle this situation in a future release.

A composition of a stoichiometric phase must be given with seven correct digits:

```
CREATE 1 1
CH-ST PH LIQ A2B=FIX 1
SET-COND P=1E5 X(LIQ,B)=0.2
EXPERIMENT T=992:5
SET-ALT-COND X(A2B,B)=.6666667
```

Experimental equilibria with phases with status ENTERED or DORMANT will be ignored by the alternate mode.

The alternate mode should be used at the beginning of an assessment only, before reasonable model parameters have been determined. When it is possible to calculate the experimental equilibria in the normal way, the alternate mode should be turned off.

7.3.10 *The tricks and treats*

Each assessor will develop his personal relation to PARROT because it is such a rich item of software with many unique features. However, there are some common tricks that it may be useful to know even before the user has developed a more intimate relation with PARROT.

1. If you have trouble, use initially as few experimental data as possible to get a reasonable overall fit. In particular, the invariant equilibria are useful. You can also use metastable invariant equilibria that can be estimated by excluding some phases. These estimated equilibria should be excluded from the final optimization!
2. If you have 100 activity measurements and 10 points from the phase diagram, you may have to decrease the weightings of the data from activity experiments.
3. It may be practical to exclude some or all intermediate phases initially and just optimize the liquid and the terminal phases (for the pure components); see section 6.2.4.4. In later stages of the optimization, it can be interesting to calculate the metastable phase diagram with just these phases in order to check that the metastable solubility lines do not have any strange kinks or turns.
4. Phases with miscibility gaps, stable or metastable, are often problematic; see section 6.2.7. Try to keep control of the gap by use of some real or estimated experimental information. Unfortunately, PARROT cannot calculate the top of a

miscibility gap as a single experimental equilibrium. Use the stability function QF to have control of any unwanted (or wanted) miscibility gaps.

5. Phases with order–disorder transformations are also problematic. They quite often require manual input of initial values and sometimes the ordered state may disappear during the optimization. It may help to add an experimental criterion that controls the state of order; see section 7.3.7.11.
6. The alternate mode, section 7.3.9, is a new feature in PARROT, so no-one has much experience with it. It should be used only to find an initial set of values for the adjustable variables that can be used to make it possible to calculate the experimental equilibria in the normal mode. It may also be used to get an initial set of variables for an intermediate phase that was initially excluded, keeping the values of already-fitted variables fixed.
7. When the liquid and some solution phases have been fitted reasonably well, the optimizing variables describing these phases can be set fixed and the variables for the intermediary phases optimized.
8. Check continually that the optimized variables fall within reasonable ranges. When a variable starts to change by several orders of magnitude in the later stages of an optimization, one must reconsider the experimental weightings or check whether too many variables are being used.
9. Phases that appear in wrong places can be handled with the DGM experiment; see section 7.3.7.9.
10. A final optimization with all variables and all selected experimental information, with the appropriate weightings, should be done. This file should be saved as documentation of the assessment.

7.3.11 Time taken for stages of an assessment using PARROT

A rough and maybe very personal estimate of the time spent on various stages of an assessment is

25% on collecting experimental data and creating, correcting, and updating the POP, setup, and dataplot files;
25% on "optimizing the weights" to find the correct balance among various types of experimental information and selecting a critical set of experimental information;
25% on optimizing the adjustable variables; and
25% on writing the report.

Quite often one has to go back and reoptimize when one finds that a selection or decision made during the optimization cannot be explained or defended in the report.

7.4 Final remarks

Most of the information here can also be found in chapter 6, but repetition is one way of learning important facts. It is rarely the case that the assessor finishes an assessment with the feeling that it cannot be improved. It is usually money or time that determines when the assessment is finished.

7.4.1 *Conflicting data*

In some cases one may find measurements of the same quantity that are widely different. All such data should be entered into the experimental data file unless there are obvious reasons (impurity of samples, for example) why some set can be rejected. However, during the optimization one should not include two conflicting sets at the same time but rather use only one at a time together with the rest of the data. This follows from a simple rule formulated by Bo Jansson, the creator of PARROT: *if you have two conflicting datasets then either one or the other may be correct or both may be wrong.* A fitted curve in between is most certainly wrong. It is to be hoped that the optimization can clarify which dataset is most in agreement with the other information on the system.

There are also cases in which conflicting data cannot be detected directly. For example, it might not be possible to reconcile activity data with solubility data from the phase diagram. This is indicated by large errors in the fit when both datasets are included. Thus it may be necessary to try to optimize with some datasets excluded in order to find these inconsistencies. For more discussion on this, see section 6.4.2.

7.4.2 *The number of adjustable coefficients*

Usually an assessment is considered better the fewer adjustable coefficients are needed to get the same fit, although it does not matter much whether 12 or 13 coefficients were needed. However, if one can get almost the same fit with 8 instead of 12 coefficients, then the assessment with 8 can be considered a superior fit. More of this topic is discussed in section 6.3. However, in an assessment one may put different weightings on different kinds of information and it is very difficult to compare assessments.

It is important to use a small number of coefficients also for the reason that the assessment will be used for extrapolations to higher-order systems. From experience it has been found that the fewer coefficients are used for binary systems the fewer problems occur with higher-order systems.

7.4.3 *Analysis of results*

In section 6.6 the points to check are listed, including the analysis of the final result, which should include the following.

- A satisfactory description of the critical set of experiments.
- A satisfactory description of data not included in the critical set.
- A reasonable set of values of the adjustable coefficents.
- Reasonable extrapolations of the thermodynamic properties, also to higher-order systems.
- A comparison with the results obtained using other critical sets or models.

7.4.4 *Rounding off of coefficients*

At the end of the optimization all optimized coefficients have several digits that must be truncated. There is a simple method by which to round them off when dealing with metallic systems, by keeping as many digits as will give less than $1\,\mathrm{J\,mol^{-1}}$ difference at 1000 K. For aqueous systems or systems assessed for very different temperature ranges, one may have to use other criteria. Rounding off that gives differences larger than $1\,\mathrm{J\,mol^{-1}}$ may give detectable changes in the phase diagram or for other experimental data.

If the coefficients have many digits, this can give an impression of high accuracy, which fact has led to another philosophy for rounding off. In PARROT one starts this rounding off by considering the optimized variable with the highest RSD, which is usually larger than 0.1. This means that this variable has only one significant digit. One may thus round it off to just one digit and trailing zeros. However, this will change the sum of errors and the remaining variables must be reoptimized with the rounded variable kept fixed. One should reach almost the same reduced sum of errors with the remaining variables; otherwise the variable set fixed was not the best one to round off.

When rounding off the first variable has been successful, one may continue to round off the new variable with the largest RSD in the remaining set by setting it fixed and reoptimizing the rest. The RSD for the remaining variables will decrease for each variable set fixed and, if $0.1 > \mathrm{RSD} > 0.01$, the variable has two significant digits and so on. This can continue until there is just one variable left. The final sum of errors should not deviate significantly from the initial one obtained when all variables were optimized. The rounded variables are easier to handle than those just rounded with many non-zero digits.

7.4.5 *The SGTE raw data format for experimental data*

The SGTE has decided on a format for storing experimental data for use in an assessment program. All software that is used for such a purpose should be able to read and write a file in this format. It is important that all decisions made by the assessor on the quality of the data and the final selection are reflected in this file, also those made interactively. In this way it is possible to reproduce the assessment at a later time, possibly together with new experimental data or to test a new thermodynamic model or new software. This format is available on the website for this book.

8 Creating thermodynamic databases

In the previous chapters it has been shown how to obtain the best possible agreement between thermodynamic models and experimental data using adjustable model parameters for binary and ternary systems. Even if each such assessment can be very important by itself, the main purpose of these assessments is to provide the building blocks of multicomponent thermodynamic databases. This objective must be considered when performing an assessment because it imposes some restrictions on the assessment of the individual system and on the possibilities of adjusting data and models to new experimental data. Such problems will be discussed in this chapter, together with the general concepts concerning thermodynamic databases.

Experience has shown that thermodynamic databases based on a limited number of ternary assessments, all centered around a "base" element like Fe or Al, can give reliable extrapolations to multicomponent alloys based on that element. This means that the database can be used to calculate the amounts of phases, their compositions, and transformation temperatures and that the calculated values have an accuracy close to that of an experimental measurement. Such databases are a very valuable tool for planning new experimental work in alloy development, since detailed experimental investigations of multicomponent systems are very expensive to perform. It is important that the databases are based on ternary assessments, not just binaries, because the mutual solubilities in binary phases must be described, otherwise the extrapolations are not reliable. There are also many ternary compounds that must be in the database. The number of quaternary compounds, however, is much smaller.

Another factor that makes ternary assessments important is that they can reveal that many binary assessments require modification when used in ternary assessments. There can be many reasons for this, the most important being that the experimental information on a binary system is scattered and insufficient. In such a case many sets of model parameters can reproduce the available data in a binary system with equal accuracy, but these parameters will give different extrapolations into ternary systems. Extrapolations from a binary system to several ternary systems should be carried out and the results compared before a binary assessment is accepted.

8.1 Unary data

The data for the pure elements must be the same in all assessments in which the element appears in order to make it possible to merge these into a database. This may be considered a trivial statement, but, since assessments may be performed by different research groups and over a period of several years, there is a need for international agreement on pure-element data. In particular, data for the metastable modifications of the element are critical because these are essential for describing solubilities. The assessment of solution phases requires data for the "end members" of the models and, in many cases, such "end members" are not stable, for example Cr in the fcc structure and Au in the bcc structure. Data for stable and metastable modifications of the elements, i.e. lattice stabilities, see section 1.2, have been collected by Dinsdale (1991) and these are the recommended values. Updated versions of this collection are available on several websites, for example http://www.sgte.org.

The accepted set of unary data will inevitably be subject to changes when new data become available, but any attempt to improve the description of the data for the pure elements in an assessment will make this assessment incompatible with other assessments performed using the old value and thus not suitable for merging with an existing database. A thermodynamic database consists of a large number of separate assessments and requires several years' development. The initial selection of unary data must be maintained even if it is later evident that some values for the pure elements could be improved. Changing an important item of unary data usually means the start of a new database.

It is important that the development of new data for unaries does not stop. The 1995 Ringberg meeting (Aldinger *et al.* 1995) was designated to set the foundation for a completely new set of such data. When this work has been completed, it will be necessary to reassess all binary and ternary systems. For a period of time assessments of new systems may have to be done using both the new and the old set of unary data. Such reassessments may occur even more frequently in the future and it is desirable to develop an automatic and continuous reassessment procedure.

8.2 Model compatibility

When combining two assessments it is essential that a phase that forms or may form a continuous solution from one binary system to another is described with the same model in both assessments. This is obvious for the terminal solution phases, but some intermediate phases may also exist across a system. It is then important that the models describing the binary systems make it possible to combine these to give a single Gibbs-energy description. The naming of such phases is a practical problem.

A simple case is a phase with fcc lattice with and without a sublattice for interstitials. It is unproblematic to combine a model for a substitutional phase in the Fe–Cr system with an fcc lattice with the austenite phase in the Fe–C system which has a B1 structure with Fe and C on two interwoven fcc lattices. One just adds the interstitial sublattice to the model for fcc in the Fe–Cr system with the vacancy as the only constituent. In the ternary Fe–Cr–C system this phase will then form a reciprocal solution with the model

(Cr, Fe) (C, Va). This will create a new "end member" belonging to the Cr–C binary system, which has to be estimated (no phase with A1 or B1 structure is stable in the Cr–C binary system).

Phases with the simple fcc, bcc, and hcp lattices appear in many systems, but also many other phases with more complicated lattices, such as σ and spinel phases, are not uncommon. It is important that they are modeled in the same way in all systems. Deciding whether two phases in different systems should be combined and, if so, what model to select is not trivial. In the Cr–Si, Fe–Si, and Mn–Si systems there are phases with the general stoichiometry M_3Si. However, only Fe_3Si and Mn_3Si are similar; Cr_3Si has a different structure. A great help in identifying phases that are the same is the Strukturbericht notation, but it is not available for many phases.

Many phases should be treated as the same even if they do not form a continuous stable solution from one end to the other because other phases are more stable. It is very important to identify the phases in different assessments that should be treated as the same, because this has a large influence on the extrapolations to higher-order systems.

The recommended modeling of phases with an order–disorder transition is to partition the Gibbs energy into a disordered part and an ordered part as described in section 5.8.4.1. That makes it easier to combine the ordered phase with phases from other systems in which only the disordered part may be stable. For example, the B2 phase in the Al–Ni system should be combined with the A2 phase in the Cr–Ni system. The B2 phase with partitioning of the Gibbs energy has one part describing the disordered part (which has an A2 structure) and one part describing the contribution due to ordering. One can then simply add the parameters from the A2 phase in the Cr–Ni system to the A2 part in the Al–Ni system and keep the additional parameters for the ordered part as zero. Great care must be taken when merging phases that have an order–disorder transition, since it is easy to make mistakes that make the disordered state become unstable at all temperatures.

Sometimes the combination of two assessments will require reassessments of a system in order to achieve the required model compatibility. It is **strongly** recommended that one select models that are compatible with the systems with which the assessment should be combined later. Otherwise, one should be prepared to reassess all systems for which the "wrong" model has been used.

8.3 Experimental databases

In each assessment the assessor must create a file with experimental data used in the optimization. In BINGSS it is the *.dat and in PARROT it is the *.POP file, as described in chapter 7. In the experimental datafile each item of experimental information extracted from the literature should be referenced correctly and all transformations or corrections made from the original publication documented. In some cases the assessor may have used theoretical information, either from publications, which then should be referenced, or introduced as "fictitious" experimental information, for example to avoid a solution phase appearing in a region where there is no information indicating that this phase should be stable. This should be documented in the experimental datafile also. Each item of experimental information has an uncertainty that normally should have been provided by the experimentalist, but often the assessor has to modify this uncertainty and the assessor can also assign a weight to each experimental value. The final weightings depend on all experimental values used in the assessment, as discussed in more detail in chapter 9.

The experimental datafile represents a large amount of work and it is important that it is kept for future use, for example when new experimental data become available or when a model for a phase is changed, both of which cases may necessitate a reassessment. However, even with a well-documented experimental datafile, it is not trivial to carry out a reassessment. The original assessment may have included conflicting datasets and the original assessor may have selected one of these. The new information may indicate that this selection was wrong and then a completely new assessment is required because all weightings for various experimental data must be carefully revised.

The experimental datafiles should be kept on a public internet site and be made available to anyone who is interested in carrying out a reassessment using a new model for a phase or a new set of unary data, for example. The authors of this book will provide such an internet site, from which assessors may upload their experimental datafile or download previous files. For testing new software for assessments, it is also interesting to have access to previous experimental datafiles. Testing new models, in particular, must be done using identical experimental data; otherwise it is impossible to make proper comparisons. The experimental data files are useful also when assessing a ternary or higher-order system when it may be necessary also to vary model parameters for some of the lower-order systems. As noted above, changes for a lower-order system may require further modifications of other higher-order systems dependent on that system.

8.4 Naming of phases

In a thermodynamic database it is desirable to have a unique name for a phase that can extend into a multicomponent system, even if that phase may have a different name in each subsystem. A simple example of the problem can be found in the Ca–Mg–Fe–O system, where pure CaO is called lime, pure MgO is called periclase, and both of these form a continuous solution with the wustite phase in the Fe–O system, all having the NaCl (B1) structure type. In a database the parameters for these three phases must be stored together with a single name. At present the name "halite" has been selected for this because halite is a kind of "generic" name for all phases with the NaCl structure type.

For metallic systems the phases are usually given greek letters as names, starting with α for the low-temperature form, but α-Fe and α-Cu do not even have the same lattice. For binary systems, phase names like "(Fe, Ni)" are used for extended solution phases, but on adding carbon, the same solution phase may be called "austenite." In a database all parameters for a phase with the same structure type and the same Gibbs-energy function must be stored with the same phase name.

In a database for steel, the phase name *ferrite* can be used for the bcc structure and *austenite* for the fcc structure. However, in an aluminum database the fcc structure type must not be called austenite and one should try to find a phase name that is independent of the composition of the phase. In diverse fields of science like geochemistry and metallurgy the same phase has often been assigned different names, but it would be helpful, in particular for students and other non-experts, if one could agree to use the same name for the same phase for many applications.

In the assessments of a binary, ternary, or higher-order system one should thus try to select phase names that are related to the structure rather than to a particular composition. Some simple structures based on bcc, fcc, and hcp lattices appear in many systems and, although these notations actually characterize lattice types only, rather than a structure or a phase, it is generally accepted to use fcc, bcc, and hcp as phase names. The use of these phase names in a database can be extended to include also phases with more complex structures, but that are modeled in such a way that the simple structure is included as a special case. For example, the interstitial solution of carbon in austenite can be regarded as an fcc phase even if it has a B1 structure and might be called "halite." The TiC phase has the same structure type and, in the Fe–Ti–C system, the Fe-rich phase with fcc lattice with interstitial solution of C and substitutional solution of Ti can be modeled as the same phase as the Ti-rich phase with a high fraction of interstitial C and a small fraction of substitutional Fe. This model is actually an example of a reciprocal solution, in this case (Fe, Ti)(C, Va); see section 5.8.1.

To go further, one must first review the various ways in which a structure type can be designated. There are many ways to express the crystalline properties, such as in terms of *Pearson symbols*, the *space group*, and the *prototype phase*; see section 2.2.3.2. All of them are related to the crystal symmetry needed for identification of X-ray information on a phase with given composition. This is not sufficient or even correct for a thermodynamic identification of the phase, for example the space group $Fm\bar{3}m$ includes both the A1 and the $D0_3$ structure.

The *Strukturbericht* designation has the advantage that the nomenclature has no relation to the actual structure. It consists of a letter followed by a number and possibly superscripts and subscripts. The letter A is used for phases for pure elements (although A15 is a mistake), B for binary compounds with equal stoichiometry, C for compounds with stoichiometry 2 : 1, etc. After C the letter has less obvious meaning. There is no independent authority assigning these any longer, but the notation is popular and various scientists are inventing "Strukturbericht"-like notations for new phases.

It is generally advisable to use the Strukturbericht, if it is known, as a prefix or suffix to a more "application-oriented" phase name. Many phases have no Strukturbericht designation and in that case the *prototype phase* can be used instead. In the Strukturbericht designation there are subscripts and sometimes superscripts. The various parts of a phase name are usually connected by "_" (underscore), and for simplicity that can be used also to separate subscripts or superscripts in the Strukturbericht designation, for example L1_2 for the $L1_2$ phase. This is still a matter of discussion though; in this book several styles of phase names are used since that is the current situation.

It still remains to consider phase names when the same thermodynamic model is used to describe several structure types, for example A1, $L1_2$, and $L1_0$ as discussed in section 5.8.4. Such a "structure family" is usually based on a simple disordered lattice such as fcc, bcc, or hcp and one may use this as the first part of a name. In the existing databases there are ordered structures like B2, $D0_3$, and $L2_1$ based on the bcc lattice, $L1_2$ and $L1_0$ based on the fcc lattice, and $D0_{19}$ and B19 based on the hcp lattice. With a four-sublattice model it is possible to treat both $L1_2$ and $L1_0$ ordering with a single Gibbs-energy model, but, to include also $D0_{22}$ ordering, eight sublattices are needed, which is not realistic at present, so it must be treated as a separate phase. Using the disordered lattice as the first part and the structure with the most complex ordering as the last part, the name of an fcc phase with four sublattices for ordering is thus fcc_L1_02.

The partitioning of the Gibbs energy of a phase with B2 structure into a disordered part (with a substitutional lattice) and an ordered part (with two sublattices), as described in section 5.8.4.1, makes it possible to use the phase name bcc_A2 for the parameters describing the disordered part and the phase name bcc_B2 for the ordered part. When a user asks the database for the parameters of a system in which there is no ordered phase with B2 structure, the parameters for the ordering may be ignored and parameters for the disordered phase only are retrieved. However, on retrieving data for a system in which the B2 phase may be stable, as in the Al–Ni–Cr system, it may happen that, when the user calculates an equilibrium in which a stable phase called bcc_B2 appears, it may actually be disordered A2. It is possible to implement in the software the use of different names of phases depending on the composition or the state of order of a phase, but it may be even more confusing for the user if a phase suddenly is replaced by another just because it has undergone ordering. An interstitial sublattice for carbon and nitrogen can be added both to the ordered and to the disordered model without changing the ordering model or the use of a single Gibbs-energy function both for ordered and for disordered phases.

Phases with more complicated structures that are not ordered superstructures of simpler structures sometimes have accepted names like the Laves phase, spinel, and M23 carbide. It is recommended that such names be retained unless they are ambiguous. In the case of Laves phases, which have the general formula A_2B, there are three different structures with the Strukturbericht names C14, C15, and C36 and it is recommended that these be used as prefixes, for example C14_Laves phase.

A phase that has a fixed composition and does not dissolve any other elements should normally keep its accepted name, provided that there is little or no solubility of other elements in higher-order systems. Some structure information should be added, give a name like graphite_A9 or cristobalite_C9. If there is no accepted name, the recommendation is to use the stoichiometric formula, possibly prefixed with some structural information, like D0_I_MO2B5. Note that some stoichiometric phases can be stable with different structure types at different temperatures.

Some compounds and intermetallic phases with solubilities have traditional names like σ and μ, and, although the use of greek names like α, β, etc. is discouraged, names like sigma_D8_B and mu_D8_5 are acceptable because they have been used for just one unique structure. For carbides, names like M23C6 and M7C3 are established. Here M stands for metallic atoms like Cr, Fe, and Mo, and C can actually be replaced by N or B, but one should give the Strukturbericht designation as a suffix, thus M23C6_D8_4 and M7C3_D10_1. For oxides, names like periclase_B1 (MgO), corundum_D5_1 (Al_2O_3), and spinel_H1_1 (Al_2MgO_4) for the respective structure types are recommended. For salt systems, the phase name halite_B1 is recommended for NaCl and all phases with the same lattice. Note that halite, periclase, and an interstitial solution of carbon in fcc titanium all have the same Strukturbericht designation, but there is no possibility of forming a continuous solution between these extremes and thus they can be treated as different phases.

The liquid phase usually has complete solubility of all elements and should thus be treated as a single phase in all systems. In some systems the liquid has a large miscibility gap and it may be tempting to treat, for example, the metallic liquid as a phase other than an oxide or sulfide liquid and use different models for these. However, this should be discouraged, since it is possible to change from a metallic liquid composition to almost any other liquid composition by adding appropriate elements without passing any phase interface. The amorphous phase can be treated as a supercooled liquid if an appropriate glass-transition model is implemented, otherwise it should be treated as a separate phase.

Aqueous solutions and polymers may be treated as separate phases because it is not possible to form a continuous solution with metallic or oxide liquids.

8.5 From assessments to databases

A database is a merged set of thermodynamic assessments of binary, ternary, and higher-order systems. To create a database, one must first collect all the necessary assessments. Each such assessment usually represents a complete description of a system and there is thus considerable redundancy if each assessment is stored separately. It is nonetheless important to have them separate in order to check that the unary data are identical, that compatible models are used for the same phases, and that the parameters really describe the system as expected.

When the assessments have been checked, the phase names should be unified. For this there is specific software, *tdbmerger*, by M. Jacobs (unpublished, 1999, available from the website of this book), which can merge database files for individual assessments written with the Thermo-Calc package and performs several checks during the merging. This software allows selection of data and some interactive modifications of the data in order to make data compatible. However, if different unaries or incompatible models have been used in two assessments, that means that one of the assessments must be revised.

The main reason for keeping individual assessments on separate files and using merging software is to simplify the updating of a database. If the database manager at a later stage wants to replace a binary assessment with a new, better, one, that is virtually impossible unless the database can be merged again using the new binary data.

In some cases it may be possible to transform one model into a more general one without reassessment. For example, a metallic liquid treated as a substitutional solution, (A, B), may be changed by manual editing into an ionic liquid model, $(A^{a+}, B^{b+})_P(Va)_Q$, and merged with other assessments of ionic systems, but the tdbmerger software cannot make this change by itself.

Each thermodynamic software system will store its databases in a format unique to that software. However, there is a definition by the SGTE for a *database-transfer* format that is close to the TDB database format used by *tdbmerger*. The intention is that it should be possible to translate the SGTE transfer format to and from any specific software-database format. The SGTE transfer format is described on the website.

The effort involved in combining published assessments to produce a validated database is considerable and should not be confused with a simple merging of assessments.

8.5.1 Merging of assessments

Output from BINGSS or other software used for an assessment must first be converted to the Thermo-Calc format. For BINGSS, conversion software, coe2ges, is available on the website of this book. The LIST-DATA command in the GES module of Thermo-Calc can generate a database file with option N.

The output from this LIST-DATA command will normally contain some V variables that have been optimized. The software *rvbv*, available on the website of this book, will replace the V variables by their final values.

8.5.2 Unassessed parameters

On merging assessments, a large number of *unassessed* parameters may appear. Such parameters can be, for example, ternary interactions, but also carbides and intermetallic phases like a Laves phase may cause problems. If the Laves phase exists in two binaries and is modeled as $(A, B)_2(A, B)$ and $(A, C)_2(A, C)$, this phase will, in the database, be modeled as $(A, B, C)_2(A, B, C)$ and the parameters $^\circ G_{\mathrm{B:C}}$ and $^\circ G_{\mathrm{C:B}}$ will be listed as unassessed. By default, such parameters are treated as zero in most software, but that might not be a good assumption.

All unassessed parameters can be listed using the GES module and the database manager has to decide on a value of these or choose to leave them unassessed as a warning to the users of the database.

Many such unassessed parameters can be assigned an estimated value by the database manager because usually they have very little influence on any stable equilibrium calculated with the database. The estimated parameters are usually far from any stable composition of the phase and have no influence on the calculated equilibria. However, the manager should be aware of the fact that the database is sometimes used also to extrapolate to metastable states by suspending the stable phases and in such cases the estimates may be very critical because the composition of the phase can then be far from the stability range. Thus a clear reference must be given for each estimated parameter as well as for the assessed parameters.

8.5.3 Missing parameters

Another problem that may appear on merging assessments is that phases ignored in the assessment of a binary system may appear as stable in that system after the merger. For example, the Cu–Fe system has no hcp-A3 phase and, in a normal assessment of the

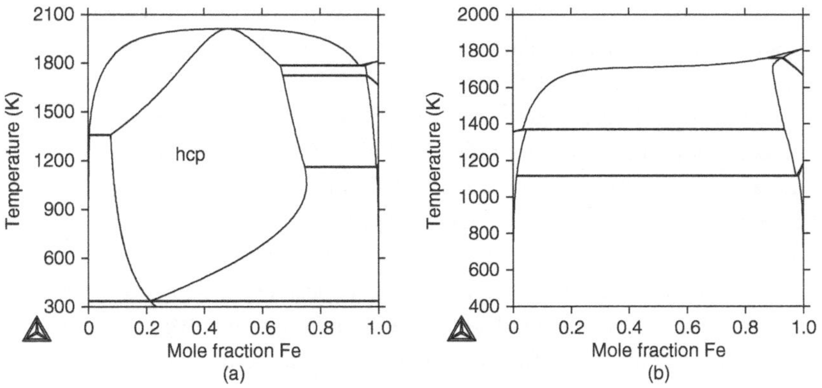

Figure 8.1 What may happen when a binary is added to a database. (a) The Cu–Fe system with an ideal modeling for the hcp phase. (b) The Cu–Fe system with the interaction for the hcp phase as for the fcc phase.

Cu–Fe system, no interaction parameter is assessed for the hcp-A3 phase. However, if this assessment is merged with systems in which Cu and Fe dissolve in an hcp-A3 phase, for example, Cu–Zn and Fe–Zn, the database may be used to calculate a ternary section, Cu–Fe–Zn. The binary interaction in hcp-A3 between Cu and Fe will be zero, and, since there is a positive interaction between Cu and Fe in the liquid, fcc-A1, and bcc-A2 phases, one will find that the hcp-A3 phase becomes a stable phase along the Cu–Fe binary, even with very little Zn present. Unless the hcp-A3 phase is explicitly suspended, this phase will also appear in the binary Cu–Fe system; see Fig. 8.1(a). The correct Cu–Fe phase diagram is shown in Fig. 8.1(b). It is the database manager's responsibility to check all assessed systems after the database has been merged and ensure that there are no surprises. Some software may notify the user of such missing parameters, but, in a multicomponent database, there is a very large number of those and it is a tedious problem to handle these for the user.

This is something to consider whenever one is adding a binary assessment to a database. The important phases are the phases which normally exhibit large solubilities, like liquid, fcc, bcc, and hcp. For the liquid there will always be some parameters and, if hcp is missing but fcc has been assessed, one should copy the same parameters in hcp as in fcc. If there is neither fcc nor hcp in a binary, one could set the regular-solution parameter for these equal to that for the liquid or the bcc phase, whichever seems the most reasonable at low temperature. If the bcc phase is not stable in the binary, its regular-solution parameter can be set the same as that of the liquid.

8.5.4 *Validation of the database*

Checking that the assessed systems can be recalculated from the complete database without errors and that no new phases appear is a first step in the validation, but it does not guarantee that errors do not appear on extrapolating to higher-order systems. The database manager must thus have experimental information on a number of multicomponent systems that he uses to check that the extrapolations from the assessed systems are reasonable. These checks can be used to define the valid composition range for each component in the database.

It may be interesting to learn from an example to do with compiling a thermodynamic database for steels. The first SGTE solution database released in 1989 could not describe the austenite/ferrite equilibria at 1150 °C for duplex stainless steels correctly, although it contained the best available assessments of the most important ternary and, in some cases, quaternary systems. These steels contain high amounts of Cr, Ni, and Mo (22, 5, and 3 mass percent, respectively) and all ternaries, Fe–Cr–Ni, Fe–Cr–Mo, Fe–Ni–Mo, and Cr–Ni–Mo, were assessed. The problem could not be solved by adjusting parameters for alloying elements with smaller amounts. Instead, these ternaries were investigated and the problem could be solved when it was realized that the fcc-A1 phase was being treated as ideal in the Cr–Mo system, because it is not stable in that binary. By introducing the same positive interaction in the fcc-A1 phase in the Cr–Mo system as for the bcc-A2 phase and reassessing the Fe–Cr–Mo system, it was possible to get reasonable descriptions of the duplex stainless steels.

8.6 Database management and updating

A database requires constant updating and the *tdbmerger* software of Jacobs was designed for this. Updating can be of two kinds, adding new assessments or replacing existing assessments. In order to simplify the replacement, all individual assessments used to create a database must be kept on separate files all the time. The tdbmerger software should merge these files each time the database is updated and it is then easy to replace one assessment with another simply by selecting the new file for that assessment. Replacing an old assessment or adding a new one can affect the extrapolations to higher-order systems and this should be checked by the manager.

The estimated parameters that the database manager has added to the database should be kept in a special "add-on" file that is merged last.

8.6.1 Cancellation of errors

It is important to consider that adding a binary that was missing from the original database does not necessarily improve the database. It may actually make the extrapolations in the database worse. One reason for this is that the manager may have added estimations of missing parameters to the previous version of the database in order to compensate for the missing binary. Revision of such estimated parameters can be very difficult.

8.6.2 Documentation of a database

Since a database is likely to be used for longer than a single manager can keep it updated, it is important that the manager keep sufficient documentation of the systems in the database and of the test points he uses for the extrapolations. This makes it possible for the next manager to continue the work on updating and extending the database.

For commercially available thermodynamic databases, a user should always request a list of references of all assessed systems in the database. Some databases can provide a list of references for each system retrieved from the database, which makes it easier to verify that the database is up to date. Databases should be expected to be useful for many years with proper updating, but it is important to know the origin of all parameters in a database. There is a facility to reference each parameter to either a paper or any other documentation in the database format used by Thermo-Calc, namely the so-called "TDB format," which is very similar to the "SGTE format for database transfer." The documentation of both of these is available from the appropriate websites.

8.7 Existing thermodynamic databases

Several databases are available and they are usually classified in terms of their main component (for example, Al databases), a special physical property (semiconductors database, superconductors database) or a class of materials (steels database, Ni-based-superalloys database), or even a special application (solders database). More details about existing databases and the software required in order to use them are given in a special issue of *Calphad* (2002, **26**, pp. 141–312).

8.7.1 Referencing a database

When a database is used for a publication, it should be properly referenced with its name, supplier, and year and version identification if possible. The reference for a database is more like describing an instrument used for a measurement than referencing an article or book. Stating, for example, "SGTE solution database" is not sufficient because there are many such databases. Referencing software is not appropriate because most software can be used with several different databases. If the database supplier has a website, that should be included in the reference.

8.8 Mobility databases

Although this whole book is devoted to thermodynamics, the fact is that computational thermodynamics is most important for simulating phase transformations. The thermodynamic description of the phase gives important quantities also in the metastable ranges of the phases, such as the chemical potentials of the components and the thermodynamic factor for diffusion. In addition, one must have mobilities of the elements in the various phases, but that quantity usually varies more smoothly than does the diffusion coefficient. The main problem is that there is so little experimental data, in most cases one may know only the self-diffusion or tracer diffusion. The mobility is often closely related to the bulk modulus, which can easily be calculated with *ab initio* techniques; this is a possibility for estimating mobilities.

There is a special version of PARROT for assessment of mobilities, either using diffusion data or by fitting directly to measured concentration profiles. These assessments are dependent on the thermodynamic descriptions, of course, since they are used to calculate the thermodynamic factors and chemical potentials. The model for mobilities used in PARROT is described in Campbell (2005) and is based on mobilities of each component in each single-component phase. There are a few commercial mobility databases available; see, for example, http://www.thermocalc.com.

8.9 Nano-materials

One new area of thermodynamic applications is to nano-materials. Real materials often consist of several crystalline phases spatially arranged. Between the phases there are phase interfaces where the composition and crystal lattice change. In single-phase regions of the materials there are grain boundaries between grains that have the same crystal lattice but varying orientation of the crystal axes. Each grain is a "single crystal" and its size is typically of the order of 100 nm to 100 μm. The fact that almost all inorganic materials are "polycrystalline" is very important because that is the reason for their isotropic mechanical properties. Only for a few applications, such as turbine blades, is there a reason to take advantage of the fact that a single crystal has different mechanical properties in different directions.

Thermodynamics normally applies to "bulk" polycrystalline materials when the surface properties are several orders of magnitude less important than the bulk properties. The phase boundaries in phase diagrams represent equilibria between two phases separated by a planar interface. If the interface is curved, there may be a pressure term to be added even to the "bulk" thermodynamics in order to get the correct equilibrium composition; this is discussed in detail by Hillert *et al.* (1998).

The interface region between two different crystalline grains has a larger density of defects than do the inner regions of the grains and the atoms at the interface can have different coordination numbers from those they have in the bulk. These factors make the interface properties different from those of the "perfect" crystal by a factor that can reach 30%. This fact has inspired the idea of creating materials mainly formed by interfaces in order to discover properties different from those of a perfect crystal. Reducing the size of the grains to nanometers makes the number of atoms in the grain comparable to the number of atoms at the interface and, as a consequence, the macroscopic properties of the material are no longer governed by the interaction of atoms inside a crystal. These so-called "nano-materials" can be in a metastable equilibrium or even in a "non-equilibrium" state (Bustamante *et al.* 2005) and their properties will be different from those of the equilibrium state of materials that has been treated in this book. In other cases equilibrium thermodynamics can still be applied and an addition to the Gibbs thermodynamics of small systems can be found in the book by Hill (1994).

8.9.1 *Surfaces in materials*

Although this book deals only with bulk thermodynamics of materials, it is important to mention some relations to surface properties. The grains in a polycrystalline material usually form at solidification or when the material is recrystallized, as described in any textbook on materials science. The internal surfaces like phase interfaces and grain boundaries are important for many macroscopic properties, but also for phase transformations in materials. Grain boundaries can move and, with increasing temperature, the grains may become very large, since this makes the surface area smaller and minimizes the energy. It has been known for a long time that mechanical properties are usually better for materials with small grains and much effort has been devoted to stabilizing the grain size against changes in temperature, for example by pinning the grain boundaries by placement of particles.

For phase transformations the relative stability of the different phases at the phase interface is the most important factor determining the interface movement. Usually both the crystal lattice and the composition change across a phase interface and thus bulk diffusion is also needed in order for the interface to move. Even during phase transformations, it is a useful assumption that one has "local equilibrium" at the phase interface, which means that the compositions at the interface are given by an equilibrium tie-line in the phase diagram. The movement of the interface and the bulk diffusion must balance to maintain the equilibrium compositions for the two phases at the interface, taking into account that the actual tie-line may change with time in multicomponent systems. This is used, for example, in the DICTRA software (Andersson *et al.* 2002); see also section 8.10.2. When the elements have very different mobilities, this "local equilibrium" may be replaced by a "para-equilibrium" assumption, as discussed by Hillert *et al.* (1998). For rapidly moving interfaces, one may use the "solute-drag" theory, see section 8.10.3, to describe the transition from local equilibrium to massive transformation. All of these models for phase transformations depend on a good description of the bulk thermodynamics.

8.9.2 *Nucleation in materials*

When a new phase wants to form, there is a nucleation stage during which the surface energy of the new phase is just as important as, or more important than, the bulk energy. Classical nucleation theory gives the critical radius of a spherical particle of the new phase as

$$r^* = \frac{2\sigma V_\mathrm{m}}{\Delta G_\mathrm{m}} \qquad (8.1)$$

where σ is the surface energy, V_m the molar volume of the new phase, and ΔG_m the difference in energy between the new and old phases, also known as the "driving force" (see section 2.3.6).

Several models for estimating the surface energy, σ, from the bulk thermodynamic data have been proposed. Since this energy depends on many local factors, such as lattice orientations and segregation of solute toward the boundary, such models can give only a rough average value.

8.10 Examples using databases

Many examples of the use of a thermodynamic database can be found in the literature, for example Hack (1996), Kattner *et al.* (1996), and Saunders and Miodownik (1998). Here, only a few examples are given.

8.10.1 Multicomponent phase-diagram calculations

The classical example of the use of a thermodynamic database is the calculation of the phase diagram for a multicomponent system. An isopleth for a high-speed steel with metal composition 4% Cr, 9% Mo, 1.5% W, 1% V, 8% Co and the rest Fe for varying carbon content is shown in Fig. 8.2.

The reliability of any line in such a diagram calculated from a validated database within its range of applicability is normally the same as for an experimental determination. Since the calculation may take a minute whereas the experiment may take a month or

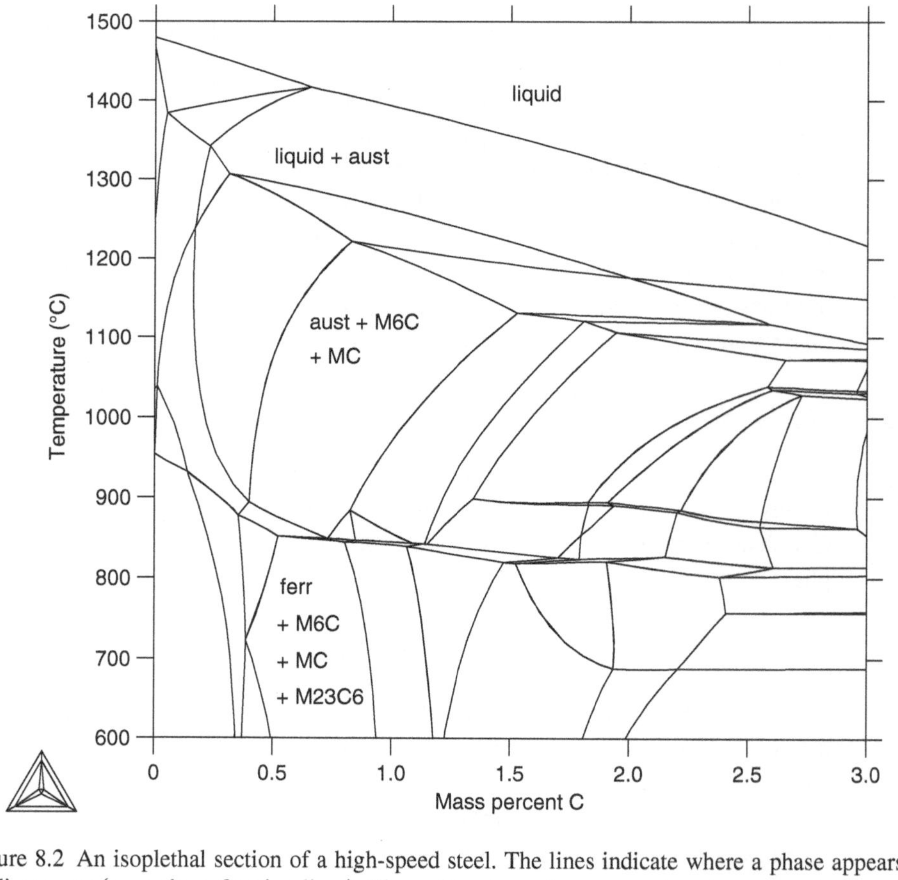

Figure 8.2 An isoplethal section of a high-speed steel. The lines indicate where a phase appears or disappears (zero-phase-fraction lines). The stable phases in some regions have been indicated; to write all of them on the figure would be too confusing.

longer, the gain in speed for alloy design is obvious. Of course, any conclusions drawn from a calculation must be verified experimentally, but one can reduce the number of experiments needed significantly. In addition, the calculation can be used also to obtain chemical potentials, heats of transformations, and amounts and compositions of individual phases for any temperature and overall composition.

The equilibrium phase diagram will normally deviate from the behavior of any normal material that has been processed too quickly to obtain full equilibrium. Below 800 °C diffusion, except of carbon, is usually too slow to produce the equilibrium set of phases.

8.10.2 *Simulation of phase transformations*

A thermodynamic database is developed to describe the stable state of a system, but it provides the possibility of extrapolating outside the stability range because the modeling of each phase is usually done for the whole system. This is a very important feature when

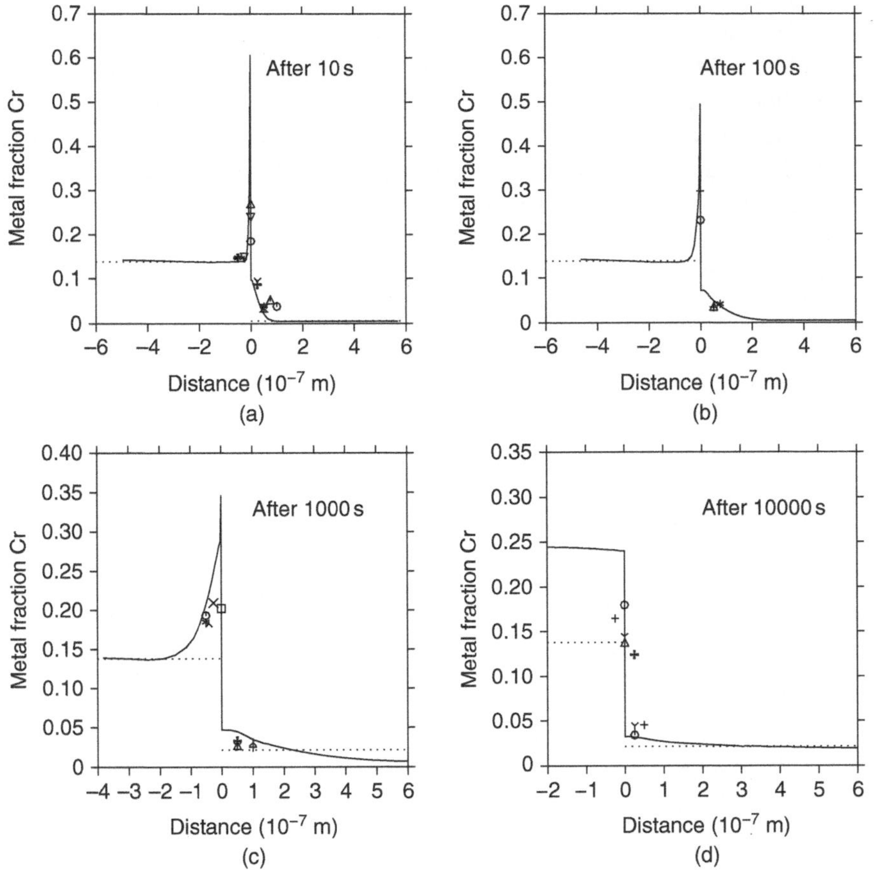

Figure 8.3 Concentration profiles of Cr across the interface after various times. The origin of the horizontal axis is always at the interface.

there is an interest in describing phase transformations since it provides the possibility of calculating driving forces for nucleation and the thermodynamic factor for diffusion also outside the stable equilibrium. In order to simulate processes and other transformations together with kinetic data for mobilities, accurate values of such thermodynamic properties are essential.

The DICTRA software (Andersson *et al.* 2002) was developed to handle phase transformations, including diffusion in multicomponent alloys, which can be simulated along one spatial coordinate. The application selected here is the dissolution of a spherical cementite particle in an Fe–Cr–C matrix. The cementite particle was formed at low temperature in equilibrium with ferrite in order to have an equilibrium partitioning of Cr and C. The material was then heated to 1183 K, the ferrite transformed to austenite, and the particle started to dissolve in the austenite matrix. In Figs. 8.3(a)–(d) the concentration profiles of Cr around the interface after various times are shown. At the interface it is assumed that one has equilibrium partitioning of Cr and C, but since C can move much faster, there is a very sharp concentration profile for Cr. Measurements of the concentration profile from Liu *et al.* (1991) are shown together with the calculated profile.

The interface between the cementite and austenite cannot move until the Cr concentration in the cementite particle begins to become more uniform. In Fig. 8.4 the volume fraction of the particle is plotted versus time. The experimental points at short times are very uncertain and deviate from the calculation, but the simulation predicts correctly the amount of carbide after a long time. The dissolution of a cementite particle in a binary Fe–C alloy takes less than a second, as a comparison. The Cr content of the alloy thus has the effect of slowing down the transformation and that is correctly predicted by the simulation. The simulation is based on independent thermodynamic

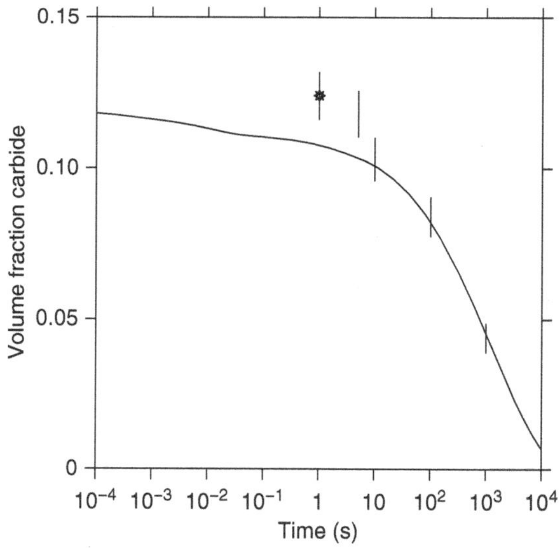

Figure 8.4 The volume fraction of carbide.

assessments of the Fe–Cr–C system and mobilities of Cr and C in austenite and cementite. The experimental data shown in the figures have not been used to fit the simulation.

8.10.3 Solute-drag simulations

The solute-drag model was developed to explain how grain-boundary mobilities could depend on the concentration of solutes at the grain boundary (Lücke and Detert 1957, Cahn 1962, Hillert and Sundman 1976). This has been extended to phase interfaces (Hillert and Sundman 1977) and, more recently, to multicomponent systems (Hillert and Schalin 2000). The interesting feature of the solute-drag theory is that it gives a continuous curve of the Gibbs energy across the phase interface between two different phases as shown for a binary system in Fig. 8.5.

The dotted curve in Fig. 8.5 represents the Gibbs energy at any position across the interface and is simply given by the equation

$$G_{\rm m}^{\rm I} = \chi G_{\rm m}^{\rm Liquid} + (1-\chi) G_{\rm m}^{\rm fcc} \tag{8.2}$$

where χ is the length coordinate across the interface, going from zero to one. At $\chi = 0$ one has only fcc with the composition $x_{\rm B}^{\rm fcc/I}$ and at $\chi = 1$ one has only liquid with the composition $x_{\rm B}^{\rm I/Liquid}$. In between the composition varies between these limits and the Gibbs energies of the two phases can be calculated from the thermodynamic models using the same composition for the same value of the coordinate χ. Note that the equilibrium compositions for the fcc and liquid phases are given by the common tangent. When the interface is moving the composition at the interface of the growing phase can differ from the equilibrium composition, but the composition at the interface of the shrinking

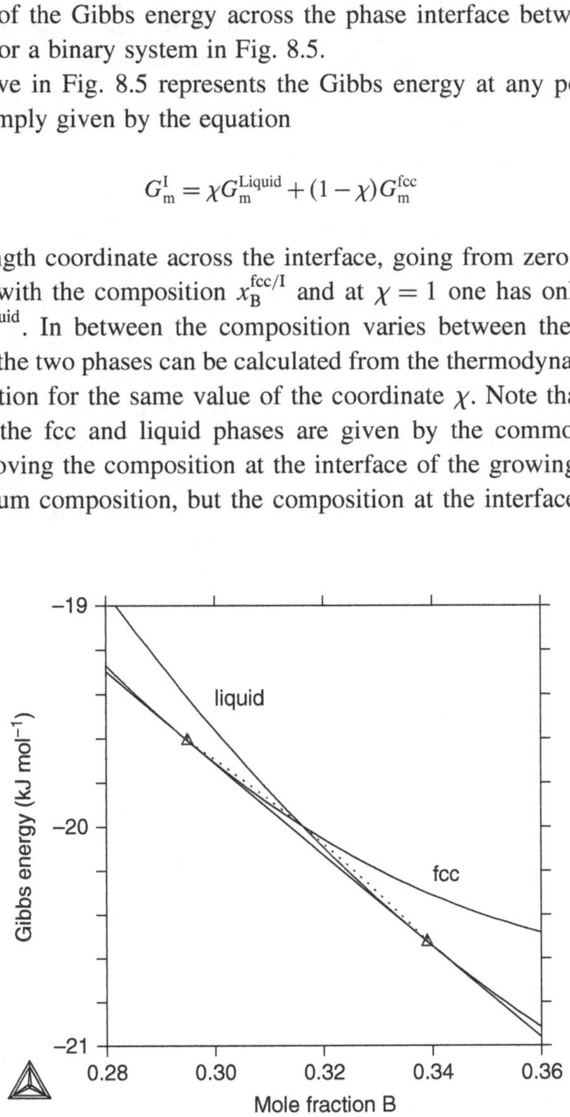

Figure 8.5 Gibbs-energy curves related to an interface between the liquid and fcc phases.

phase must always be inside the two-phase region. These compositions as well as the composition profile across the interface can be solved using a differential equation as described in Hillert and Sundman (1977).

8.10.4 *The phase-field method*

The simulation of phase transformations in two or three dimensions is based on the phase-field method, since it is impossible to treat directly the movement of "sharp" individual phase interfaces in two and three dimensions in the same way as is done in DICTRA for movement in one dimension. Instead a grid is imposed and the amounts of phases at each grid point are calculated by the phase-field method, which is based on thermodynamics and kinetic data, i.e., a "diffuse" interface. The first simulations using the phase-field method had very simple thermodynamic descriptions and a single continuous Gibbs-energy function across the composition range. This would apply only to ordering transformations in which the ordered and disordered phases have the same Gibbs-energy function. The phase-field method combined with realistic Calphad databases has recently been applied to a variety of phase transformations (Grafe *et al.* 2000, Warnken *et al.* 2002, Qin and Wallach 2003, Loginova *et al.* 2004, Zhu *et al.* 2004, Böttger *et al.* 2006).

Figures 8.6 to 8.8(d) show a simulation of equiaxial solidification of an Al alloy using MICRESS phase-field software (http://www.micress.de), the TQ interface of Thermo-Calc (http://www.thermocalc.com) and the COST-507 light-alloy database (Ansara 1998a). The alloy contains $x_{Mg} = 5.53$ at%, $x_{Mn} = 0.401$ at%, and the rest is Al. The figures show the formation of the fcc, Al_6Mn, and AlMg-β solid phases from the liquid.

In Fig. 8.6 the cooling curve of the alloy is shown. In the first part of the curve dendrites of Al are formed. The start of the solidification of Al dendrites is magnified in Fig. 8.6(b), which shows that a slight undercooling is necessary to nucleate the solid

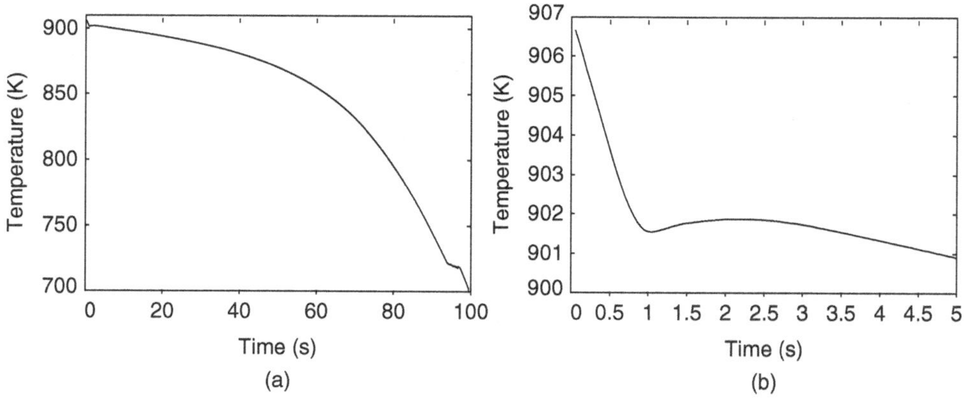

Figure 8.6 The temperature–time curves for solidification of an Al–Mg–Mn alloy at constant heat flux. In (b) a magnification of the first part of the curve in (a) is shown. Courtesy of Bernd Böttger.

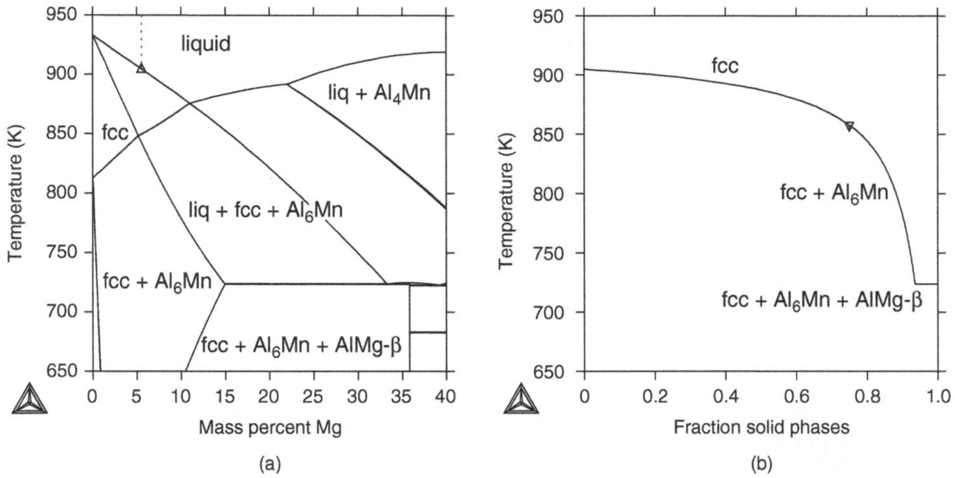

Figure 8.7 The phase diagram for Al–0.4Mn–Mg is shown in (a). A curve of the fraction of solid versus temperature from a Scheil-type solidification simulation is shown in (b).

phase. After most of the alloy has solidified, the Al_6Mn starts to form and finally also AlMg-β. This four-phase equilibrium is an invariant reaction in the ternary system and the temperature remains constant.

The cooling curves can be related to the calculated phase diagram in Fig. 8.7(a) from Ansara (1998a) and compared with a "Scheil–Gulliver" (SG)-type solidification (Gulliver 1913, Scheil 1942) calculation shown in Fig. 8.7(b). The alloy composition is shown as a dotted line in Fig. 8.7(a). When it starts forming fcc (Al dendrites), the liquid composition will move toward higher Mg content until the Al_6Mn phase starts to form also. That temperature is marked in Fig. 8.7(b). In the SG simulation no diffusion takes place in the solid phases and the liquid is assumed to be homogeneous.

The following figures are from simulations obtained using the phase-field method and show the development of the microstructure in two dimensions. In Fig. 8.8(a) the temperature is just below the liquidus temperature and the formation of Al dendrites has just started. The compositions of the components are indicated by the grayscale. In Fig. 8.8(b) the dendrites have grown and one can notice a gradient in composition from the center to the interface with the liquid.

In Fig. 8.8(c) the precipitation of Al_6Mn can be seen to have occurred inside the liquid phase, but this cannot be noticed on the cooling curve because the latent heat is very small. Finally, in Fig. 8.8(d) the dissolution of the Al_6Mn phase has started and the final liquid transforms to fcc and AlMg-β.

In the phase-field method the surface energies and interface mobilities play an important role. Such quantities are even more difficult to assess than the thermodynamic and mobility data. Often one can only vary these in a simulation model until one finds agreement with a real microstructure, but using validated thermodynamic databases makes it easier to find realistic values for these quantities.

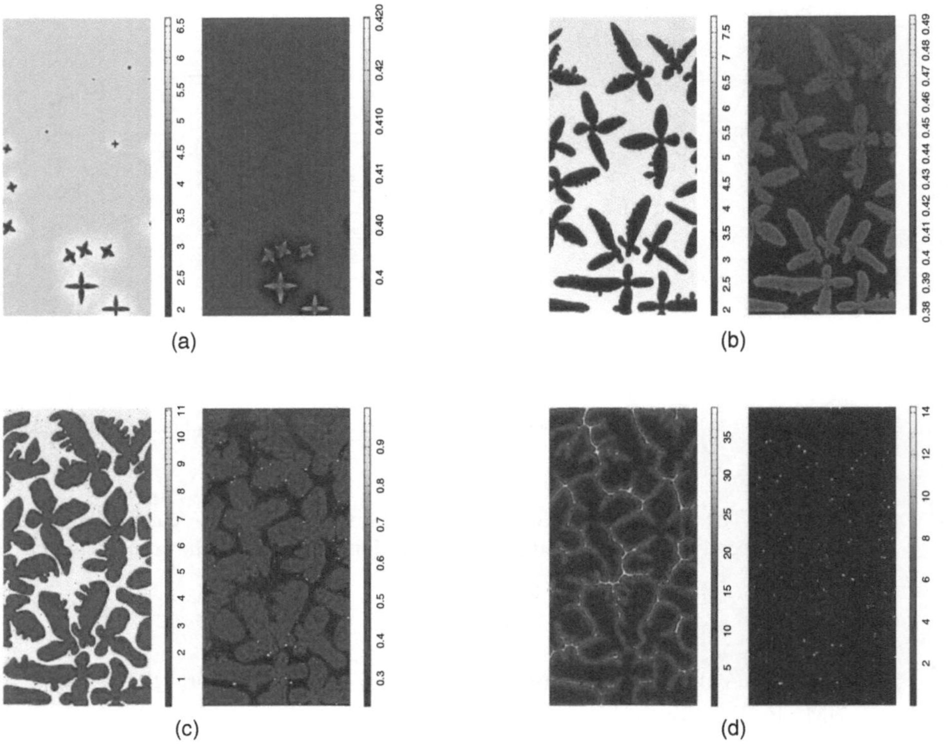

Figure 8.8 Steps in the phase-field-method simulation of the solidification. (a) Temperature just below liquidus after 1 s, the Al dendrites have nucleated. (b) The Al dendrites have grown after 16 s. (c) The dendrites after 40 s and the Al_6Mn phase has nucleated in the liquid. (d) The material is now completely solidified. The composition is displayed as a grayscale. The formation of the second solid phase in the liquid is shown in (c). In (d) the alloy is completely solid. Courtesy of Bernd Böttger.

However, there are many problems still to solve, such as that of which value of the Gibbs energy to use across a phase interface. Various approximate methods have been applied in various forms of software, but, unfortunately, there is still little interest in developing this further because the thermodynamic coupling slows the software down considerably and many software users have more interest in producing nice graphs than in realistic microstructures.

The technique from the multi-component solute-drag theory (Odqvist *et al.* 2003) which is used to connect the Gibbs energies across the interface is applicable also in the phase-field method. One can simply calculate the Gibbs energy at any point in the interface as the weighted average of the Gibbs energies of the phases involved, calculated for the same composition in all phases and using the amount of the phases as weighting:

$$G_{\mathrm{m}}^{\mathrm{I}} = \sum_{\alpha} m^{\alpha} G_{\mathrm{m}}^{\alpha}(T, P, x_i) \tag{8.3}$$

Some of the phases may be ordered and then the Gibbs energies for those phases must be calculated using the site fractions with the constraint that the mole fractions are the same as for the other phases.

If any phase has a stoichiometric constraint so it does not have a Gibbs energy value for all compositions, one can either assess the phase with another model with extended (metastable) solubility or assume some reasonable value of the missing Gibbs energy.

9 Case studies

The systems described here are real assessments, most of which have been published and the reference is given; but the descriptions here include some of the mistakes made when solving the problems leading to the publication. Such things are never included in the final publication. Discussing such problems does not mean that the assessment technique described here is bad or wrong, only that learning from mistakes is the only way to become a successful assessor, in the same way as many mistakes are inevitably made before one can learn how to be a good experimentalist.

9.1 A complete assessment of the Cu–Mg system

The Cu–Mg system published by Coughanowr *et al.* (1991), shown in Fig. 9.2 later, is very simple but offers some interesting examples of modeling. Assessments with two different software packages will also be discussed.

9.1.1 *Physical and experimental criteria for solution model selection*

There are five phases in the system, the liquid phase, the Cu phase with fcc lattice with some solubility of Mg, the Mg phase with hcp lattice and hardly any solubility of Cu, and two intermetallic phases:

- $CuMg_2$, a stoichiometric phase, and
- Cu_2Mg, with some range of homogeneity, having the cubic Laves-phase structure, C15 in the Strukturbericht notation.

9.1.1.1 The Laves phase Cu_2Mg

The range of homogeneity of the Laves phase is very well determined experimentally; it deviates on both sides from the ideal composition of 66.7% Cu. The cubic Laves-phase structural type has the following crystallographically equivalent atomic positions:

Pearson symbol:	cF24				
Space group:	Fd3m				
8 Mg:	8(a)	1/8, 1/8, 1/8;	7/8, 7/8, 7/8		
16 Cu:	16(d)	1/2, 1/2, 1/2;	1/2, 1/4, 1/4;	1/4, 1/2, 1/4;	1/4, 1/4, 1/2

The physical background that determines the formation of Laves phases is close packing of atoms of two different sizes. In a hard-sphere packing model the ideal ratio of the atomic radii is $r_A/r_B = 1.225$; in practice it is known that the ratio of pure element radii can vary from 1.05 to 1.68 in various systems. Although the ideal crystal structure has been very well studied, there is no systematic evidence giving a hint about the species occupancies for both sublattices on the Cu- and Mg-richer sides of the off-stoichiometric ranges.

In order to model this phase using the CEF, some assumptions should be made. According to the crystal structure, the description should present exactly two sublattices, the first for the 16(d) Cu atoms and the second for the 8(a) Mg atoms.

In order to describe the experimentally determined range of solubility, anti-site defects will be used, i.e., Cu atoms will occupy sites in the Mg sublattice and Mg atoms will occupy sites in the Cu sublattice. Note that this is just a model for the crystallographic occupancy, since the true nature of defects enabling the compositional range of stability of the phase has not been determined experimentally (e.g., by measurements of the lattice parameter versus x) or even theoretically. The model is then given by $(\mathrm{Cu, Mg})_2(\mathrm{Mg, Cu})$.

This assumption is not demonstrably true, but, since the stabilization of the Laves phase tolerates some deviation of the atomic radii of the components at the ideal composition, one can imagine an "effective" Cu radius that includes some Mg atoms and vice versa. In the small range of solubility of this phase, this should be a reasonable approximation.

The existence of vacancies in both sublattices would also allow the deviation from stoichiometry; however, in a close-packed structure, the vacancy probability is not expected to be large. The CEF for the selected crystallographic model is given by

$$
\begin{aligned}
G^{\phi}(T,x) - H^{\mathrm{SER}}(298\,\mathrm{K}) = {} & G^{\phi}_{\mathrm{Cu}_p:\mathrm{Mg}_q} \cdot (1-y') \cdot (1-y'') \\
& + G^{\phi}_{\mathrm{Cu}_p:\mathrm{Cu}_q} \cdot (1-y') \cdot y'' \\
& + G^{\phi}_{\mathrm{Mg}_p:\mathrm{Mg}_q} \cdot y' \cdot (1-y'') \\
& + G^{\phi}_{\mathrm{Mg}_p:\mathrm{Cu}_q} \cdot y' \cdot y'' \\
& + R \cdot T \cdot \{p \cdot [(1-y') \cdot \ln(1-y') + y' \cdot \ln(y')] \\
& \qquad + q \cdot [(1-y'') \cdot \ln(1-y'') + y'' \cdot \ln(y'')]\} \\
& + {}^{0}L^{\phi}_{\mathrm{Cu,Mg:Mg}} \cdot (1-y') \cdot y' \cdot (1-y'') \\
& + {}^{0}L^{\phi}_{\mathrm{Cu,Mg:Cu}} \cdot (1-y') \cdot y' \cdot y''
\end{aligned}
$$

$$
\begin{aligned}
&+ {}^{0}L^{\phi}_{\mathrm{Cu:Mg,Cu}} \cdot (1-y') \cdot (1-y'') \cdot y'' \\
&+ {}^{0}L^{\phi}_{\mathrm{Mg:Mg,Cu}} \cdot y' \cdot (1-y'') \cdot y'' \\
&+ {}^{1}L^{\phi}_{\mathrm{Cu,Mg:Mg}} \cdot (1-y') \cdot y' \cdot (1-2y') \cdot (1-y'') \\
&+ {}^{1}L^{\phi}_{\mathrm{Cu,Mg:Cu}} \cdot (1-y') \cdot y' \cdot (1-2y') \cdot y'' \\
&+ {}^{1}L^{\phi}_{\mathrm{Cu:Mg,Cu}} \cdot (1-y') \cdot (1-y'') \cdot y'' \cdot (1-2y'') \\
&+ {}^{1}L^{\phi}_{\mathrm{Mg:Mg,Cu}} \cdot y' \cdot (1-y'') \cdot y'' \cdot (1-2y'') \\
&+ \cdots
\end{aligned}
$$

This is the expression for the Gibbs energy of one mole of the phase (three moles of atoms). The first term in this summation expresses the Gibbs energy of the ideal compound. The second and third terms are the Gibbs energies for pure Cu and pure Mg, respectively, in a metastable state with the Laves structure. The fourth term describes the Gibbs energy of the metastable Laves phase containing only anti-structure (anti-site) atoms.

9.1.1.2 Lattice stability

Since the stable pure Cu phase has the fcc-A1 structure, the end member Cu_2Cu in a Laves-phase state is not stable. The same is true for the end member Mg_2Mg. One should have a value that expresses the difference between the Gibbs energy of the stable phase of the pure element phase and the Gibbs energy of these end members.

These differences are not determined experimentally and can be provided by band-structure total-energy calculations, *ab initio*, if the C15 structure is dynamically stable for the pure elements. If there are no values for these differences, a reasonable value should be chosen. This value must be selected carefully, since it must be used for all binary or higher-order systems with Cu or Mg dissolving in a cubic (C15) Laves phase.

In the published assessment of the Cu–Mg system by Coughanowr *et al.* (1991) these differences, the **lattice stabilities**, were obtained as an extrapolation of the Gibbs energy of the Laves phase to the pure elements. However, in a further update of the system in order for it to be used in the COST507 Al light-alloys database, a value of $5000\,\mathrm{J\,mol^{-1}}$ of atoms relative to the SER was arbitrarily chosen.

The selection of the adjustable model parameters for all the phases present in this system is well discussed in the original paper. The same assumptions were used in the more recent update of this description used in the COST507 database.

The files containing the original description in the Lukas program format (cumg.coe) are given in the website directory related to this chapter. The present COST507 description is also available in Thermo-Calc format (cumg.tdb).

The original experimental data (cumg.dat) file in Lukas format was translated into the Thermo-Calc format (cumg.pop) using the dat2pop.exe software (ftp account). The obtained pop file was used in the following optimization using PARROT.

In Fig. 9.1 the enthalpies of the fictitious states of the Laves-phase model for various compositions are shown. The stable composition is Cu_2Mg.

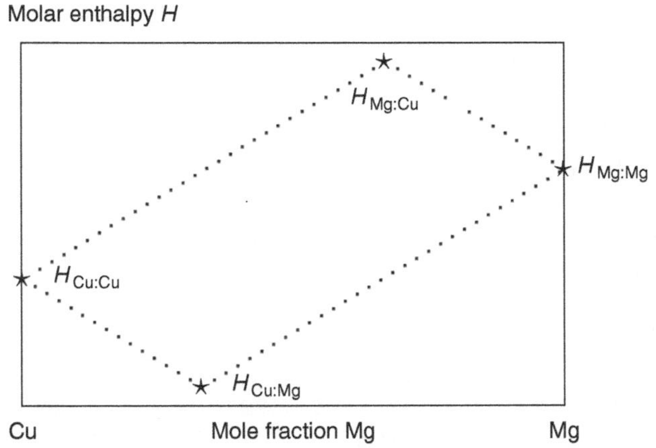

Figure 9.1 A schematic figure of the enthalpies of the Laves phase.

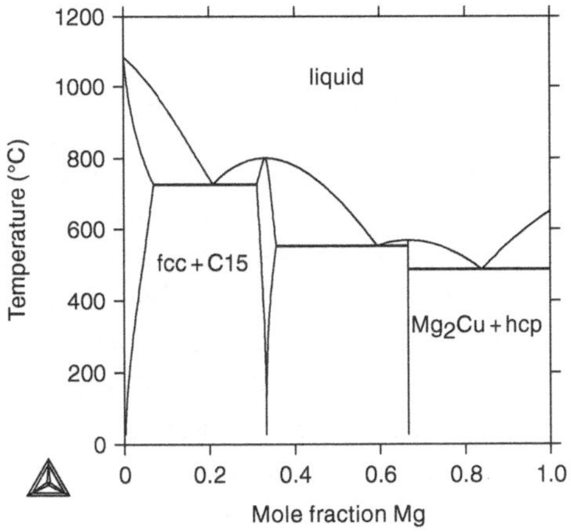

Figure 9.2 The calculated phase diagram for the Cu–Mg system.

In Fig. 9.2 the calculated Cu–Mg phase diagram is shown. In Fig. 9.3(a) the calculated Gibbs energies at 1000 K are shown. In the original assessment the hcp was treated without solubility of Cu since no experimental solubilities were available. When the system was added to a database, an ad-hoc interaction parameter was added to describe a low solubility. This was, probably by mistake, set to be positive, which creates a metastable miscibility gap in hcp. According to the recommendations in section 8.5.3, the hcp phase should have an interaction parameter equal or similar to that of the fcc phase.

In Fig. 9.3(b) the calculated enthalpies at 1000 K are shown. The miscibility gap in hcp is explained in connection with Fig. 9.3(a). The enthalpy of the Laves phase has a "V"-shaped enthalpy due to the anti-site defects used. In fact, the model used here for the Laves phase is a Wagner–Schottky model as described in section 5.8.2.3.

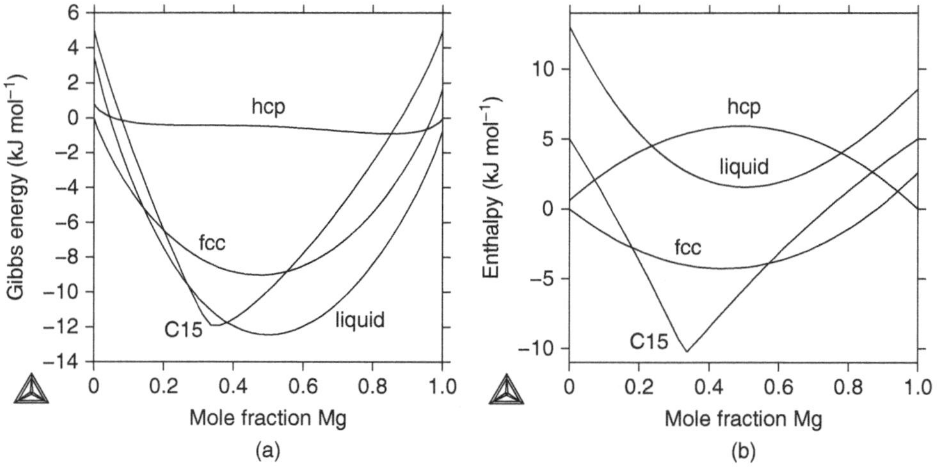

Figure 9.3 The calculated Gibbs energies (a) and enthalpies (b) for the Cu–Mg system, both at 1000 K.

9.1.2 *Assessment showing features of PARROT*

The Cu–Mg system assessed by PARROT is now described as a comparison with the use of BINGSS software for the same problem. The choice of models and the experimental data are described above. The choice of optimizer is mainly dependent on the availability of someone who can answer questions, but one of the advantages of PARROT is that it can treat ternary and higher-order systems; also it is extremely flexible.

It must be emphasized that BINGSS, PARROT, and other software for assessment of thermodynamic data can never be used as a substitute for an understanding of thermodynamics and phase diagrams. They are just tools that make it possible to handle large amounts of diverse types of data when trying to model a system. They are indispensable for the creation of large validated databases.

9.1.2.1 The setup file and experimental data file

As described in section 7.3 on software, two files should be prepared before the assessment starts. One is a file with data for the elements, phases, and models (the "setup" file). The other contains the experimental data (the "POP" file). These files may need changes during the assessment and at the end they should reflect the final choices of models and experimental data. The original files for BINGSS were converted using the conversion software and, to make it easier to understand, the file has now been edited and put into a form suitable for interactive use in PARROT.

It is also advisable to have a file with the experimental data in a format that is suitable for use in the graphical post-processor. It is often easier to see differences on a diagram. It is also useful to have some additional macro files to calculate and plot results together with the experimental data.

The setup file is a macro file that is run using the **macro** command in Thermo-Calc. It creates a "work" file in PARROT with the elements, phases, and models with parameters to optimize. The experimental data are then **compiled**:

```
tc
SYS: macro cumg
....output deleted
PARROT:compile
input file: cumg
....output deleted
PARROT:
```

After this, the user continues to run PARROT interactively in order to obtain the best set of parameters to describe the experiments. For description and explanation of the various commands, please read section 7.3 about PARROT or the PARROT user's guide. It is the responsibility of the user to select the best experimental information and determine what to keep if there are conflicting experimental data. One should not keep two experimental datasets that are in conflict because that will confuse the optimizer. With PARROT it is possible to change the selection interactively at any time if all data are in the POP file. The use of LABEL and COMMENT in the POP file to identify different datasets is recommended, since that simplifies later selection.

9.1.2.2 The alternate mode

In order to minimize the error between the experimental data and the corresponding values calculated from the models, it must be possible to calculate the experimental equilibria. Since the parameters in the thermodynamic models that should be assessed are initially zero, it might not be possible to calculate some equilibria corresponding to experimental points.

PARROT has a special mode called *alternate* to find initial values of model parameters. The reason for having an alternate mode is to handle initial values of model parameters using an alternate equilibrium calculation as described in section 7.1.1. Experimental equilibria involving two or more phases might not exist if the start values of the parameters are zero. The alternate mode in PARROT is further described in section 7.3.7.3. When the alternate mode has found a set of values for the model parameters that makes it possible to calculate the experimental equilibria using normal equilibrium calculations, one should turn off the alternate mode.

The number of experimental data is very large for this system. To make it easier to explain using PARROT, one can first try to fit just the single-phase data in the liquid phase and the invariant equilibria. To select experimental data, one uses the **set-weight** command in the EDIT module. In the EDIT module the experimental data must first be read from the work file. Setting the weightings to zero for all equilibria and then to 1 for the equilibria with label ABA means that one will use only the experimental data with that label:

```
PARROT: edit
EDIT: read 1
EDIT: set-weight 0 first-last
```

```
EDIT: set-weight 1 aba
EDIT: select-equil first
EDIT:compute-all
EDIT: save
EDIT: back
PARROT: opt 0
PARROT: list-result,,,
```

After having set the weightings, the first equilibrium was selected and all equilibria were calculated to check that there were no errors. Finally, all changes were saved back to the file. The first time the equilibria are calculated usually reveals some errors in the input POP file. These should be corrected by editing the POP file, of course, and the compilation and calculation run again.

It is essential to use SAVE in the EDIT module in order to make any changes there permanent. The command BACK goes back to the PARROT module and it is advisable to calculate all equilibria once again by activating the command OPT 0. The command LIST-RESULT will write a condensed output of all experimental data and their currently calculated values; see Table 7.3. It may be necessary to go to the EDIT module again many times and correct or modify the initial compositions or weightings of the experimental equilibria before the first real optimization in order to have a good set of experimental data to optimize.

9.1.2.3 The first diagram

When the optimization with the alternate mode has converged, it is possible to calculate the phase diagram and other thermodynamic properties and compare the simulation graphically with experimental data. This is done by running a macro file for this in the PARROT module. Of course, it is possible to do such calculations interactively, but, since they may be done very often, it is convenient to have the necessary commands on macro files.

The calculated diagram shown in Fig. 9.4(a) together with experimental data has the main features correct and further optimization will most certainly lead to a correct diagram, maybe after adjusting some of the weightings of the experimental data. However, in Fig. 9.4(b), which was optimized with the same experimental data but a slightly different set of parameters, the fcc phase has become much too stable and one has to change strategy a little in order to achieve good results. The recommended method is to optimize fewer parameters initially and maybe only two or three of the phases, for example the liquid and the terminal phases.

One may wonder what the optimizer is doing when it produces a diagram like Fig. 9.4(b), but actually it has done quite a good job of fitting the liquid and solid compounds as shown in Fig. 9.4(c), which was calculated with the same set of parameters as Fig. 9.4(b) but with the fcc phase suspended. In the experimental data file (POP file) the equilibria are usually created with just the known phases set stable, with all others suspended. During the optimization, other phases more stable than those in the POP file may appear. It is possible to add other phases as DORMANT, for example, and set as

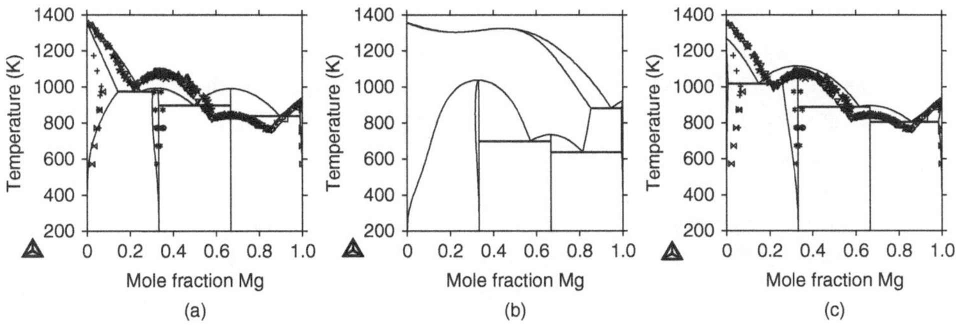

Figure 9.4 The first calculated phase diagram after optimizing with PARROT using only alternate mode. In (a) the features of the diagram are correct and, with further optimization using normal mode, the congruent-melting temperatures will change and the liquidus curves will be correct. However, often the first diagram calculated during an optimization is far from correct. As an example, in (b) a Cu–Mg diagram in which the fcc phase has extended far across the system is shown. In (c) the metastable diagram without fcc for the same set of parameters as in (b) is shown; see the text for an explanation.

an experimental criterion that their driving forces should be negative. Such things are typically done interactively when necessary. The number of experimental points for fcc is fewer than for the compounds and, depending on the parameters selected to be optimized, results like those in Figs. 9.4(b) and (c) are not unusual until one has found the most reasonable set of parameters that should be included in the optimization.

9.1.2.4 The normal mode

When the optimization with the alternate mode has converged, the user must switch off the alternate mode. After turning off the alternate mode, he must try to calculate all experimental equilibria in the EDIT module. This is done by setting the weightings of all experimental data to unity. Some equilibria may fail to converge, but it is not necessary at this stage to include all of the experimental data. The most important experimental equilibria to be included are the invariants, such as eutectics and congruent transformations. If some invariant cannot be calculated, it may be better to exclude optimization of some parameters of the corresponding phases and try to optimize them later when a good representation of the other phases has been obtained. Maybe the alternate mode will have to be used again to obtain initial values for such parameters.

One should not switch back and forth between alternate and normal mode. The alternate mode should be used only to find initial values for parameters so that normal-mode calculations can be used.

When starting to use the normal mode it is important to rescale the variables. The RESCALE command will set the current values as initial values and scaling factors. This makes it easier for the optimizing software to vary them to obtain the best fit. The complementary command RECOVER will overwrite the current values with the initial values and thus make it possible to restore the previous set of parameters. Note that this will not restore the compositions and temperatures for the experiments and it is usually necessary to calculate all equilibria in the EDIT module after a RECOVER command, and maybe also correct problems with convergence. For this reason it is advisable to make copies of the PAR file now and again, to make it easier to start again from an earlier point when the optimization fails.

At this stage it is also important to look carefully at the error for each experiment and try to identify conflicting and inconsistent information. A selection of the consistent data must be made, otherwise the optimization will not be able to give any reasonable result. The selection of data is made in the EDIT module using the weightings. It may also be necessary to add information that was not originally included in the POP file. For example, a phase may appear in the wrong part of a phase diagram. A convenient way to suppress such a phase is to add the lines

```
CHANGE-STATUS PHASE phase = DORMANT
EXPERIMENT DGM(phase)< -0.01:.001
```

These lines can be added to some existing experimental information for the composition and temperature range where the phase appears. The state variable DGM(phase) is the driving force for "phase" and, if this is negative, the phase is unstable. With this additional information, PARROT will try to change the parameters of all phases to achieve this. Such changes of the experimental information can be made directly in the POP file.

It is also possible to change the set of model parameters to be optimized in order to find a better solution. In general, the fit is improved the more parameters are added but is better the fewer parameters are used. Below will be described how to determine whether too many parameters have been used. Changing the set of parameters may require a return to the alternate mode in order to determine initial values. It is advisable to have copies of the PAR files that represent the best set of parameter values for each set tried because it is not always easy to reproduce a fit that has once been obtained but for which the PAR file has been lost. The current PAR file, selected by the SET-STORE-FILE command in the PARROT module, is continuously updated to the last optimized set. If the user wants to retain a copy of the present results before continuing, he must copy the PAR file using commands in the operating system of the file manager. When the user has several such PAR files, he may compare the results using various selections of experimental data and parameters.

In the assessment of the Cu–Mg system, the result of experiment 205 was found to be inconsistent with the other liquidus information, for example equilibrium 297, so its weighting was set to zero and the parameters were optimized again.

9.1.2.5 Large numerical values

Optimizations should continue until PARROT thinks it has found a minimum. If the weighting and selection of the experimental information is not done carefully, it sometimes happens that the optimized parameters reach very large values. The user must then again carefully examine his experimental data and their weightings. He may also have to reduce the number of parameters to avoid having large numerical values of parameters compensating for each other. For example, if the temperature range of data for a phase is very small, the temperature dependence sometimes appears to be very large. That happens typically if the temperature dependence is not determined by the experimental information. Then one has to set a constraint relating temperature-dependent and -independent coefficients of the parameter in Eq. (5.13) or Eq. (5.66), as suggested in section 6.2.1.3. Another way to avoid this is to use the same temperature-dependence coefficient for several phases. Such tricks should be used only when there is a theoretical foundation.

9.1.2.6 The final steps

When PARROT has reached a minimum for the selected set of experimental data and parameters and the user is satisfied with this, or fed up with the system and software, there are still some final checks that should be made. The criterion for reaching a minimum, which may be a local minimum, is that the optimization terminates with as many iterations as there are parameters to be optimized.

The first check may be to ascertain that the parameter values are reasonable. It may be very difficult for a beginner to decide, but, as a general rule, enthalpies should not exceed a few times $100\,\mathrm{kJ\,mol^{-1}}$ and entropies should not exceed few times $10\,\mathrm{J\,mol^{-1}\,K^{-1}}$. One may use the experimental information on enthalpies of mixing or formation to estimate a reasonable range of values. If several coefficients in the excess parameters are used, these should not increase with the increasing degree.

Another thing to check is whether the number of parameters used is reasonable. Since the sum of errors usually decreases with the number of parameters, it may be tempting to use as many as possible. There are several problems with having many parameters, though. One problem is that the values of the parameters are less well determined or significant the more parameters are used. The significance of the value of a parameter is given by the column labeled REL.STAND.DEV, which is shown for each parameter in the LIST-RESULT and LIST-ALL-VARIABLES commands.

In order for values in this column to be significant, the user must have rescaled the variables before the last OPTIMIZE command and the parameter values must be almost the same as the initial values and scaling factors. This is necessary in order for values in the final column, "REL.STAND.DEV" (relative standard deviation), to be significant. Study the values in this column carefully. The value in this column shows how significant the parameter value is. One may change the parameter within the range of this standard deviation and the reduced sum of errors will then change only by unity. Parameters with large REL.STAND.DEV may be set to zero and the optimization re-run with fewer parameters will still give almost the same result.

In the Cu–Mg case the REL.STAND.DEV is around 0.8 for one of the parameters in the $CuMg_2$ phase. One may thus change this parameter value by 80% and still not change the sum of errors significantly. In this case it is not possible to put it to zero, but maybe it could be rounded off to a value like 2 to indicate that it is not well determined. The other parameters must then be optimized again to obtain the best fit with this new fixed

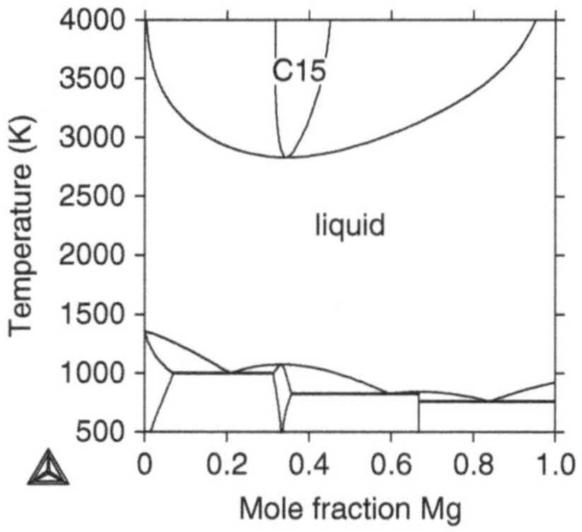

Figure 9.5 The calculated phase diagram for the Cu–Mg system extending to high temperatures.

parameter value. The values of REL.STAND.DEV for these parameters now generally tend to be smaller as they become more well determined. One may continue this rounding off of parameter values, but this is a matter of taste for the assessor.

9.1.2.7 Why constant updating is needed

After all the efforts put into this assessment by experts of types 1 and 3, a recent use of this system revealed some problems. A calculation of the system at much higher temperatures than had been considered during the assessment showed that the Laves phase becomes stable above 3000 K as shown in Fig. 9.5.

This is due to the fact that the heat-capacity expression of the Laves phase can be assessed only up to the melting temperature. The term $a_5 \cdot T^3$ of the G description contributes as $6a_5 \cdot T^2$ to C_p, leading to impossible values at very high temperatures. The simplest solution is to set breaks in the parameter descriptions of the end members of the Cu_2Mg model slightly above the melting temperature (e.g. 1100 K) and continue with descriptions using only a_0, a_1, and a_2, corresponding to constant C_p.

9.2 Checking metastable diagrams: the Ag–Al system

This binary system presents interesting features with which to illustrate many of the comments in the methodology chapter. The Al and Ag phases are both terminal fcc solid solutions that present extended ranges of homogeneity. These terminal solid solutions are modeled as the same phase. Two intermediate solution phases are stable in this system; they are disordered and their crystal structures are different from those of the terminal solutions: one phase is an hcp and the other is a bcc phase.

When modeling phases, one should take into account that not only the stable calculated phase diagram but also metastable diagrams should be reproduced by the calculations. Having reasonable metastable descriptions *increases the quality of extrapolations to higher-order systems.*

The stable phase diagram assessed by Lim *et al.* (1995) is shown in Fig. 9.6 together with several metastable diagrams with only selected phases. In Fig. 9.6(b) it is shown that the fcc phase has a metastable miscibility gap. In Fig. 9.6(f) the stable phase diagram is shown overlayed with the metastable solubility lines from the previous diagrams.

When a phase is stable only within a small composition range but the model extends across the whole system one must take care to ensure that the phase behaves reasonably well. In Fig. 9.6(d) the metastable fcc + hcp diagram is shown and its shape is strange. The reason is that the excess parameters in the hcp model have been fitted to the small stable composition range. However, hcp and fcc are very similar; both have 12 nearest neighbors and the regular-solution parameter should be almost the same for hcp and fcc since it depends mainly on the first-nearest-neighbor interaction. Higher-order coefficients in the RK series may be different, but as a rule of thumb one should have the same regular-solution parameter in fcc as in hcp. However, this "rule" is quite recent and one will normally not find that it is used in any existing assessment. The rule was created to estimate parameters in unstable phases in order to allow extrapolations to ternary systems. For example, see the hcp description for

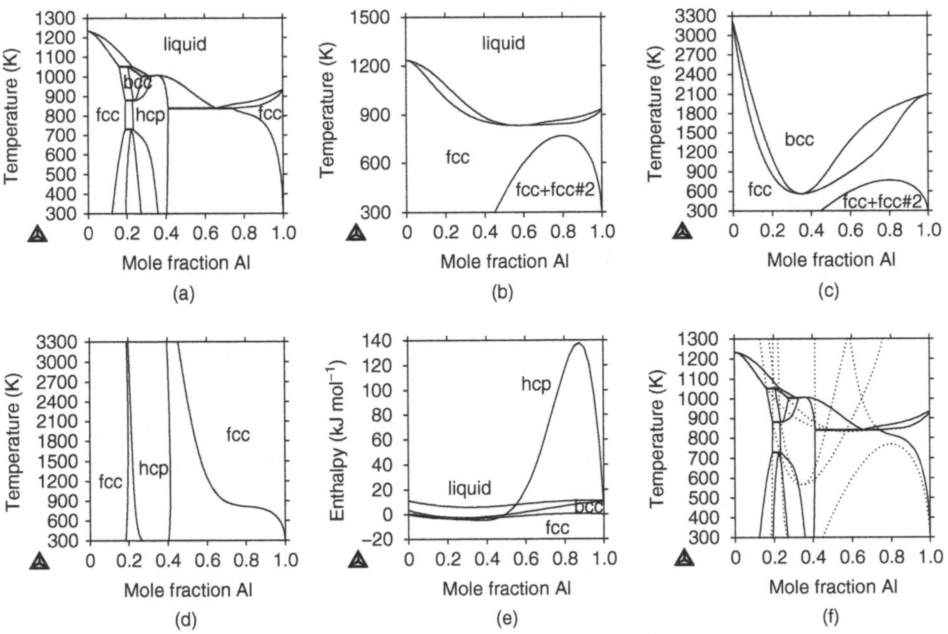

Figure 9.6 Various stable and metastable phase diagrams and the enthalpies of mixing for the Ag–Al system. (a) The stable phase diagram for the Ag–Al system. (b) The metastable phase diagram for the Ag–Al system with only liquid and fcc. (c) The metastable phase diagram for the Ag–Al system with fcc and bcc. (d) The metastable phase diagram for the Ag–Al system with fcc and hcp. (e) The enthalpy of the liquid, fcc A1, bcc A2 and hcp A3 phases at 873 K in the Ag–Al system. (f) The stable phase diagram for the Ag–Al system with overlayed metastable solubility curves.

the Cu–Fe system mentioned in section 8.5.3 and shown in Fig. 8.1(b). Since fcc and hcp are so similar, it is natural to set their excess parameters equal. For bcc it may be more adequate to set them equal to the excess parameters for the liquid. Since the phases in the Ag–Al system are typical Hume-Rothery phases, differences in G_m between these phases may mainly be due to the contribution of the electron gas to G_m (see section 6.2.8).

The calculated enthalpies for the various phases at 873 K are shown in Fig. 9.6(e). They are all reasonable except the high positive enthalpy of hcp A3 on the Al-rich side. This does not matter in the binary system, but, if a ternary element were added, forming a stable hcp A3 phase at lower Ag contents it would be very difficult to fit any ternary parameters for hcp. In the ternary Ag–Al–Zn system hcp has a continuous range of solubility from the Ag–Al binary system to Ag–Zn with a Zn mole fraction between 0.7 and 0.85 (Köster *et al.* 1964). Thus far no-one has attempted a thermodynamic assessment of this ternary system. The lack of experimental data on the extension of the other binary phases into the ternary system is another possible reason for why this has not yet been done.

9.3 The Re–W σ phase refit using first-principles data

The σ phase has the Strukturbericht designation $D8_b$ and a complex structure with five different crystallographic sites. It is shown in Fig. 5.22.

The coordination number (CN) for sublattices 1 and 4 is 12, i.e., the same as for fcc, so these sublattices are preferred by elements that have fcc lattices as pure elements. The second sublattice has CN = 15 and is preferred by bcc-type elements such as V, Cr, and W. The remaining two sublattices have CN = 14 and have more-random occupancy. Assuming that Re and W can occupy any sublattice, one has the CEF model

$$(\mathrm{Re},\mathrm{W})_2(\mathrm{Re},\mathrm{W})_4(\mathrm{Re},\mathrm{W})_8(\mathrm{Re},\mathrm{W})_8(\mathrm{Re},\mathrm{W})_8 \tag{9.1}$$

This gives 32 end members for the σ phase in the Re–W system. Their 0-K Gibbs energies were calculated *ab initio* by Berne *et al.* (2001). They are listed in Table 9.1 and plotted in Fig. 9.7(a). The figure is reprinted with permission from Fries and Sundman (2002). In the figure the Gibbs-energy curve calculated from the CEF model using the tabulated end-member values is also given.

It is worth noting that the primary *ab initio* data should always be included in a publication, not only fitted CVM cluster energies. It is the primary calculated or experimental data that are needed in the Calphad method.

The CEF model using the *ab initio* data was constructed by Fries and Sundman (2002) in order to demonstrate that the difference between the CVM and the CEF is very small for intermetallic compounds that never undergo disordering. In Fig. 9.7(b) the difference between the site fractions calculated from the CVM and from the CEF is shown. With a CEF model it is easy to calculate also other quantities, such as the

Table 9.1 *The* ab initio *calculated energies for the 32 configurations of Re and W, also translated into J mol*$^{-1}$ *used as CEF end-member energies (see also Fig. 9.7(a))*

Lattice occupation 1	2	3	4	5	Energy (meV/atom)	x_W	CEF parameter	J mol^{-1}
Re	Re	Re	Re	Re	0	0	$^\circ G_{Re:Re:Re:Re:Re}$	0
W	Re	Re	Re	Re	5.9598	0.0667	$^\circ G_{W:Re:Re:Re:Re}$	17 349
Re	W	Re	Re	Re	−41.8254	0.1333	$^\circ G_{Re:W:Re:Re:Re}$	−121 754
W	W	Re	Re	Re	−34.5563	0.2	$^\circ G_{W:W:Re:Re:Re}$	−100 593
Re	Re	W	Re	Re	−58.8566	0.2667	$^\circ G_{Re:Re:W:Re:Re}$	−171 332
W	Re	W	Re	Re	−50.4266	0.3333	$^\circ G_{W:Re:W:Re:Re}$	−146 792
Re	W	W	Re	Re	−85.7745	0.4	$^\circ G_{Re:W:W:Re:Re}$	−249 690
W	W	W	Re	Re	−70.2813	0.4667	$^\circ G_{W:W:W:Re:Re}$	−204 589
Re	Re	Re	W	Re	16.0299	0.2667	$^\circ G_{Re:Re:Re:W:Re}$	46 663
W	Re	Re	W	Re	31.0186	0.3333	$^\circ G_{W:Re:Re:W:Re}$	90 295
Re	W	Re	W	Re	−20.5734	0.4	$^\circ G_{Re:W:Re:W:Re}$	−59 889
W	W	Re	W	Re	−0.9511	0.4667	$^\circ G_{W:W:Re:W:Re}$	−2 769
Re	Re	W	W	Re	8.7176	0.5333	$^\circ G_{Re:Re:W:W:Re}$	25 377
W	Re	W	W	Re	25.7323	0.6	$^\circ G_{W:Re:W:W:Re}$	74907
Re	W	W	W	Re	−16.6055	0.6667	$^\circ G_{Re:W:W:W:Re}$	−48 339
W	W	W	W	Re	5.7752	0.7333	$^\circ G_{W:W:W:W:Re}$	16812
Re	Re	Re	Re	W	−44.2292	0.2666	$^\circ G_{Re:Re:Re:Re:W}$	−128 751
W	Re	Re	Re	W	−29.1608	0.3333	$^\circ G_{W:Re:Re:Re:W}$	−84 887
Re	W	Re	Re	W	−86.4593	0.4	$^\circ G_{Re:W:Re:Re:W}$	−251 683
W	W	Re	Re	W	−67.4605	0.4667	$^\circ G_{W:W:Re:Re:W}$	−196 378
Re	Re	W	Re	W	−79.9138	0.5333	$^\circ G_{Re:Re:W:Re:W}$	−232 629
W	Re	W	Re	W	−61.4605	0.6	$^\circ G_{W:Re:W:Re:W}$	−178 911
Re	W	W	Re	W	−107.0972	0.6667	$^\circ G_{Re:W:W:Re:W}$	−311 760
W	W	W	Re	W	−85.0240	0.7333	$^\circ G_{W:W:W:Re:W}$	−247 505
Re	Re	Re	W	W	0.3771	0.5333	$^\circ G_{Re:Re:Re:W:W}$	1 098
W	Re	Re	W	W	21.6766	0.6	$^\circ G_{W:Re:Re:W:W}$	63101
Re	W	Re	W	W	−31.6385	0.6667	$^\circ G_{Re:W:Re:W:W}$	−92 100
W	W	Re	W	W	−7.8781	0.7333	$^\circ G_{W:W:Re:W:W}$	−22 933
Re	Re	W	W	W	6.7427	0.8	$^\circ G_{Re:Re:W:W:W}$	19 628
W	Re	W	W	W	24.2583	0.8667	$^\circ G_{W:Re:W:W:W}$	70 616
Re	W	W	W	W	−27.5870	0.9333	$^\circ G_{Re:W:W:W:W}$	−80 306
W	W	W	W	W	0	1	$^\circ G_{W:W:W:W:W}$	0

variations of site fractions and configurational heat capacity with temperature, as shown in Fig. 9.8.

In the paper by Fries and Sundman (2002) the new CEF model for the σ phase was combined with an existing assessment of the whole phase diagram from Liu and Chang (2000) to obtain a complete description of the whole system. In order to obtain fit to the other phases, it was necessary to adjust the stability of the σ phase for the pure elements and a parameter describing the disordered substitutional regular solution was introduced for σ according to Eq. (5.161). The reason why a regular-solution parameter was needed may be that, in an assessment without experimental enthalpy data, only differences between values of G_m of the phases are needed in order to calculate the phase-diagram data. Thus an enthalpy contribution common to the Gibbs energies of all phases in the system cannot be found by the assessor.

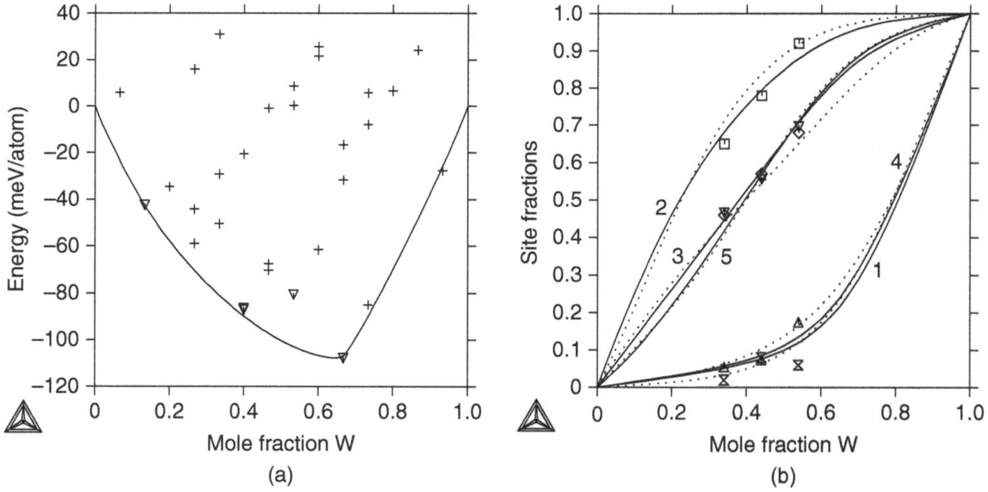

Figure 9.7 The symbols in (a) are the *ab initio* calculated energies and the curve was calculated using a model based on the CEF. The triangular points represent those end members with Re in sublattices 1 and 4 and W in sublattice 2, i.e., those which were expected to have the lowest energy. In (b) site occupancies in the σ phase, calculated for 1500 K, are compared with experimental data and with curves calculated from a CVM model by Berne *et al.* (2001) (dotted). Copyright 2002, reprinted with permission from the American Physical Society.

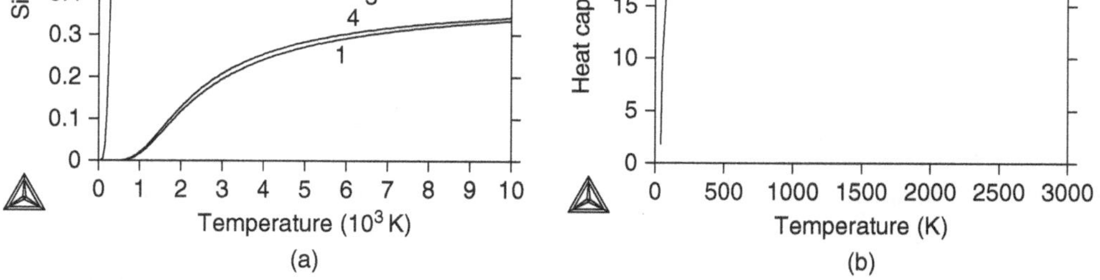

Figure 9.8 In (a) the site fractions on the various sublattices for $x_W = 0.4$ are plotted as a function of temperature up to 10 000 K and in (b) the contribution to the heat capacity from the configuration is plotted up to 3000 K. Note that the phase never undergoes disordering and that the rapid change in the slopes of the fractions around 200 K gives a bump in the heat-capacity curve. Reprinted with permission from the American Physical Society.

9.4 A complete binary system: Ca–Mg

The optimization described here was published by Agarwal *et al.* (1995).

9.4.1 *Experimental data from the literature*

A literature review and critical evaluation was given by Nayeb-Hashemi and Clark (1987). Most of the arguments given there could be used to select the more reliable one from among the contradictory sets of experimental results in Table 9.2 (see section 6.1.2). Since the least square of errors is a good indicator of the best fit only if the errors are distributed randomly and are small, contradictory experimental values must be judged before the optimization and values suffering from systematic errors must be removed (section 2.4.1). The following selections could be done before any calculation.

The liquidus data of Haughton (1937), Vosskühler (1937), and Klemm and Dinkelacker (1947) agree fairly well, whereas those of Baar (1911) and Paris (1934) deviate significantly (Fig. 9.9). Since contradictory data cannot both be correct, the latter data, being the older ones, were rejected in the optimization. The (Mg) solvus data published by these authors disagree so heavily (see Fig. 9.12 later) that a Gaussian distribution of errors cannot be assumed. Only the data of Vosskühler (1937) and Burke (1955) agree well enough for the scatter to be interpreted as randomly distributed errors. Therefore only these values were used, not those of Haughton (1937), Nowotny *et al.* (1940), and Bulian and Fahrenhorst (1946). For the temperatures of the invariant equilibria the values selected by Nayeb-Hashemi and Clark (1987) were accepted as the result of a critical judging of various measurements representing data that must have a unique value. The following values were used for the optimization: congruent-melting temperature of $CaMg_2$, 986 K; of Mg-rich eutectic, 789.5 K; and of Ca-rich eutectic, 719 K.

The thermodynamic values agree well, except the enthalpies of mixing of Sommer *et al.* (1977) and the enthalpy of formation of $CaMg_2$ of Smith and Smythe (1959). The values of Sommer *et al.* (1977) disagree significantly with later values from the same laboratory (Agarwal *et al.*, 1995) and the authors of the later paper assume the earlier data to suffer from systematic errors. The values of Sommer *et al.* (1977) were therefore not used. The values of Agarwal *et al.* (1995) were treated and plotted (see Fig. 9.13 later) as partial enthalpies of Ca in liquid (see section 4.1.1.1, on mixing calorimetry). The enthalpies of formation of $CaMg_2$ at 298 K of King and Kleppa (1964) and Davison and Smith (1968) agree well (-13.4 and $-13.1\,\mathrm{kJ\,mol^{-1}}$, respectively, referred to moles of atoms), but the bomb-calorimetry measurement of Smith and Smythe (1959) ($-19.5 \pm 13\,\mathrm{kJ\,mol^{-1}}$) deviates significantly. The latter was characterized as only tentative by the authors themselves; thus it was not used. From the heat-content values of Agarwal *et al.* (1995) (see Fig. 9.15 later) three points in the vicinity of the melting temperature were not used, since the possibility of some partial melting there, which might give some discrepancy between the measured value and its interpretation, cannot be excluded. All other values given in Table 9.3 were used for the optimization.

Table 9.2 *A summary of experimental phase-diagram data on the Ca–Mg system*

Reference	Experimental method	Equilibria	Ranges x_{Mg}	Ranges T (K)
Baar (1911)	Thermal analysis	Liquidus	0.000–1.000	750–1100
Paris (1934)	Thermal analysis	Liquidus	0.125–0.988	750–1000
Haughton (1937)	Thermal analysis	Liquidus	0.667–0.970	790–1000
		Mg-rich eutectic	0.895	790
	Metallography	(Mg) solvus	0.987–0.997	570–790
Vosskühler (1937)	Thermal analysis	Liquidus	0.578–0.977	800–1000
		Mg-rich eutectic	0.894	789
	Metallography	(Mg) solvus	0.995–0.999	570–790
Nowotny *et al.* (1940)	Metallography, Hardness, X-ray	(Mg) solvus	0.992–0.995	570–780
Bulian and Fahrenhorst (1946)	Electrical resistivity	(Mg) solvus	0.992–0.995	670–790
Klemm and Dinkelacker (1947)	Thermal analysis	Liquidus	0.350–0.839	790–990
		Ca-rich eutectic	0.270	718
Burke (1955)	Metallography	(Mg) solvus	0.994–0.999	640–790

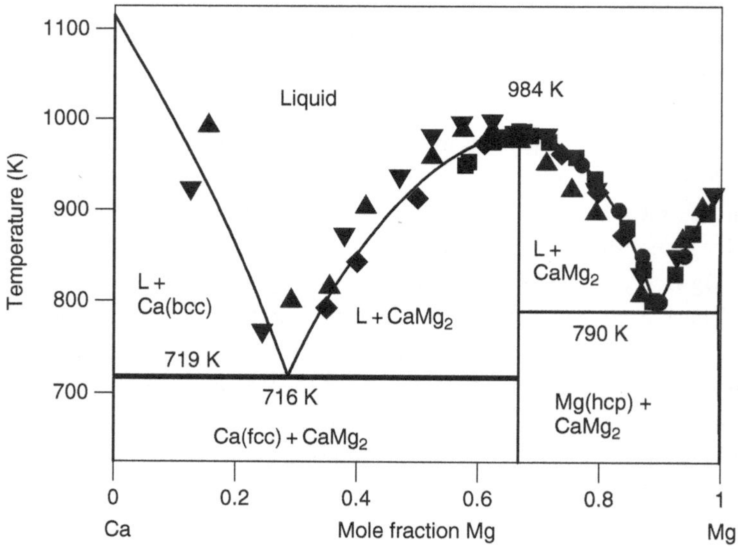

Figure 9.9 The optimized phase diagram of the Ca–Mg system. Data from ● Haughton (1937), ■ Vosskühler (1937), and ♦ Klemm and Dinkelacker (1947) were used. Data from ▲ Baar (1911) and ▼ Paris (1934) were not used.

9.4.2 *Selecting the models*

Five phases have to be modeled, the liquid, the fcc and bcc phases of pure Ca, the hcp phase of pure Mg, and the intermediate Laves phase $CaMg_2$. For both Ca modifications and for the intermediate phase the ranges of homogeneity are very small and not

Table 9.3 *A summary of experimental thermodynamic data on the Ca–Mg system*

Reference	Experimental method	Quantity	Ranges	
			x_{Mg}	T (K)
Smith and Smythe (1959)	Mg vapor pressure	$\mu_{Mg}^{CaMg_2+Ca} - {}^{\circ}\mu_{Mg}$	0.300–0.667	298–770
	Bomb calorimetry	$\Delta^{f}H^{CaMg_2}(298\,K)$	0.667	298
Chiotti *et al.* (1964)	Equilibrium with $H_2 + CaH_2$	$\mu_{Ca}^{CaMg_2+Mg} - {}^{\circ}\mu_{Ca}$	0.667–0.800	298–773
King and Kleppa (1964)	Sn solution calorimetry	$\Delta^{f}H^{CaMg_2}(298\,K)$	0.667	298
Gartner (1965)	Reaction calorimetry	$\Delta^{f}H^{CaMg_2}(298\,K)$	0.667	532
Mashorets and Puchkov (1965)	Mg vapor pressure	$\mu_{Mg}^{liq} - {}^{\circ}\mu_{Mg}^{liq}$	0.158–0.962	1200
Chiotti *et al.* (1966)	Adiabatic calorimetry	$\Delta^{fus}H^{CaMg_2}$	0.667	987
Davison and Smith (1968)	Acid solution calorim.	$\Delta^{f}H^{CaMg_2}(298\,K)$	0.667	298
Sommer (1979)	Mg vapor pressure	$\mu_{Mg}^{liq} - {}^{\circ}\mu_{Mg}^{liq}$	0.050–0.920	1010
Agarwal *et al.* (1995)	Mixing calorimetry	ΔH_{Ca}^{liq}	0.557–0.964	1023
	Drop calorimetry	$H(T) - H(298\,K)$	0.667	752–1170

known quantitatively, therefore they are ignored and these three phases are modeled as stoichiometric phases, describing the Gibbs energy G as a function of temperature T only.

The enthalpy of mixing versus mole fraction of the liquid phase exhibits roughly parabolic behavior, indicating that the bonding is predominantly metallic. Therefore the RK formalism is adequate for modeling the Gibbs energy of the liquid.

The solubility of Ca in the hcp Mg solid solution was measured several times. Since it is small, (Mg) may be treated as a Henrian solution, which is described by the regular-solution formula, Eq. (5.65), with a single parameter ${}^{0}L_{ij} = a_0 + a_1 T$, using the fixed value for the unary Ca(hcp) parameter proposed by Dinsdale (1991).

9.4.3 *Selecting the adjustable parameters*

The unary parameters must be identical in all systems containing the same component and must not be changed in the optimization of a binary system. The Ca(bcc) and Ca(fcc) phases are thus totally defined and their descriptions do not contain adjustable parameters. For these unary parameters the SGTE values compiled by Dinsdale (1991) were adopted.

For the Gibbs energy of the $CaMg_2$ phase as a function of temperature $G^{CaMg_2}(T) - H_{Ca}^{SER}(T_0) - 2H_{Mg}^{SER}(T_0)$ with $T_0 = 298\,K$ three coefficients of Eq. (5.2) can be defined independently because the $H(T) - H(T_0)$ measurements of Agarwal *et al.* (1995) describe well enough the heat capacity given by the third coefficient. The enthalpy of formation (King and Kleppa 1964, Davison and Smith 1968) essentially determines the first coefficient. The μ_{Mg} values in the two-phase field Ca + $CaMg_2$ are directly related to the Gibbs energy of $CaMg_2$ and thus give a description of G^{CaMg_2} defining independently the second coefficient. The same is true for μ_{Mg} in the two-phase field Mg + $CaMg_2$, since the deviation due to non-zero solubility of Ca in (Mg) is far below significance. The heat-content values are not smooth enough to allow differentiation in order to get the heat capacity C_p as a function of T and define a fourth coefficient. Therefore the fourth and fifth coefficients were estimated using partially the idea of Kubaschewski and Ünal (1977). These authors proposed estimating C_p by assuming the fixed value 30.3 J mol^{-1} K^{-1} at the melting temperature and using the fixed fifth coefficient $E = 21\,000$ J K mol^{-1} in the expression

$$C_p = -C - 2D \cdot T - 2E \cdot T^{-2} \tag{9.2}$$

Both numerical values are referred to one mole of atoms. The determination of the two coefficients C and D cannot be done by estimating C_p at 298 K, as was done in the paper of Kubaschewski and Ünal (1977), since the method given there requires a tendency toward ionic structure. Therefore here they were adjusted during the optimization, keeping C_p as 30.3 J mol^{-1} K^{-1} at the melting temperature of 986 K. This procedure seems to be better than usage of the Kopp–Neumann rule assuming $C_p^{CaMg_2}$ to be the sum of $\frac{1}{3}C_{p\ Ca} + \frac{2}{3}C_{p\ Mg}$. C_p after this definition would exhibit several kinks due to the transformation and melting of Ca and Mg, which, however, have no reason to appear in C_p of the $CaMg_2$ phase. The term $-2E \cdot T^{-2}$ may be regarded as representing the most significant term of a Taylor-series expansion of the difference between a Debye function and a constant C_p after Dulong and Petit. Thus the method used by Agarwal *et al.* (1995) is compatible with the "Ringberg recommendation" (Chase *et al.* 1995) that one should express C_p as the sum of a Debye function and a linear function of T.

For the Mg(hcp) solid solution the only binary parameter is the regular-solution parameter. It is adjusted using the solubility data of this phase (Vosskühler 1937, Burke 1955). Since these solubilities are known over a significant range of temperature, as explained in section 6.3, it is possible to adjust the regular-solution parameter as function of temperature using two coefficients A and B of the linear function ${}^0L^{hcp}_{Ca,Mg} = A + B \cdot T$.

The RK description of the excess Gibbs energy of the liquid phase needs at least the two parameters ${}^0L^{liq}_{Ca,Mg}$ and ${}^1L^{liq}_{Ca,Mg}$ of Eq. (5.65), since the enthalpy of mixing ${}^{mix}H^{liq}(T)$ (Agarwal *et al.* 1995) is clearly asymmetrical. Since the enthalpy (Agarwal *et al.* 1995) and Gibbs-energy data (Mashorets and Puchkov 1965, Sommer 1979) of the liquid are measured independently, it is possible to adjust these parameters as (linear) functions of temperature, ${}^0L^{liq}_{Ca,Mg} = {}^0A + {}^0B \cdot T$ and ${}^1L^{liq}_{Ca,Mg} = {}^1A + {}^1B \cdot T$. The question of whether a third parameter is useful can be answered by doing the optimization with two and then with three parameters and comparing the results. From that Agarwal *et al.* (1995) decided to use a third parameter.

9.4.4 Optimization

The selected coefficients were adjusted to the experimental values using the program BINGSS. The optimization was started with all adjustable coefficients zero, except C

Table 9.4 *Adjusted coefficients of the Ca–Mg system (to express G in* J mol^{-1} *of atoms)*

Phase	Parameter	A_ν or A	B_ν or B	C	$D \cdot 10^3$	E
Liquid	${}^0L^{\text{liq}}_{\text{Ca,Mg}}$	−32 322.4	16.7211			
	${}^1L^{\text{liq}}_{\text{Ca,Mg}}$	60.3	6.5490			
	${}^2L^{\text{liq}}_{\text{Ca,Mg}}$	−5 742.3	2.7596			
Mg(hcp)	${}^0L^{\text{hcp}}_{\text{Ca,Mg}}$	−9 183.2	16.9810			
$CaMg_2$	$G^{\text{CaMg}_2} - H^{\text{SER}}_{\text{Ca}} - 2H^{\text{SER}}_{\text{Mg}}$	−67 874.7	466.5126	−82.720 14	−4.762 23	630 000

of the term $C \cdot T \cdot \ln(T)$ for the $CaMg_2$ phase, which was set to 92.1987 to give $C_p = 90.9\,\text{J mol}^{-1}\,\text{K}^{-1}$ at 986 K after Eq. (9.2) (1 mole = 3 moles of atoms). By imposition of constraints (section 7.2), the coefficient C was forced to change by −1972 for unity change of D in this equation, in order to keep $C_p(986\,\text{K})$ constant.

All values mentioned in Tables 9.2 and 9.3 were put into the file camg.dat, but, as explained above, the values from Baar (1911), Paris (1934), Haughton (1937), Nowotny *et al.* (1940), Bulian and Fahrenhorst (1946), and Sommer *et al.* (1977) and the $\Delta^{\text{f}}H^{\text{CaMg}_2}$ value of Smith and Smythe (1959) were excluded from the calculation.

The calculation was started with the variable "IVERS" set equal to 1 (see sections 7.2 and 7.1.1). The Marquardt parameter became 10.0 after ten calculations in the first run. After this run, the mean-squared error dropped significantly in relation to the definition of the dimensionless errors in Eq. (2.59). The calculation was continued with "IVERS" equal to 2, i.e., using common tangents, the Marquardt parameter 10.0 and eight calculations. After four more steps with "IVERS" equal to 2, the Marquardt parameter became 10^{-9} and, after two more calculations, the mean-squared errors remained constant. The resulting parameters were taken as the optimized dataset. They are given in Table 9.4.

9.4.5 *Results*

Diagrams calculated with these parameters are compared with the experimental data in Figs. 9.9–9.14. There is a small disagreement for the heat of melting, the enthalpy of formation of $CaMg_2$, and the enthalpy of mixing of the liquid. Because the enthalpy is a state function (section 2.1.7), the enthalpy of formation plus the heat content plus the enthalpy of melting must be the same as the heat content of the liquid pure elements plus the enthalpy of mixing of the liquid with an Mg content of 66.7 mol%. The disagreement, although it is visible in Fig. 9.15, is fairly well within the accuracy of the measurements.

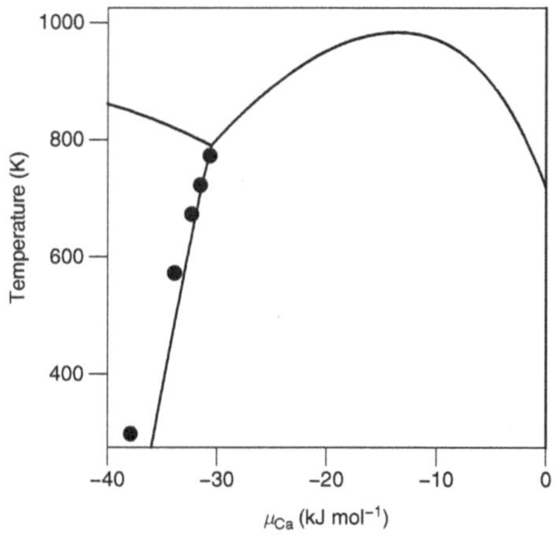

Figure 9.10 The chemical potential of Ca in the two-phase equilibria of the Ca–Mg system, with data from Chiotti *et al.* (1964).

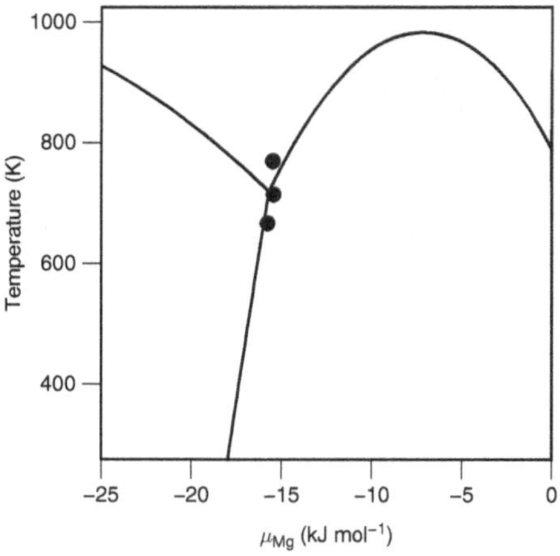

Figure 9.11 The chemical potential of Mg in the two-phase equilibria of the Ca–Mg system, with data from Smith and Smythe (1959).

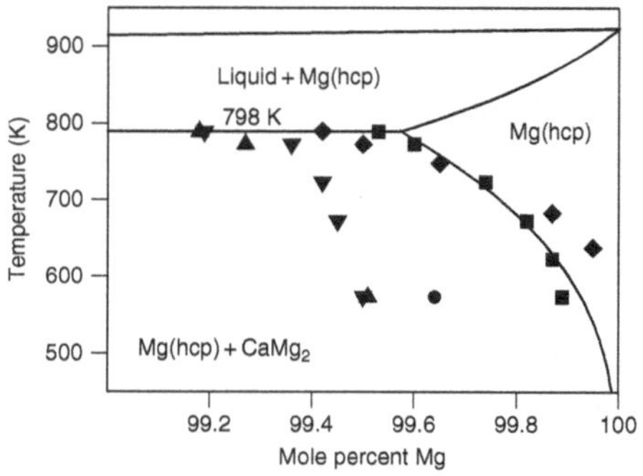

Figure 9.12 The Mg-rich edge of the Ca–Mg phase diagram. Data from ■ Vosskühler (1937) and ♦ Burke (1955) were used. Data from • Haughton (1937), ▲ Nowotny *et al.* (1940), and ▼ Bulian and Fahrenhorst (1946) were not used.

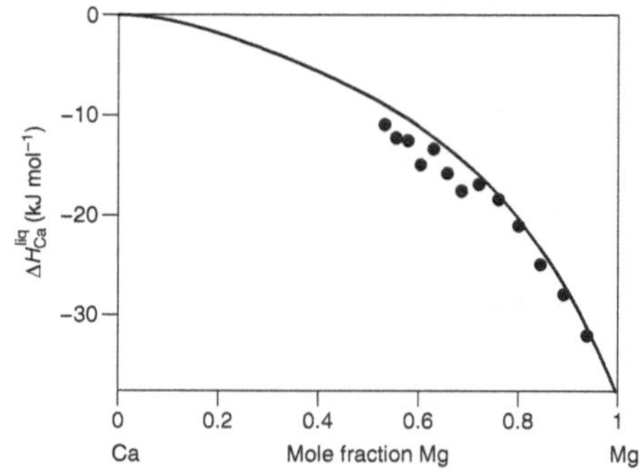

Figure 9.13 The partial enthalpy of Ca at 1032 K, with data from Agarwal *et al.* (1995).

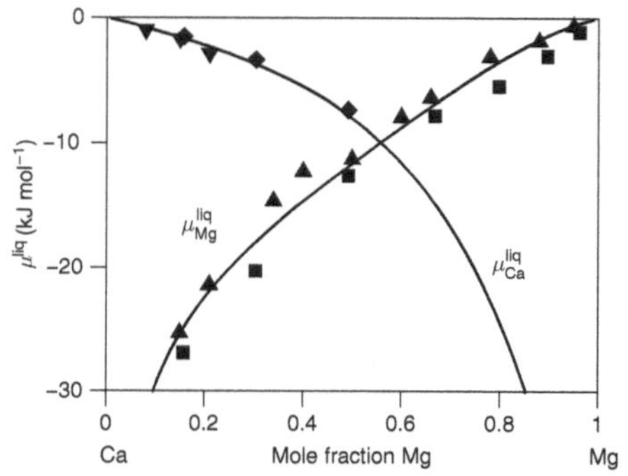

Figure 9.14 The chemical potential of Ca and Mg in liquid Ca–Mg alloys. The data are for Ca ♦ and Mg ■ from Mashorets and Puchkov (1965), and for Ca ▼ and Mg ▲ from Sommer (1979).

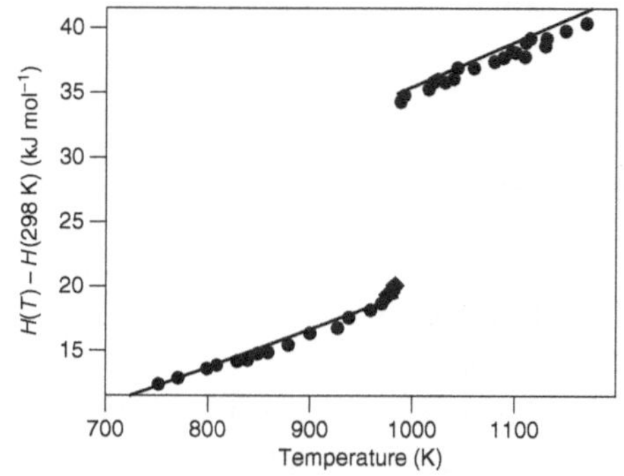

Figure 9.15 The heat content of $CaMg_2$, with data Agarwal *et al.* (1995), • used and ▲ not used.

9.5 Modeling the γ–γ′ phases: the Al–Ni system

When assessing a system with phases that can undergo disordering, like the B2 and $L1_2$ phases, it is strongly recommended that a model that includes the disordered states be used, i.e., that the same Gibbs-energy function can be used also for A2 and A1, respectively. The models recommended for ordered phases are described in section 5.8.4. The Calphad models usually do not include an explicit dependence on short-range order (SRO) as in the CVM, for example, because the contribution to the Gibbs energy from SRO is usually much smaller than many others. As described in the chapter on modeling, it is possible to describe the topological features of fcc ordering with a Bragg–Williams-type configurational entropy.

9.5.1 Order–disorder reactions

The stable phase diagram for Al–Ni is shown in Fig. 9.16(a). There is both an ordered phase (B2) based on the bcc lattice (A2) and an ordered phase ($L1_2$) based on the fcc lattice (A1). Al and Ni are both fcc as pure elements and the bcc phase (A2) is not stable in this system. This system is of great technological interest because superalloys are based on the A1 + $L1_2$ equilibria, so it has been assessed several times using various models. Superalloys contain from five to eight additional elements and thus it is important to use a model that can easily be extended to multicomponent systems and also can calculate equilibria and chemical potentials rapidly enough for it to be used in simulations of phase transformations using phase-field methods.

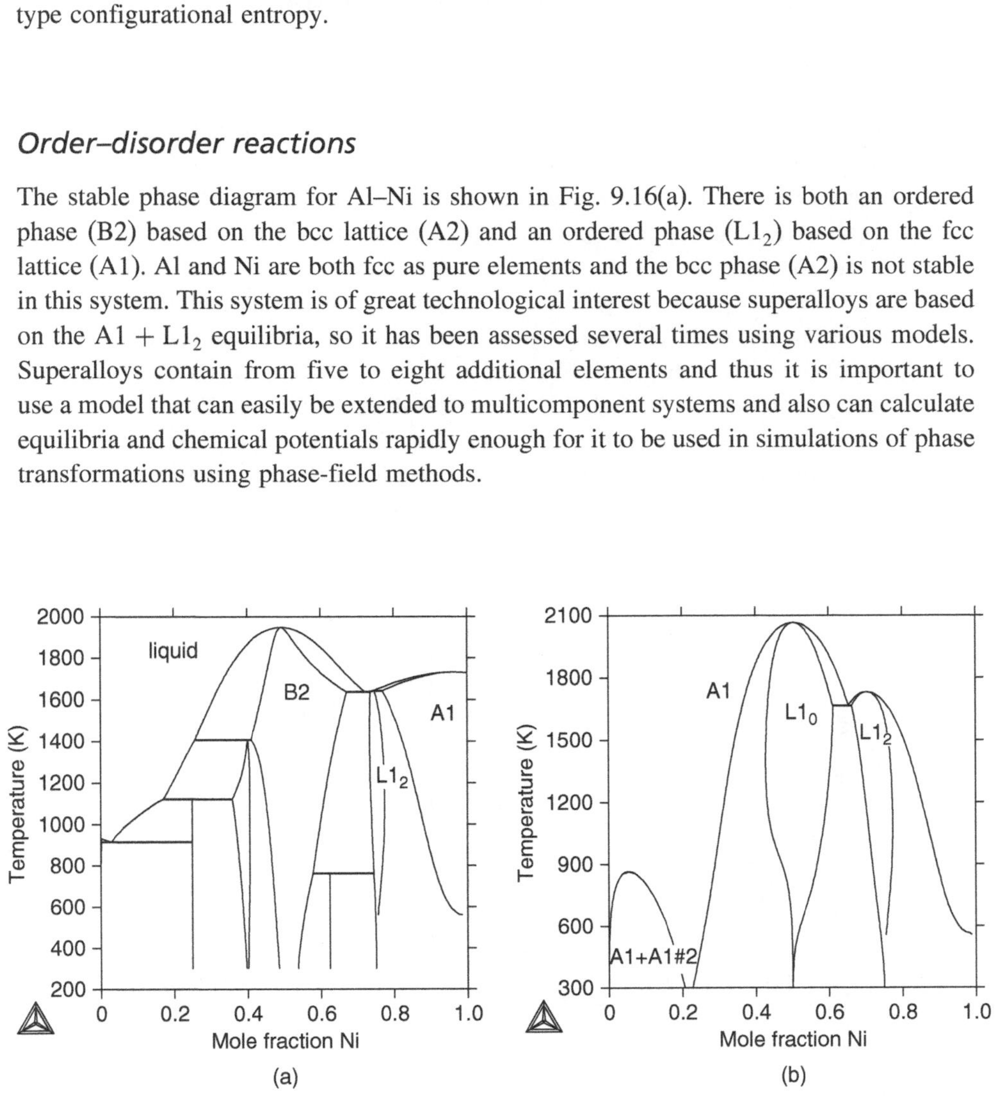

Figure 9.16 Calculated stable (a) and metastable (b) phase diagrams for the Al–Ni system, from an assessment by Ansara *et al.* (1997b) and Sundman (unpublished work, 2002).

By suspending all phases except the fcc, one can calculate a metastable phase diagram for Al–Ni with only ordered and disordered states of the fcc phase. Such a phase diagram is shown in Fig. 9.16(b). In order to perform an assessment of these phases, one can use first-principles calculations of the energies for the various **end members** in the model, i.e., $°G_{Al_3Ni}$, $°G_{AlNi}$, and $°G_{AlNi_3}$. The stable parts of the phase diagram must also be reproduced correctly, of course. It is interesting to note that on the Al-rich side there is a miscibility gap rather than an ordered $L1_2$ phase.

9.5.2 The driving force and thermodynamic factor

In a simulation of phase transformations, one must know the chemical potentials outside the stability range of the phase. That is provided by the Gibbs-energy function, which has been assessed for the whole composition range, not just for the stability range. In Fig. 9.17(a) the Gibbs-energy functions for the disordered A1, ordered $L1_2$, and metastable $L1_0$ ordered phases are shown. The $L1_0$ ordered phase is metastable because another ordered phase, B2, is more stable, but the Gibbs energy for B2 is not shown.

The diffusion follows gradients in the chemical potentials and the diffusion coefficient is dependent on the thermodynamic factor (also known as the stability function), which is obtained from the second derivatives of the Gibbs energy. The values of the thermodynamic factor as functions of composition are shown in Fig. 9.17(b), for the stable $L1_2$ phase as well as for the metastable $L1_0$ phase at $T = 1273$ K. This figure illustrates the abrupt changes of these quantities at the ideal ordering composition. In Fig. 9.18 a simulation using diffusion in a superalloy is shown.

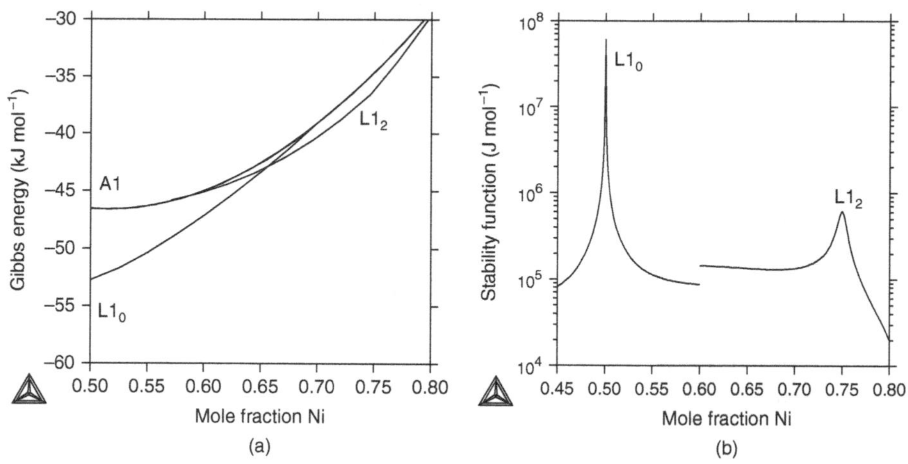

Figure 9.17 Thermodynamic functions for the fcc phase in the Al–Ni system. (a) The Gibbs energy of the fcc and ordered fcc based phases. (b) The stability function for the ordered phases.

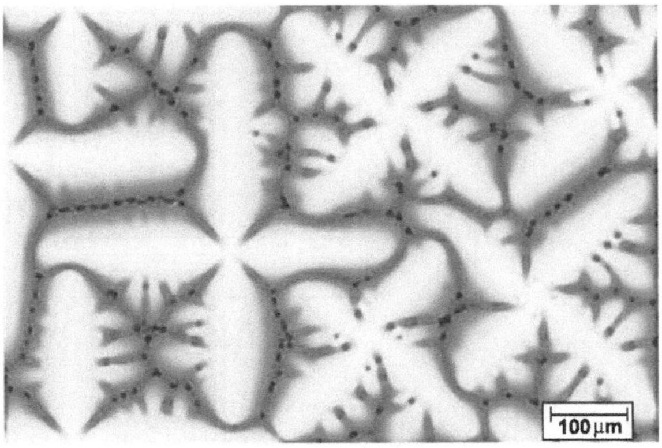

Figure 9.18 A simulated microstructure for an Ni-based five-component model alloy (Warnken *et al.* 2002) obtained using an assessed thermodynamic database. The simulation was done using the MICRESS code for the phase-field model http://www.micress.de. It reproduces the horizontal section of a directionally solidified alloy, showing γ dendrites and interdendritic γ' particles. Courtesy of N. Warken.

9.6 Assessment of a ternary oxide system

This case study will describe the first steps of an assessment of the quasiternary system Al_2O_3–CaO–SiO_2. This is a fairly well established system with some particular modeling problems. It will be used to show some special features of the Thermo-Calc software, including

- how to handle quasiternary (and quasibinary) systems,
- how to control a ternary miscibility gap, and
- how to select ternary parameters.

The quasibinary assessments were taken from previous publications: Al_2O_3–CaO from Hallstedt (1990), Al_2O_3–SiO_2 from Wang and Sundman (unpublished work, 1992), and CaO–SiO_2 from Hillert and Sundman (1990). The phase diagrams calculated from these assessments are shown in Figs. 9.19(a), (b), and (c), respectively.

9.6.1 *The quasiternary system*

The term "quasi" means that one of the components is redundant, insofar as its amount cannot vary independently of those of the others. For example, the amount of oxygen is fixed when the amounts of the other elements have been set. Of course, no real system has this property, but it is a useful approximation. The Gibbs phase rule states that there should be one condition for each component plus two for temperature and pressure. In a "quasi" system the last degree of freedom can be removed by imposing the condition that the activity of the "redundant" component is set to an arbitrary value. However, one must be careful to ensure that the modeling of all the phases in the system does not allow the composition to vary outside the "quasi" limits. If this is not the case, the equilibrium results are not independent of the selected activity.

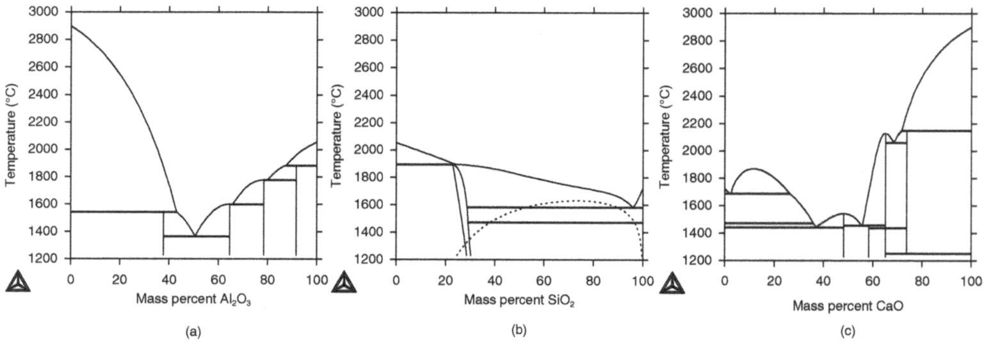

Figure 9.19 The three quasibinary phase diagrams: (a) CaO–Al_2O_3, (b) Al_2O_3–SiO_2 (the dotted line is a metastable miscibility gap in the liquid), and (c) SiO_2–CaO.

9.6.2 *The liquid model*

The partially ionic two-sublattice liquid model described in section 5.9.4 was used for the liquid phase. This is a very flexible model, which can take into account many kinds of ions and other species. In the original model, both CaO and Al_2O_3 were treated as "basic oxides," i.e., they give away oxygen and form Ca^{2+} and Al^{3+} cations and O^{2-} anions in the liquid:

$$(Ca^{2+}, Al^{3+})_P(O^{2-}, SiO_4^{4-}, SiO_2)_Q \quad (9.3)$$

whereas SiO_2 is treated as an "acid oxide" because it takes up oxygen, creating the complex SiO_4^{4-}. This model has a reciprocal subsystem,

$$(Ca^{2+}, Al^{3+})_P(O^{2-}, SiO_4^{4-})_Q \quad (9.4)$$

and it was found to be impossible to control the appearance of a reciprocal miscibility gap in this subsystem, see section 5.8.1.2, without using too many parameters. No such reciprocal miscibility gap is found in the real system. One reason for the problem is that there existed no experimental data to enable the determination of values for the "end member" $Al_4(SiO_4)_3$.

The model was then simplified by treating Al_2O_3 as "amphoteric," i.e., it neither gives up oxygen nor forms any complexes with oxygen, and thus can be modeled as a neutral species. The size of this species was taken to be half of Al_2O_3, i.e., $AlO_{3/2}$, because too large a molecule would change the entropy in the liquid and using the $AlO_{3/2}$ molecule made the model $(Ca^{2+})_P(O^{2-}, AlO_{3/2})_Q$ identical to the previous ionic model $(Ca^{2+}, Al^{3+})_P(O^{2-})_Q$. The ternary model was thus

$$(Ca^{2+})_P(O^{2-}, SiO_4^{4-}, AlO_{3/2}, SiO_2)_Q \quad (9.5)$$

Since there are no Al^{3+} ions in the first sublattice, there is no longer any reciprocal miscibility gap. In both sublattices only a single ionic species is left. Because they have the same charge, their amounts must always be equal. Owing to this condition, the model can be further simplified, without any change in the mathematical formulation, to

$$(\mathrm{AlO}_{3/2}, \mathrm{CaO}, \mathrm{Ca_2SiO_4}, \mathrm{SiO_2}) \tag{9.6}$$

and this is a model for an associate solution as discussed in section 5.9.4.3.

Owing to the problems with this model discussed below, a recent assessment of the Al_2O_3–CaO–SiO_2 system by Mao *et al.* (2006) treated the Al_2O_3 oxide as an acid by introducing the anion AlO_2^{1-} into the liquid. This new constituent removes most of the problems on extrapolating into the quasiternary system, but it is anyway useful to learn about the problems with the previous model since such problems can occur with any model or system.

9.6.3 The ternary extrapolation

On putting the three quasibinary assessments together, extrapolating to the ternary system, and calculating the isothermal phase diagram at 2000 K, one obtains Fig. 9.20(a). In this diagram there is a small liquid miscibility gap in the SiO_2 corner arising from the quasibinary CaO–SiO_2 system. Additionally there is a large miscibility gap in the liquid in the central part of the diagram.

This miscibility gap originates from modeling the Ca_2SiO_4 complex in the liquid with a very large stability. The assessed parameters for interaction between CaO and Al_2O_3 and between CaO and SiO_2 are very negative, whereas that for interaction between Al_2O_3 and SiO_2 is weakly positive at 2000 K. On starting from a liquid with composition $x_{\mathrm{CaO}} = 0.6$, $x_{\mathrm{SiO_2}} = 0.4$ and adding Al_2O_3, the liquid splits up into one rich in Ca_2SiO_4 and another that is rich in Al_2O_3. Since this miscibility gap does not appear in experimental investigations, the first objective in the assessment of ternary parameters is to suppress

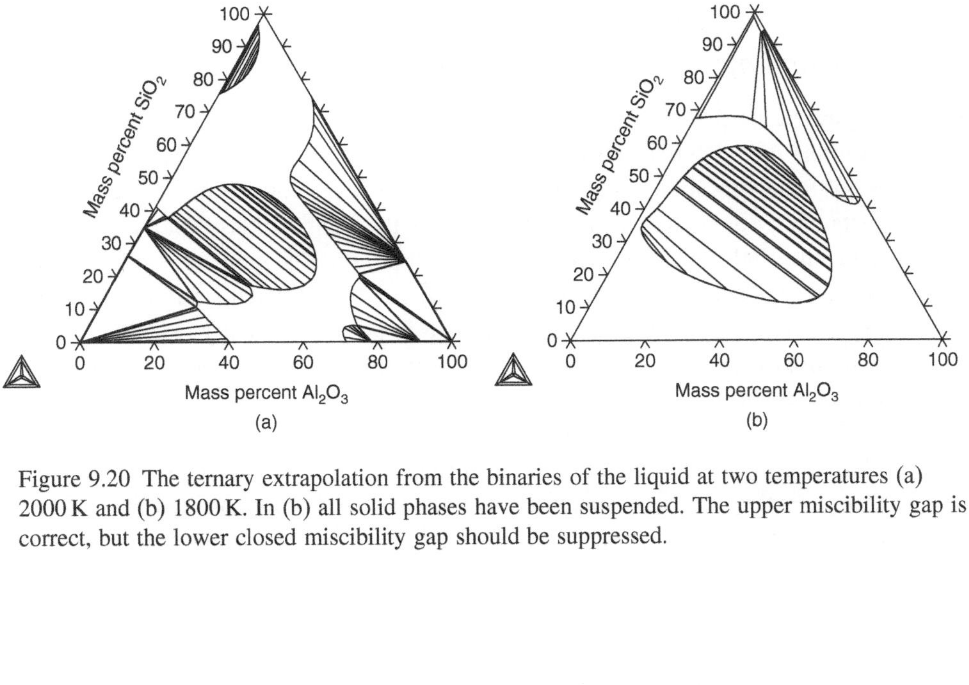

Figure 9.20 The ternary extrapolation from the binaries of the liquid at two temperatures (a) 2000 K and (b) 1800 K. In (b) all solid phases have been suspended. The upper miscibility gap is correct, but the lower closed miscibility gap should be suppressed.

the appearance of this miscibility gap below the stable liquidus temperature. This is not the only possibility; in some cases it may instead be reasonable to reassess the binary subsystems in order to avoid the appearance of ternary miscibility gaps, which are due to the very different interaction parameters in the binaries, as was done in the most-recent assessment (Mao *et al.* 2006).

In Fig. 9.20(b) the miscibility gaps at 1800 K, are shown; in this calculation all solid phases have been suspended. In the top corner the stable miscibility gap in the $CaO–SiO_2$ binary extends toward the binary $Al_2O_3–SiO_2$ edge, where it is metastable with respect to liquid + solid SiO_2 equilibria.

9.6.4 *Control of the ternary miscibility gap*

It is difficult to calculate whether the composition of a phase is inside a miscibility gap in a ternary system, but it is easy to ascertain whether the composition is inside the spinodal by calculating the stability function

$$\det\begin{pmatrix} \dfrac{\partial\mu_1}{\partial x_1} & \dfrac{\partial\mu_1}{\partial x_2} \\ \dfrac{\partial\mu_2}{\partial x_1} & \dfrac{\partial\mu_2}{\partial x_2} \end{pmatrix}$$

Inside the spinodal this determinant is negative; along the spinodal it is equal to zero. The smallest eigenvalue of this stability function can be obtained in Thermo-Calc by use of the function QF(phase name) and its value at various points in the liquid phase can be calculated. In PARROT one can create equilibria at the relevant compositions and temperatures and use "EXPERIMENT QF(ION_LIQ)>0.1:.001," which means that there will be a contribution to the sum of errors if this inequality is not fulfilled, which will force the adjustable variables to change during the optimization in order to suppress the miscibility gap.

9.6.5 *Experimental liquidus projection*

Figure 9.21 shows an assessed liquidus-surface phase diagram, reprinted with permission from the ACerS-NIST Phase Equilibria CDROM V.3.1 (http://www.ceramics.org). There is a stable miscibility gap only close to the SiO_2 corner. In the central part of the system there are two ternary compounds with the compositions $(CaO)_2.Al_2O_3.SiO_2$, called gehlenite, and $CaO.Al_2O_3.(SiO_2)_2$, called anorthite. The gehlenite will be called melilite because it has the same structure as $(MgO)_2.Al_2O_3.SiO_2$ and melilite is the generic name of this structure. Both melilite and anorthite melt congruently and their melting points will be used in the experimental file. There are also experimental thermochemical data for the compounds.

From the liquidus surface, one may use the invariant equilibria with the liquid in equilibrium with three solid phases as experimental points. This is not strictly correct because experimentally one has determined only points on the liquidus and this information has been graphically interpolated and extrapolated to provide isotherms, monovariant lines,

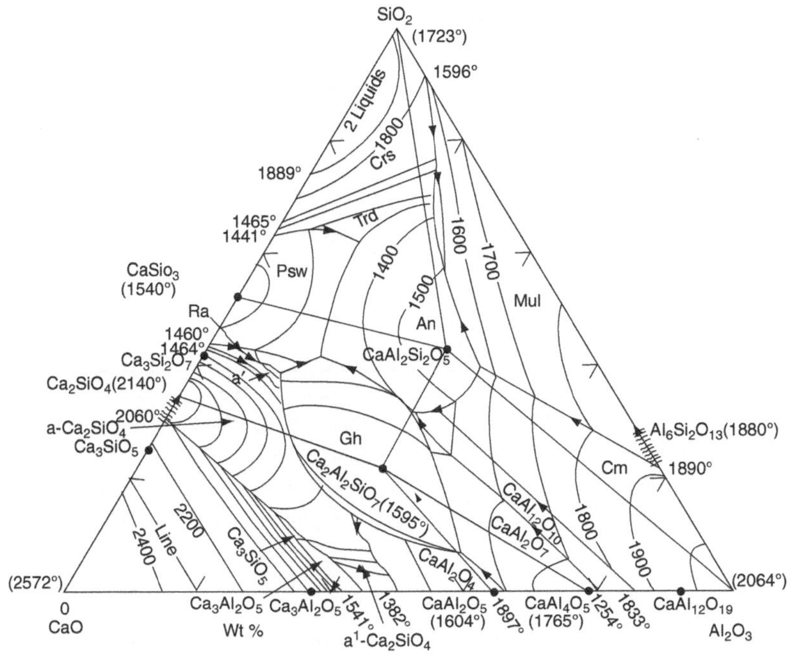

Figure 9.21 The experimental liquidus surface in the Al_2O_3–CaO–SiO_2 system. Reprinted with permission from ACerS-NIST Phase Equilibria CDROM V.3.1 (http://www.ceramics.org).

and invariant equilibria. In a proper assessment one should go back to measure all these experimental liquidus points because a graphical extrapolation might not be accurate, but, for the sake of simplicity, in this case study we will just use the invariant equilibria from the phase diagram.

There are also some measurements of the activities of the various oxide species. With this information, one can create an experimental data file for the system.

9.6.6 *The interaction parameters*

In the above selected model for the liquid there are four constituents, the three "end members," $AlO_{3/2}$, CaO, and SiO_2, and the complex Ca_2SiO_4. There is no corresponding complex for Al–Si–O because $AlO_{3/2}$ is treated as a neutral species in the liquid.

The interaction parameters involving the end members were assessed when the binaries were modeled, and this also includes the interaction parameters involving the complex and the CaO and SiO_2 end members. However, there is a new "quasibinary" interaction between Ca_2SiO_4 and $AlO_{3/2}$ that can be assessed inside the ternary system since it will give zero contribution to the three initial quasibinary systems. In fact, the interaction between Ca_2SiO_4 and $AlO_{3/2}$ is the most important one to suppress the miscibility gap in the central part. Figure 9.22 shows the miscibility gaps at 1800 K, calculated using the parameter

$$L_{AlO_{3/2},Ca_2SiO_4} = -15\,000 \tag{9.7}$$

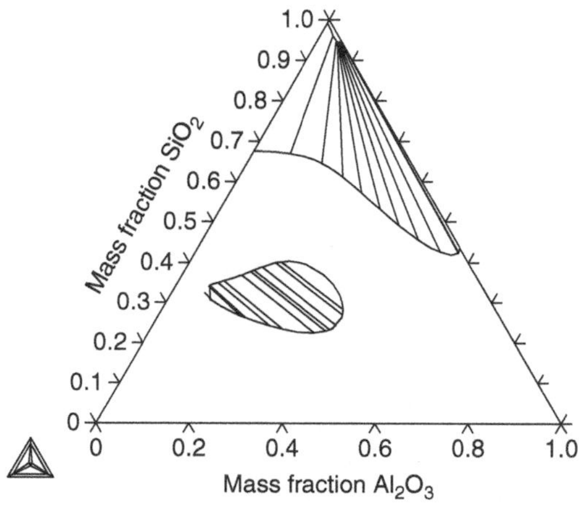

Figure 9.22 The liquid miscibility gap at 1800 K with all solid phases suspended and using the quasibinary parameter $L_{Ca_2SiO_4,Al_2O_3} = -15\,000\,J\,mol^{-1}$. Compare this with Fig. 9.20(b).

The parameter should be negative in order to stabilize the liquid along the line from Ca_2SiO_4 to Al_2O_3. It is possible to model this quasibinary parameter in a composition-dependent way by using an RK series, in which each parameter ${}^{\nu}L$ can be temperature-dependent.

There are three ternary parameters, each of which involves three different constituents:

$$L_{AlO_{3/2},CaO,SiO_2} \tag{9.8}$$

$$L_{AlO_{3/2},Ca_2SiO_4,SiO_2} \tag{9.9}$$

$$L_{AlO_{3/2},CaO,Ca_2SiO_4} \tag{9.10}$$

All three of these interaction parameters may be both temperature-dependent and composition-dependent according to Eq. (5.68).

As pointed out before, one needs some parameters to suppress the miscibility gap and, by testing, one can find that the quasibinary parameter has the largest effect. Checking the output from the PARROT module shows that the largest error occurs close to the SiO_2 corner where the experimental temperature for the invariant equilibrium of liquid + SiO_2 + mullite + pseudo-wollastonite is much lower than the calculated value. The very steep liquidus from the Al_2O_3–SiO_2 side that is found experimentally indicates that either there are more species contributing to the ideal entropy of mixing than considered in the model, which would be difficult to model, or the excess parameters need to give larger contributions. The ternary parameter having the largest effect in the SiO_2 corner is $L_{AlO_{3/2},Ca_2SiO_4,SiO_2}$ and it is actually necessary to model this parameter in a composition-dependent way. Making the quasibinary parameter temperature-dependent significantly improves the assessment. Finally, the optimization was done using five ternary coefficients.

9.7 Some notes on a ternary assessment, the Cr–Fe–Ni system

The POP file with experimental data for the Cr–Fe–Ni system is from an assessment done many years ago by Hillert and Qui (1990). It contains many types of experimental data: tie-lines between fcc, bcc, and liquid; liquidus temperatures; activities and activity coefficients; heats of mixing; and partition coefficients.

Most of the experimental information is just the same as one would have in a binary assessment. Since it is a ternary system, one must have one more condition for each equilibrium. Ternary tie-lines can have two experimental compositions in each phase and one of these must be used in a SET-CONDITION while the others are used as EXPERIMENT. As the condition one should select the experimental composition that has the smallest estimated uncertainty, but the problem is not so critical because there is the possibility of setting an estimated error also for a condition. One may thus give

```
SET-CONDITION P=1E5 T=1473 X(fcc,NI)=.05:.01
EXPERIMENT X(fcc,CR)=.22:.01
or
SET-CONDITION P=1E5 T=1473 X(fcc,CR)=.22:.01
EXPERIMENT X(fcc,NI)=.05:.01
```

and the optimizer will give the same result.

For some activities obtained by measuring the ratio of ion currents of two elements, an "instrument constant" relating the two ion currents was optimized with a special feature available in PARROT. The measured quantity, $a_{\mathrm{Fe}}/a_{\mathrm{Ni}}$ for example, was multiplied by the instrument constant, VAL1 in the POP file, and VAL1 was allowed to be varied by PARROT until the best fit was found. This is achieved by the IMPORT VAL1#11 line in the POP file. The command IMPORT will connect a variable in the POP file, VAL1 in this case, with an optimizing variable, V11 in this case. The value after the hash sign "#" specifies to which V-variable it is connected. V11 can either be set constant or allowed to be an optimizing variable. The same technique can be used also in other cases, for example if several samples are kept at the same carbon activity but the activity itself is not well determined. The activity in several experimental equilibria can thus be set to the same value but allowed to be varied by PARROT to get the best fit.

Data for the binary systems Fe–Cr, Cr–Ni, and Fe–Ni can be extracted from the SGTE binary database, for example. All phases, except the σ phase, are solution phases and one may add ternary interaction parameters to all three or just for the liquid. A ternary parameter can either be just an $a + bT$ function or composition-dependent. The special form for composition dependence derived by Hillert and used in Thermo-Calc requires that one use three composition-dependent parameters. The principle is that each of these three parameters will have the most influence in one of the corners of the ternary system. A possible set of commands to enter the ternary parameters is

```
ENTER-PARAM L(LIQUID,CR,FE,NI;0) 298.15 V1+V4*T
ENTER-PARAM L(LIQUID,CR,FE,NI;1) 298.15 V1+V2+V4*T
ENTER-PARAM L(LIQUID,CR,FE,NI;2) 298.15 V1+V3+V4*T
```

If one has just the first parameter, ending with ";0," that would represent the composition-independent parameter. If all three parameters are equal, that would also mean that there is no composition dependence. If the assessment gives small values for V2 and V3, it is advisable to use just a single composition-independent ternary parameter. This is the reason for using V1 and V4 as optimizing variables for all three parameters. There will rarely be enough data to evaluate the temperature dependences of three ternary parameters.

9.7.1 *Assessing binary and ternary data together*

A common problem is that one may wish to assess binary and ternary data together. This is possible in PARROT and one may add binary information to the same POP file as for the ternary or compile several POP files together into PARROT. When more than one POP file is used, one must answer NO to the third question of the COMPILE command when the second or later POP file is compiled. For example,

```
PARROT: compile
INPUT FILE: abc.POP
LIST-FILE /SCREEN/:
INITIATE STORE FILE /Y/:
...output...

PARROT: compile
INPUT FILE: ab.POP
LIST-FILE /SCREEN/:
INITIATE STORE FILE /Y/: N
...output...
```

If one answers Y (the default) to the prompt INITIATE STORE FILE, any experimental data from previously compiled POP files will be deleted and the first experimental data will be placed in the first block in the PARROT/EDIT workspace. If there is a FLUSH command in the POP file, that will continue storing experimental data in the second block. The second compile command when the POP file for the ab system has been compiled and when the prompt INITIATE STORE FILE has been answered N or NO will store the experimental data on this file in the next block, keeping whatever has been compiled in the previous blocks. In the EDIT module, one must use the READ command to select the block with equilibria.

Note that a binary POP file may require some modification before it can be used in a ternary assessment. The CREATE-NEW-EQUILIBRIUM command has a second argument that specifies the initialization code. Normally this initialization code is unity (1) and that means that all components in the system will be entered and all phases suspended initially. One may then explicitly set the phases FIXED, ENTERED, or DORMANT. For a binary file compiled with a ternary system, one should change the initialization code to zero (0) because that means that all components and phases are suspended initially.

Thus one should change from the binary file

```
CREATE-NEW 1 1
CH-ST PH fcc bcc=FIX 1
SET-COND T=1273 p=1E5
```

to

```
CREATE-NEW 1 0
CH-ST COMP A B=ENT
CH-ST PH fcc bcc=FIX 1
SET-COND T=1273 p=1E5
```

or

```
CREATE-NEW 1 1
CH-ST COMP C=SUS
CH-ST PH fcc bcc=FIX 1
SET-COND T=1273 p=1E5
```

However, the latter is less safe. It is always possible to have problems with equilibria with one or more components suspended. Even when one enters only a few, the suspended components may appear with non-zero fractions. There is a fix to this, which may have to be executed explicitly for each troublesome equilibrium. By simply setting the component first entered and then suspended, the Gibbs Energy System (GES) will explicitly set the fractions to zero for the suspended components.

There may also be trouble with phases that dissolve the suspended component on a sublattice together with one of the other components. For example, if one has a phase with the model (A)(B, C) and suspends component C, then the phase appears as stoichiometric in the AB system and for stoichiometric phases the calculation of G is different when one calls GES to calculate it. Deeper inside GES there will be an error when GES discovers the suspended component. This bug may be fixed in a later version, but at present one may have to enter the same phase twice, both as (A)(B) and as (A)(B,C), with the same coefficients for the parameter $^\circ G_{\mathrm{A:B}}$.

Appendix – websites

Bilbao crystallographic server, http://www.cryst.ehu.es
Crystal Lattice Structures website, http://cst-www.nrl.navy.mil/lattice
International Union of Crystallography, http://www.iucr.org
MICRESS phase field software, http://www.micress.de
SGTE website, http://www.sgte.org
Sluiter enthalpy database, http://www-lab.imr.edu/~marcel/enthalpy/enthlp.html
Thermo-Calc Software AB, http://www.thermocalc.com
Widom alloy database, http://alloy.phys.cmu.edu/

References

Abe T. and Sundman B. (2003) *Calphad* **27** 403–408

Abrikosov I. A., Niklasson A. M. N., Simak S. I., Johansson B., and Ruban A. V. (1996) *Phys. Rev. Lett.* **76** 4203–4206

Agarwal R., Fries S. G., Lukas H. L. *et al.* (1992) *Z. Metallkd.* **83** 216–223

Agarwal R., Lee J. J., Lukas H. L., and Sommer F. (1995) *Z. Metallkd.* **86** 103–108

Ågren J., Cheynet B., Clavaguera-Mora M. T. *et al.* (1995) *Calphad* **19** 449–480

Aldinger F., Sundman B., and Seifert H.-J. (Eds.) (1995) "The Ringberg Workshop 1995 on Pure Element Data," *Calphad* **19** 433–575

Aldinger F., Sundman B., and Seifert H.-J. (Eds.) (1997) "The Ringberg Workshop 1996 on Solution Modelling," *Calphad* **21** 139–287

(2000) "The Ringberg Workshop 1997 on Applications of Thermodynamic Data," *Calphad* **24** 15–94

(2001) "The 4th Ringberg Workshop," *Z. Metallkd.* **92** 510–562

Aldinger F., Sundman B., and Zinkevich M. (Eds.) (2007) "The Ringberg Workshop 2005," *Calphad* **31** 2–74

Allendorf M. D. (2006) http://www.ca.sandia.gov/HiTempThermo (gas data from *ab initio* methods)

Andersson J.-O. (1985) *Int. J. Thermophys.* **4** 411–419

Andersson J.-O. and Sundman B. (1987) *Calphad* **11** 83–92

Andersson J.-O., Helander T., Höglund L., Shi P., and Sundman B. (2002) *Calphad* **26** 273–312

Ansara I. (Ed.) (1998a) *COST 507 Light Alloy Database*, Luxembourg: Office of Official Publications of the European Community

(1998b) Al–B in *COST 507* pp. 15–17

(1999) SGTE substance database, http://www.sgte.org

(2001) *Ecole d'été de calculs thermodynamiques*, ed. Record M. C. and Tedenac J. C., Amsterdam: Elsevier, pp. 23–39

Ansara I., Bonnier E., and Mathieu J.-C. (1973) *Z. Metallkd.* **64** 258–268

Ansara I., Sundman B., and Willemin P. (1988) *Acta Metall.* **36** 977–982

Ansara I., Chatillon C., Lukas H. L. *et al.* (1994) *Calphad* **18** 177–222

Ansara I., Chart T. G., Fernández Guillermet A. *et al.* (1997a) *Calphad* **21** 171–218

Ansara I., Dupin, N., Lukas H. L., and Sundman B. (1997b) *J. Alloys Compounds* **247** 20–30

Baar N. (1911) *Z. Anorg. Allg. Chem.* **370** 352–394

Bain E. C. (1924) *Trans. AIME* **70** 25–35

Bakker H. (1998) *Enthalpies in Alloys*, Uetikon-Zürich: Trans Tech

(Bale C. W., Chartrand P., Degterov S. A. *et al.* 2002) *Calphad* **26** 189–228

Berne C., Sluiter M., Kawazoe Y., Hansen T., and Pasturel A. (2001) *Phys. Rev. B* **64** 144103-1–144103-8

Binczycka H., Uhrmacher M., Elidrissi-Moubtassim M. L., Jumas J. C., and Schaaf P. (2005) *Phys. Stat. Solidi B* **242** 1100–1107

Birks N., Meier G. H., and Pettit F. S. (2006) *Introduction to the High-temperature Oxidation of Metals*, 2nd edn., Cambridge: Cambridge University Press

de Boer F. R., Boom R., Mattens W. C. M., Miedema A. R., and Niessen A. K. (1988) *Cohesion in Metals: Transition Metal Alloys*, Amsterdam: North-Holland

Boettinger W. J. and Kattner U. R. (2002) *Metall. Mater. Trans. A* **33** 1779–1794

Bonhomme F. and Yvon K. (1996) *J. Alloys Compounds* **232** 271–273

Böttger B., Eiken J., and Steinbach I. (2006) *Acta Mater.* 2697–2704

Bradley A. J. (1951) *J. Iron Steel Inst.* 233–244

Bragg W. L and Williams E. J. (1934) *Proc. Roy. Soc.* **A145** 699–730

Bulian V. W. and Fahrenhorst E. (1946) *Z. Metallkd.* **37** 70

Burke E. C. (1955) *Trans. Metall. Soc. AIME* **203** 285–286

Burton P. B., Dupin N., Fries S. G. *et al.* (2001) *Z. Metallkd.* **92** 514–525

Buschow K. H. J. and van der Goot A. S. (1971) *Acta Crystallogr.* **B 27** 1085–1088

Bustamante C., Liphardt J., and Ritort F. (2005) *Phys. Today* **58** 43–48

Cahn J. W. (1962) *Acta Metall.* **10** 789–798

Callen H. H. (1985) *Thermodynamics and an Introduction to Thermostatistics*, New York: Wiley

Campbell C. E. (2005) *J. Phase Equilibria Diffusion* **26** 435–440

Car R. and Parrinello M. (1985) *Phys. Rev. Lett.* **55** 2471–2474

Chartrand P. and Pelton A. D. (2001) *Metall. Mater. Trans A* **32** 1397–1407

Chase M. W. Jr. (Ed.) (1998) *NIST–JANAF Thermochemical Tables*, 4th edn., Gaithersburg, MA: NIST

Chase M. W., Ansara I., Dinsdale A. T. *et al.* (1995) *Calphad* **19** 437–447

Chen Q., Hillert M., Sundman B. *et al.* (1998) *J. Electr. Mater.* **27** 961–971

Chen S.-L., Chou K.-C., and Chang Y. A. (1993) *Calphad* **17** 237–250

Chiotti P., Curtis R.W., and Woerner P.F. (1964) *J. Less-Common Metals* **7** 120–126

Chiotti P., Gartner G. J., Stevens E. R., and Saito Y. (1966) *J. Chem. Eng. Data* **11** 571–574

Colinet C. (1967) Estimation des grandeurs thermodynamiques des alliages ternaires, D.E.S., Faculté des Sciences, Université de Grenoble, France

(2002) in *Calphad and Alloy Thermodynamics*, ed. Turchi P. E. A., Gonis A., and Shull R. D., Warrendale, PA: TMS pp. 21–60

(2003) *Intermetallics* **11** 1095–1102

Connolly J. W. D. and Williams A. R. (1983) *Phys. Rev. B* **27** 5169–5172

Connolly J. W. D. and Kerrick D. M. (1987) *Calphad* **11** 1–55

Coughanowr C. A., Ansara I., Louma R., Hämäläinen M., and Lukas H. L. (1991) *Z. Metallkd.* **82** 574–581

Daams J. L. C., Villars P., and van Vucht J. H. N. (1992) *J. Alloys and Compounds* **182** 1–33

Daams J. L. C. and Villars P. (1993) *J. Alloys Compounds* **197** 1–34
(1994) *J. Alloys Compounds* **215** 1–34

Darken L. S. (1967) *Trans. Metall. Soc. AIME* **239** 90–96

Davison J. E. and Smith J. F. (1968) *Trans. Metall. Soc. AIME* **242** 2045–2049

Dinsdale A. T. (1991) *Calphad* **15** 317–425

Dupin N. (1995) Contribution à l'évaluation thermodynamique des alliages polyconstituées à base de nickel, unpublished Ph.D. Thesis, Institut National Polytechnique, Grenoble

Durand-Charre M. (2004) *Microstructure of Steels and Cast Irons*, Berlin: Springer

van Emmerik E. P. (2005) *J. J. van Laar: A Mathematical Chemist*, Ph.D. Thesis, Delft: Delft Technical University

Eriksson G. (1971) *Acta Chem. Scand.* **25** 2651–2658

Fernández Guillermet A. (1982) *Calphad* **6** 127–140

Fernández Guillermet A. and Gustafson P. (1984) *High Temp. High Press.* **16** 591–610

Ferrier A. and Wachtel E. (1964) *C. R. Acad. Sci. (Paris)* **258** 5424–5427

Ferro R. and Saccone A. (1996) Structure of intermetallic compounds and phases, in *Physical Metallurgy*, Vol. 1, ed. Cahn R. W. and Haasen P., Amsterdam: Elsevier, Chapter 4

de Fontaine D., Fries S. G., Inden G. *et al.* (1995) *Calphad* **19** 499–536

Fries S. G. and Lukas H. L. (1993) *J. Chim. Phys.* **90** 181–187

Fries S. G. and Sundman B. (2002) *Phys. Rev. B* **66** 012203-1–012203-4

Frisk K., Dumitrescu L., Ekroth M. *et al.* (2001) *J. Phase Equilibria* **22** 645–655

Gartner G. J. (1965) Application of an adiabatic calorimeter to the determination of the heats of fusion and heats of formation of several metallic compounds, unpublished Ph.D. Thesis, Iowa State University, Ames, Iowa

Gebhardt E., Seghezzi H. D., and Dürrschnabel W. (1961) *J. Nucl. Mater.* **4** 255–268

Ghosh G. and Asta M. (2005) *J. Mater. Res.* **20** 3102–3117

Ghosh G., Delsante S., Borzone G., Asta M., and Ferro R. (2006) *Acta Mater.* **54** 4977–4997

Goodman D. A., Cahn J. W., and Bennett L. H. (1981) *Bull. Alloy Phase Diagrams* **2** 29–34

Grafe U., Böttger B., Tiaden J., and Fries S. G. (2000) *Scripta Mater.* **42** 1179–1186

Grimvall G. (1991) *Thermophysical Properties of Materials*, Amsterdam: Elsevier
(1999) *Thermophysical Properties of Materials*, Amsterdam: Elsevier, p. 318

Grover R., Getting I. C., and Kennedy G. C. (1973) *Phys. Rev. B* **7** 567–571

Grube G. (1929) *Z. Elektrochem. Angew. Phys. Chem.* **35** 315–332

Grumbach M. P. and Martin R. M. (1996) *Phys. Rev. B* **54** 15 730–15 741

Grundy A. N., Povoden E., Ivas T., and Gauckler L. J. (2006) *Calphad* **30** 33–41

Grytsiv A., Ding J. J., Rogl P. *et al.* (2003) *J. Alloys Compounds* **350** 315–359

Grytsiv A., Rogl P., Giester C., and Pomjakushin V. (2005) *Intermetallics* **13** 497–509

Gueneau C., Baichi M., Labroche D., Chatillon C., and Sundman B. (2002) *J. Nucl. Mater.* **304** 161–175

Guggenheim E. A. (1937) *Trans. Faraday Soc.* **33** 151–155

(1952) *Mixtures*, Oxford: Clarendon Press

Gulliver G. M. (1913) *J. Inst. Metals* **9** 120–157

Guye H. and Welfringer J. (1984) Modelling of the thermodynamic properties of complex metallurgical slags, in *Second International Symposium on Metallurgical Slags and Fluxes*, ed. Fine H. A. and Gaskell D. R., Warrendale, MA: TMS, pp. 357–375

Hack K. (Ed.) (1996) *The SGTE Case Book*, London: The Institute of Metals

Hafner J., Wolverton C., and Ceder G. (2006) *Mater. Res. Soc. Bull.* **31** 659–665

Hallstedt B. (1990) *J. Am. Ceram. Soc.* **73** 15–23

(1992) *J. Am. Ceram. Soc.* **76** 1497–1507

Hasebe M. and Nishizawa T. (1972) *Bull. Jap. Inst. Metals* **11** 879–891

Haughton J. L. (1937) *J. Inst. Metals* **61** 241–246

Henig E.-Th., Lukas H. L., and Petzow G. (1980) *Z. Metallkd.* **71** 398–402

Henry N. F. M. and Lonsdale K. (Eds.) (1965) *International Tables for X-Ray Crystallography*, vol. I, Symmetry Groups, 2nd edn., Birmingham: The International Union of Crystallography and the Kynoch Press

Hill T. L. (1994) *Thermodynamics of Small Systems*, New York: Dover

Hillert M. (1980) *Calphad* **4** 1–12

(1981) *Physica* **103B** 31–40

(1986) *Metall. Trans. A* **17** 1878–1879

Hillert M. and Staffanson L.-I. (1970) *Acta Chem. Scand.* **24** 3618–3626

Hillert M. and Sundman B. (1976) *Acta Metall.* **24** 731–743

(1977) *Acta Metall.* **25** 11–18

Hillert M. and Jarl M. (1978) *Calphad* **2** 227–238

Hillert M. and Qui C. (1990) *Metall. Trans. A* **21** 1673–1680

Hillert M. and Sundman B. (1990) *Calphad* **14** 111–114

Hillert M., Jansson B., Sundman B., and Ågren J. (1985) *Metall. Trans. A* **16** 261–266

(1997a) *J. Phase Equilibria* **18** 249–263

(1997b) *Calphad* **21** 143–154

(1998) *Phase Equilibria, Phase Diagrams and Phase Transformations: A Thermodynamic Basis*, Cambridge: Cambridge University Press

(2001) *J. Alloys Compounds* **320** 161–176

Hillert M. and Schalin M. (2000) *Acta Mater.* **48** 461–468

Hoehne G. W. H., Hemminger W., and Flammersheim H.-J. (1996) *Differential Scanning Calorimetry*, Berlin: Springer

Hohenberg P. and Kohn W. (1964) *Phys. Rev.* **136** 864–871

Hultgren R., Desai P. D., Hawkins D. T. *et al.* (1973) *Selected Values of the Thermodynamic Properties of the Elements*, Metals Park, OH: American Society of Metals

Hume-Rothery W., Smallman R. E., and Haworth C. W. (Hume-Rothery *et al.*1969) *The Structure of Metals and Alloys*, 5th edn., London: Institute of Metals

Inden G. (1981) *Physica* **103B** 82–100

Ising E. (1925) *Z. Phys.* **31** 253–258

Jansson Å. (1987) *TRITA-MAC 340*, Stockholm: Royal Institute of Technology
Jansson B. (1983) Computer operated methods for equilibrium calculations and evaluation of thermochemical model parameters, unpublished Ph.D. Thesis, Royal Institute of Technology, Stockholm
Jones J. P. (2002) *Adv. Imaging Electron Phys.* **125** 63–117
Joubert J. M. (2002) *Calphad* **26** 419–425
Joubert J. M. and Feutelais Y. (2002) *Calphad* **26** 427–438
Kapoor M. L., Mehrota G. M. and Froberg M. G. (1974) *Arch. Eisenhütt.* **45** 663–669
Kattner U. R. (1997) *J. Minerals, Metals Mater. Soc.* **49** 12 14–19
Kattner U. R., Boettinger W. J., and Coriell S. R. (1996) *Z. Metallkd.* **87** 522–528
Kaufman L. (1959) *Acta Metall.* **15** 575–587
(2002) Hume-Rothery and CALPHAD thermodynamics, in *Calphad and Alloy Thermodynamics*, ed. Turchi P. E. A., Gonis A., and Shull R. D., Warrendale, PA: TMS, pp. 3–19
Kaufman L. and Bernstein H. (1970) *Computer Calculations of Phase Diagrams*, New York: Academic Press
Kikuchi R. (1951) *Phys. Rev.* **81** 988–1003
King R. C. and Kleppa O. J. (1964) *Acta Metall.* **12** 87–97
Klemm W. and Dinkelacker F. (1947) *Z. Anorg. Chem.* **255** 2–12
Kohler F. (1960) *Monatsh. Chem.* **91** 738–740
Kohn W. and Sham L. J. (1965) *Phys. Rev. B* **140** A1133–A1139
Konetzki R., Schmid-Fetzer R., Watson A. *et al.* (1993) *Z. Metallkd.* **84** 569–573
Kopp H. (1864) *Ann. Chem. Pharmacie* **126** 362–372
Korzhavyi P. A., Sundman B., Selleby M., and Johansson B. (2005) *Mater. Res. Soc. Symp. Proc.* **842** S4.10.1–6
Köster W., Müller R., and Seelhorst J. (1964) *Z. Metallkd.* **55** 589–596
Kowalski M. and Spencer P. J. (1993) *J. Phase Equilibria* **14** 432–438
Kubaschewski O. and Ünal H. (1977) *High Temp. High Press.* **9** 361–365
Kubaschewski O., Evans E. L., and Alcock C. B. (1967) *Metallurgical Thermochemistry*, Oxford: Pergamon Press
Kubaschewski O., Alcock C. B., and Spencer P. J. (1993) *Materials Thermo-chemistry*, 6th revised edn., Oxford: Pergamon Press
Kumar K. C. H. and Wollants P. (2001) *J. Alloys Compounds* **320** 189–198
Kusoffsky A. and Sundman B. (1998) *J. Phys. Chem. Solids* **59** 1549–1553
Kusoffsky A., Dupin N., and Sundman B. (2001) *Calphad* **25** 549–565
van Laar J. J. (1908a) *Z. Phys. Chem.* **63** 216–253
(1908b) *Z. Phys. Chem.* **64** 257–297
Landolt-Börnstein (1999) *Landolt-Börnstein: Numerical Data and Functional Relationships in Science and Technology – New Series. Group 4: Physical Chemistry. Vol. 19: Thermodynamic Properties of Inorganic Materials. Subvolume A: Elements and Compounds*, Part 1, ed. SGTE Scientific Group Thermodata Europe, Berlin: Springer
(2002) *Landolt-Börnstein: Numerical Data and Functional Relationships in Science and Technology – New Series. Group 4: Physical Chemistry. Vol. 19: Thermodynamic Properties of Inorganic Materials. Subvolume B: Binary Systems*, Part 1, ed. SGTE Scientific Group Thermodata Europe, Berlin: Springer

Lechermann F., Fahnle M., and Sanchez J. M. (2005) *Intermetallics* **13** 1096–1109

Lee B. J. (1992) *Calphad* **16** 121–149

Lee S. W., An S. Y., Shim I. B., and Kim C. S. (2005) *J. Magnetism Magn. Mater.* **290** 231–233

Liang P., Tarfa T., Robinson J. A. *et al.* (1998) *Thermochim. Acta* **314** 87–110

Liang P., Dupin N., Fries S. G. *et al.* (2001) *Z. Metallkd.* **92** 747–756

Libowitz G. G. (1971) *Metall. Trans.* **2** 85–93

Lim S. S., Rossiter P. L., and Tibbals J. W. (1995) *Calphad* **19** 131–142

Liu Z.-K., Höglund L., Jönsson B., and Ågren J. (1991) *Metall. Trans. A* **22** 1745–1752

Liu Z.-K. and Chang Y. A. (2000) *J. Alloys Compounds* **299** 153–162

Loginova I., Ågren J., and Amberg G. (2004) *Acta Mater.* **52** 4055–4063

Lu X. G., Selleby M., and Sundman B. (2005) *Calphad* **29** 49–55

Lücke K. and Detert K. (1957) *Acta Metall.* **5** 628–637

Lukas H. L. and Fries S. G. (1992) *J. Phase Equilibria* **13** 532–541

Lukas H. L., Henig E.-Th., and Zimmermann B. *et al.* (1977) *Calphad* **1** 225–236

Lukas H. L., Weiss J., and Henig E.-Th. (1982) *Calphad* **6** 229–251

Lupis C. H. P. (1967) *Acta Metall.* **15** 265–276

Mao H. H., Selleby M., Hillert M., and Sundman B. (2006) *J. Am. Ceram. Soc.* **69** 298–308

Marquardt D. W. (1963) *J. Soc. Indust. Appl. Math.* **11** 431–441

Martin R. M. (2004) *Electronic Structure – Basic Theory and Practical Methods*, Cambridge: Cambridge University Press

Mashorets V. P. and Puchkov L. V. (1965) *J. Appl. Chem. USSR* **38** 937–940, translated from *Zh. Prikl. Khim.* **38** 1009–1013 (1965)

Mirkovic D., Gröbner J., Schmid-Fetzer R., Fabrichnaya O., and Lukas H. L. (2004) *J. Alloys Compounds* **384** 168–174

Mizutani U., Asahi R., Sato H., and Takeuchi T. (2006) *Phil. Mag.* **86** 645–654

Muggianu Y.-M., Gambino M., and Bros J.-P. (1975) *J. Chim. Phys.* **72** 83–88

Murnaghan F. D. (1944) *Proc. Nat. Acad. Sci. (USA)* **30** 244–247

Nayeb-Hashemi A. A. and Clark J. B. (1987) *Bull. Alloy Phase Diagrams* **8** 58–65

Neckel A. and Wagner S. (1969) *Monatsh. Chem.* **100** 664–670

Neumann F. (1865) *Ann. Phys. Chem.* **202** 123–142

Nowotny H., Wormnes E., and Mohrnheim A. (1940) *Z. Metallkd.* **32** 39–42

Oates W. A. and Wenzl H. (1996) *Scripta Mater.* **35** 623–627

Odqvist J., Sundman B., and Ågren J. (2003) *Acta Mater.* **51** 1035–1043

Paris R. (1934) *Publ. Sci. Techn. Minist. Air (France)* **45** 39–41

Paxton A. T., Methfessel M., and Pettifor D. G. (1997) *Proc. Roy. Soc. A* **453** 1493–1514

Pelton A. D. (2001) *Calphad* **25** 319–328

Pelton A. D. and Bale C. W. (1986) *Metall. Trans. A* **17** 1211–1215

Pelton A. D. and Blander M. (1986) *Metall. Trans. B* **4** 805–815

Pelton A. D. and Chartrand P. (2001) *Metall. Mater. Trans. A* **32** 1355–1383

Pelton A. D., Degterov S. A., Eriksson G., Robelin C., and Dessureault Y. (2000) *Metall. Mater. Trans. A* **31** 651–659

Pelton A. D., Chartrand P., and Eriksson G. (2001) *Metall. Mater. Trans. A* **32** 1409–1416

Perepezko J. H. and Wilde G. (1998) *Ber. Bunsen-Ges. Phys. Chem. Chem. Phys.* **102** 1074–1082

Pettifor D. G. (1996) *Bonding and Structure of Molecules and Solids*, Oxford: Oxford University Press

(2003) *Acta Mater.* **51** 5649–5673

Pisani C. (Ed.) (2000) *Quantum-Mechanical Ab-initio Calculations of the Properties of Crystalline Materials*, Berlin: Springer

Qin R. S. and Wallach E. R. (2003) *Acta Mater.* **51** 6199–6210

Redlich O. and Kister A. (1948) *Indust. Eng. Chem.* **40** 345–348

Sanchez J. M., de Fontaine D., and Teitler W. (1982) *Phys. Rev. B* **26** 1465–1468

Sanchez J. M., Ducastelle F., and Gratias D. (1984) *Physica A* **128** 334–350

Saunders N. and Miodownik A. P. (1998) *Calphad*, Vol. 1, ed. Cahn R. W., Oxford: Pergamon Press

Scheil E. (1942) *Z. Metallkd.* **34** 70–72

Schmalzried H. and Pelton A. D. (1973) *Ber. Bunsenges. Phys. Chem.* **77** 90–94

Schmid-Fetzer R., Andersson D., Chevalier P. Y. *et al.* (2007) *Calphad* **31** 38–52

Schubert K. (1964) *Kristallstrukturen zweikomponentiger Phasen*, Berlin: Springer.

Schubert K., Lukas H. L., Meißner H.-G., and Bhan S. (1959) *Z. Metallkd.* **50** 534–540

Seetharaman S., Mukai K., and Sichen D. (2005) *Steel Res. Int.* **76** 267–276

Seiersten M. and Tibballs J. (1993) *The Al–Fe System*, Sintef report STF28F93051, Oslo: Sintef

Shao G. (2001) *Intermetallics* **9** 1063–1068

Shockley W. (1938) *J. Chem. Phys.* **6** 130–144

Shull R. D. and Joshi A. (Eds.) (1992) *Thermal Analysis in Metallurgy*, Proceedings of the TMS Meeting in Anaheim, CA, 1991, Warrendale, MA: TMS, pp. 77–306

Skriver H. L. (1985) *Phys. Rev. B* **31** 1909–1923

Sluiter M. (2006) *Calphad* **30** 357–366

Smith J. F. and Smythe R. V. (1959) *Acta Metall.* **7** 261–267

Sommer F. (1979) *Z. Metallkd.* **70** 545–547

Sommer F., Predel B., and Aßmann D. (1977) *Z. Metallkd.* **68** 347–349

Spence J. C. H. and Tafto J. (1983) *J. Microscopy* **130** 147–154

Spencer P. J. (1998) *Thermochim. Acta* **314** 1–21

Sundman B. (1990) *Anales Fís. B* **86** 69–82

(1991a) *J. Phase Equilibria* **12** 127–140

(1991b) *Calphad* **15** 109–119

Sundman B. and Ågren J. (1981) *J. Phys. Chem. Solids* **42** 297–301

Sundman B. and Mohri T. (1990) *Z. Metallkd.* **81** 251–254

Sundman B., Jansson B., and Andersson J.-O. (1985) *Calphad* **9** 153–190

Sundman B., Fries S. G., and Oates W. A. (1998) *Calphad* **22** 335–354

(1999) *Z. Metallkd.* **90** 267–273

Sykes P. W. (1926) *Trans. ASST* **10** 839–871

Tanaka T., Gokcen N. A., and Morita Z. I. (1990) *Z. Metallkd.* **81** 49–54 and 349–353

Tanaka T., Gokcen N. A., Kumar K. C. H., Hara S., and Morita Z. I. (1996) *Z. Metallkd.* **87** 779–783

Temkin M. (1945) *Acta Phys. Chim.* **20** 411–420

Tepesch P. D., Asta M., and Ceder G. (1998) *Modelling Simul. Mater. Sci. Eng.* **6** 787–797

Toop G. W. (1965) *Trans. Metall. Soc. AIME* **233** 850–854

Turchi P. E. A., Abrikosov I. A., Burton B. *et al.* (2007) *Calphad* **31** 4–27

Vosskühler V. H. (1937) *Z. Metallkd.* **29** 236–237

Wagner C. (1952) *Thermodynamics of Alloys* Cambridge, MA: Addison-Wesley Press, Inc.

Wagner C. and Schottky W. (1930) *Z. Phys. Chem.* **11** 163–210

Warnken N., Böttger B., Ma D. *et al.* (2002) *Materials for Advanced Power Engineering*, 29 September–2 October 2002, Liège, Belgium, ed. Lecomte-Beckers J., Jülich: Forschungszentrum Jülich, part I, pp. 315–324

Westbrook J. H. and Fleischer R. L. (Eds.) (1995) *Intermetallic Compounds, Principles and Practice*, vols. I and II, New York: John Wiley & Sons

Williams D. B. and Carter C. B. (1996) *Transmission Electron Microscopy,* New York: Plenum

Zarkevich N. A. and Johnsson D. D. (2004) *Phys. Rev. Lett.* **92** 255702

Zhu J. Z., Wang T., Zhou S. H., Liu Z.-K., and Chen L. Q. (2004) *Acta Mater.* **52** 833–840

Zimmermann B. (1976) Rechnerische und experimentelle Optimierung von binären und ternären Systemen aus Ag, Bi, Pb and Tl, unpublished Ph.D. Thesis, Universität Stuttgart

Zunger A., Wei S. H., Ferreira L. G., and Bernard J. E. (1990) *Phys. Rev. Lett.* **63** 353–356

Index